Towards Sustainable
Agricultural Development

Towards Sustainable Agricultural Development

Edited by
M.D. Young

Published in association with the
Organisation for Economic Co-operation and Development

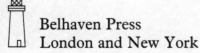

Belhaven Press
London and New York

© Organisation for Economic Cooperation and Development, 1991
© English language edition, Belhaven Press, 1991

First published in Great Britain in 1991 by
Belhaven Press (a division of Pinter Publishers),
25 Floral Street, London WC2E 9DS

British Library Cataloguing in publication data

A CIP catalogue record for this book is available from
the British Library

ISBN 1 85293 137 X

For enquiries in North America please contact
PO Box 197, Irvington, NY 10533

Library of Congress Cataloging in Publication Data

A CIP catalog record for this book is
available from the Library of Congress

Typeset by Florencetype Ltd, Kewstoke, Avon
Printed and bound by Biddles Ltd, Guildford and Kings Lynn

To
David Juckes

Contents

Preface   ix
Authors   x

Introduction: conceptual framework   1
M.D. Young

Part I   Intensive crop production and the use of agricultural chemicals

1   Impact of German intensive crop production and agricultural chemical policies in Hildesheimer Börde and Rhein-Pfalz
H. de Haen, H.F. Fink, C. Thoroe and W. Wahmhoff   9

2   The effects of Swedish price support, fertiliser and pesticide policies on the environment
K.I. Kumm   50

3   The effects of US agricultural policies on water quality and human health: opportunities to improve the targeting of current policies
R.M. Wolcott, S.M. Johnson and C.M. Long   91

Part II   Intensive animal production and the management of animal manure

4   Intensive livestock production in France and its effects on water quality in Brittany
P. Rainelli   115

5   Dutch approaches to the management of pollution from intensive livestock production
Grontmij NV   147

Part III   Dry-land farming, soil conservation and soil erosion

6   Impact of agricultural policies on soil erosion in two regions of Portugal
Alfredo Gonçalves Ferreira   175

7   Integration of agricultural and environmental policies in the United States: the case of soil erosion and soil conservation in dry-land farming
P. Crosson   189

Part IV Changing landscapes, land-use patterns and the character of rural landscapes

8   Maintaining alpine landscapes in the Austrian Tyrol: policies for the maintenance of the rural cultivated landscape in the Tyrol
     W. Puwein   221
9   Changing landscapes and land-use patterns and the quality of the rural environment in the United Kingdom
     J. Bowers and T. O'Riordan   253

Part V The impact on agriculture of pollution from other sources

10   The economic impact of air pollution on agriculture: an assessment and review
     R.M. Adams and T.D. Crocker   295
11   The impact of sewage sludge on agriculture
     M. Linster   320

12   Overview: the integration of agricultural and environmental policies
     M.D. Young   337

Index   343

Preface

This book contains a collection of studies prepared for the OECD's *ad hoc* Group on Agriculture and the Environment. Concerned about the direction of agricultural policy and the extent of environmental problems associated with many agricultural practices, they commissioned experts in eight countries to prepare studies which addressed the underlying causes of these problems. At the same time, mindful of the effects that pollution from urban sources can have on agriculture, they also commissioned two reports on the effects of environmental policies on agriculture.

The work was initiated in response to the declaration by OECD Environment Ministers that they will 'ensure that environmental considerations are taken fully into account at an early stage in the development and implementation of economic and other policies'.

All the chapters have been prepared by the *ad hoc* group's consultants with assistance from the secretariat. The views expressed in all but the last chapter are those of the authors and do not necessarily represent the views of the OECD, nor the governments of its member countries. The last chapter, however, is very similar to the Executive Summary of the OECD's report on *Agriculture and Environmental Policies: Opportunities for Integration*.

The administrators responsible for co-ordinating the work described in this book were Michael Young, Jean-Marie Debois, Christian Avérous and David Juckes. Ferenc Juhasz as secretary to the *ad hoc* group was responsible for co-ordinating the OECD's work on the integration of agricultural and environmental policies. Soon after the completion of all this work David Juckes passed away. David's tireless efforts and patience throughout the project were appreciated by us all. We dedicate the book to him.

Ferenc Juhasz
Secretary,
Ad Hoc Group on Agriculture
and Environment

Authors

R.M. Adams, Oregon State University, Corvallis, Oregon, USA

J. Bowers, School of Economic Studies, University of Leeds, Leeds, England

T.D. Crocker, University of Wyoming, Laramie, Wyoming, USA

P. Crosson, Resources for the Future, Washington, D.C., USA

J.M. Debois, Organisation for Economic Co-operation and Development, Paris, France

A.G. Ferreira, Universidade de Evora, Evora, Portugal

H.F. Fink, Institut für Agrarökonomie, University of Göttingen, Göttingen, Germany

Grontmij NV, De Bilt, The Netherlands

H. de Haen, Institut für Agrarökonomie, University of Göttingen, Göttingen, Germany

S.M. Johnson, Iowa State University of Science and Technology, Iowa, USA

D. Juckes, Organisation for Economic Co-operation and Development, Paris, France

F. Juhasz, Organisation for Economic Co-operation and Development, Paris, France

K.I. Kumm, Langsta, Sweden

M. Linster, Université de Paris XII (Paris–Val-de-Marne), Ecole Nationale des Ponts et Chaussées, Paris, France

C.M. Long, Environmental Law Institute for the Economic Studies Branch, Washington, D.C., USA

T. O'Riordan, School of Environmental Sciences, University of East Anglia, Norwich, England

W. Puwein, Klosterneuburg, Austria

P. Rainelli, Institut National de la Recherche Agronomique, Centre de Rennes, Rennes, France

C. Thoroe, Institut für Agrarökonomie, University of Göttingen, Göttingen, Germany

W. Wahmhoff, Institut für Agrarökonomie, University of Göttingen, Göttingen, Germany

R.M. Wolcott, Environmental Protection Agency, Washington, D.C., USA

M.D. Young, CSIRO Division of Wildlife and Ecology, Canberra, Australia

Introduction
conceptual framework
M.D. Young

Growing awareness and concern about the linkages between agricultural policies and practices and specific environmental problems, some of them with serious long-term implications, have focused increasing attention on the interface between agriculture and environment. This has led to a recognition of the need to better integrate agricultural policies with policies which seek to protect, preserve and enhance the environment. It is believed that by pursuing integrated policies more sustainable agricultural production systems will emerge.

The agriculture/environment integration issue can be succinctly stated as follows: to what extent do macro-agricultural policies take into account their effects on environmental quality, and to what extent do environmental policies take into account their impact on agricultural output, income and prices? Defining the problem in these terms reflects the growing recognition in both domains of the close linkages between and the potential for mutually reinforcing action by agricultural and environmental policy-makers. This is a necessary prerequisite to sustainable development. Thus it is not a question of bending agricultural policy to meet exclusively ecological goals, nor of shaping environmental policy to suit the needs of agriculture. Rather, integrated policies, by making appropriate trade-offs between the interests of both sectors, can contribute to rational, efficient and sustainable development in both an economic and an ecological sense.

The principle areas of concern leading governments to re-examine the interrelationship between agricultural and environmental policies tend to fall into five broad categories:

1 The problems of pollution and environmental degradation associated with intensive crop production.
2 The post-war trend, especially in Europe, towards intensive animal husbandry, creating its own particular environmental problems associated with animal wastes.
3 The downstream effects and lost production from soil erosion caused by extensive agricultural practices associated with dryland agriculture.
4 The changing demands and practices that have significantly altered the character of the landscapes which is leading to the development of

programmes to encourage farmers to save dwindling natural habitat and
wilderness areas for the benefit of present and future generations.
5 The impact on agriculture of pollution from other sources and non-
   agricultural activities, notably in the form of acid precipitation, water
   pollution and sewage sludge. (OECD, 1989)

Methodology

This book reports the second phase of the OECD's work on the integration of
agricultural and environmental policies. Taking a policy area approach, each
part analyses policy issues within several countries. It aims to identify general
principles which are likely to apply throughout all OECD countries.
Consequently, the examples chosen for detailed study covered as wide a
cross-section of situations as is possible.

The subject areas and country studies were organised as follows:

1 Intensive crop production/agricultural chemicals interface (Federal
   Republic of Germany, Sweden, United States, chapters 1, 2 and 3).
2 Intensive animal husbandry-animal manure interface (France, the
   Netherlands, chapters 4 and 5).
3 Dryland agriculture/soil conservation and erosion interface (Portugal,
   United States, chapters 6 and 7).
4 Changing landscape and land use patterns/rural quality interface (Austria,
   United Kingdom, chapters 8 and 9).
5 The impact on agriculture of pollution from other sources (a review of the
   literature and state of knowledge in this area, chapters 10 and 11).

Each policy area study begins with a description of the environmental,
economic and social costs and benefits of the relevant agricultural policies and
practices. Then attention is given to a set of policy issues, including the
effects of subsidies on inputs, the need to improve advisory services and the
impact of production support measures on landscape amenity. The results of
this examination are then compared with policy alternatives and practices
which could result in a better integration of agricultural and environmental
considerations.

Each study involved the following four steps:

1 The identification of the benefits and impacts which arise from different
   agricultural practices within the policy area.
2 The identification of the socio-economic, administrative and physical
   conditions which cause these impacts or which are necessary to derive
   these benefits.
3 An evaluation of the effectiveness of policy instruments which seek to
   enhance benefits and mitigate negative impacts.
4 The identification of the necessary conditions for and approaches to the
   successful implementation of policies which achieve better integration.

The principal problems and issues examined in the case studies

Intensive crop production/agricultural chemicals interface

The focus of this interface was on the effects on the quality of the environment of the fertilisers, both organic and inorganic, and pesticides used in intensive crop production. The meaning of the term 'intensive' varies from country to country, but generally such agricultural systems are characterised by irrigation and/or high rainfall and the observation that, for at least most of the year, water availability does not limit plant growth. Intensive crops include almost all horticultural crops, sugar cane, sugar beet, cotton, potatoes, maize, wheat, barley and oats.

The principal environmental problems considered in this policy area study were:

1 The effects of residual fertilizers and pesticides on human health, including those associated with spray drift from pesticides during application, and residues in food and water.
2 The effects of residual fertilisers (phosphates and nitrates) on the condition of aquatic environments (eutrophication).
3 The unintended effects of pesticides on species other than those to which they are applied.

The principal issues selected for close examination were:

1 The relationship between product price support policies and the extent of fertiliser and pesticide use.
2 The impact and effectiveness of policy instruments which seek to reduce the quantity of fertilisers and pesticides used, including charges, taxes and subsidies on inputs, quotas on inputs and production standards for fertiliser and pesticide manufacture.
3 The need for and effectiveness of regulations to prohibit the use of certain chemicals and adoption of certain application practices.
4 The effectiveness of advisory services in making agricultural practices more favourable to the environment.
5 The contribution of agricultural research to environmental policy.
6 The identification of 'non-point' sources of pollutants.

Intensive animal husbandry/animal manure interface

The focus in this policy area study was on the adverse effects that animal manure, of not properly managed, can have on the environment. These effects are most noticeable in connection with intensive animal husbandry, notably of pigs, poultry and dairy cattle.

The most significant related environmental problems include nitrate pollution of ground and surface water supplies, the effects of ammonia emissions on surrounding vegetation and their contribution to acid rain, and the accumulation of heavy metals in soils. At a local level, the emission of unpleasant odours is an environmental nuisance.

The principal issues selected for close examination in this policy area study were:

1 The effects of price support and tariff policies on the location and development of intensive animal production systems.
2 The impact and effectiveness of policy instruments (charges, taxes, subsidies, quotas and regulations) which seek to decrease environmental pollution from intensive animal husbandry.
3 The application of the 'polluter pays' principle to non-point sources of pollution.
4 The need for and effectiveness of regulations to control the quality of prepared animal feedstuffs.

Dryland agriculture/soil conservation and erosion interface

The work in this policy area recognised both the positive and the negative effects of agriculture on the environment. In particular, it examined the role for agriculture in contributing to environmental objectives through the maintenance of soil structures and productivity. The off-site benefits from desirable agricultural practices such as those which reduce silt loads in surface waters, etc., were also examined.

Dryland agriculture is characterised by the fact that moisture is often the limiting factor for plant growth. Generally, dryland agricultural systems contain either native or improved pastures grown in association with cereal crops.

The principal issues selected for close examination in this policy area study were:

1 The potential and cost effectiveness of grants and subsidies in encouraging soil conservation.
2 The effectiveness of regulations which require and/or prohibit certain agricultural practices.
3 The identification of rates of soil loss and the problem of farming activity where rates of erosion are excessive.
4 The potential of community-backed education campaigns to raise awareness and change community values.
5 The effect of capital and income tax policies on conservation.
6 The co-ordination of administrative action during policy formation and implementation.

Changing landscape and land use patterns/rural quality interface

The work in this interface focused both on the positive role of agriculture in maintaining landscapes and land use patterns and on the impact that changes in agricultural practices can have on the quality of the rural environment. Particular attention was paid to the environmental benefits derived from the role of agriculture in promoting tourism, providing wildlife habitats and maintaining aesthetically appealing landscapes. This included an examination of policies designed to encourage the development of these roles in alpine and mountainous areas. The integration of rural development and agricultural policies was also examined.

The principal issues selected for closer examination in this policy area study were:

1 The effectiveness of policy instruments in encouraging farmers to maintain a positive role in producing environmental benefits for society.
2 The effects of price support policies on landscape diversity and biological diversity.
3 The effectiveness of policy instruments which seek to reduce, redirect and/or stop the rate of change in agricultural landscapes and land use practices, including subsidies, financial incentives and integrated programmes to revitalise economically depressed regions.
4 Changing institutional arrangements to achieve the more effective integration of agricultural, regional development and environmental policies.

Impact on agriculture of pollution from other sources

In several countries pollution from other sources has led to regional decline in the quality of food production and, in other cases, the quantity of food produced. There is a growing body of literature describing such impacts on agriculture, although most of it is based on the extrapolation of the conclusions from laboratory experiments about the effects of acid rain, etc., on regional and national estimates of crop production.

In this part, a review was conducted of the literature and state of knowledge about this policy area in all OECD countries, describing and where possible quantifying the likely impact on agriculture of pollution from other sources, notably sewage sludge (chapter 10), acid rain, ambient ozone concentrations, water pollution from other sources, and rising CO_2 concentrations (chapter 11).

The final chapter develops the set of consistent themes which emerged from these studies into a series of policy recommendations. It is believed that if all countries formulated and implemented agricultural and environmental policies within the framework which emerges agricultural production would be more sustainable.

Reference

OECD (1989), *Agricultural and Environmental Policies: Opportunities for Integration*: OECD, Paris.

Part I
Intensive crop production and the use of agricultural chemicals

1. Impact of German intensive crop production and agricultural chemical policies in Hildesheimer Börde and Rhein-Pfalz

H. de Haen, H.F. Fink, C. Thoroe and W. Wahmhoff

In the Federal Republic of Germany agricultural production has increased considerably during the last twenty-five years (table 1.1). In 1985 the level of food production was about 60 per cent above that of 1960. Over the same period, labour use in agriculture decreased by roughly 60 per cent and the agricultural area diminished by 2·3 million ha (roughly 16 per cent, or 0·7 per cent per year). During the same period the use of capital increased by 60 per cent and that of intermediate inputs nearly doubled. The substantial changes in West German agriculture can be characterised by two main features: increased intensity of land use, and specialisation and concentration in agricultural production patterns. This development, supported by the European Common Agricultural Policy, involved substantial increases in farm incomes. But this growth has not continued in recent years. Severe unemployment in the economy as a whole has slowed down the process of out-migration of labour, and increasing surplus agricultural production within the European Community has narrowed the scope for income orientation.

Increasing intensity of land use is indicated by the growth of yields as well as in that of yield-improving inputs. Gross production per hectare increased by 90 per cent between 1960 and 1985 (at an average annual rate of 2·6 per cent); the use of nitrogenous fertiliser rose by 150 per cent at an average annual rate of more than 3·7 per cent; and that of pesticides increased even more. But this process of increasing the use of yield-improving inputs has slowed down during recent years. From 1980 to 1985 there was no further increase in the use of nitrogenous fertiliser, and the use of some pesticides even decreased considerably, especially that of insecticides. Yet recent growth rates of yields per hectare were only slightly lower than those of the years before. The increasing intensity of animal production has resulted in a

Table 1.1 Main indicators of agricultural development in the Federal Republic of Germany

Indicator	1960	1970	1980	1985
Agricultural area[a] (AA) (ha million)	14·3	13·6	12·2	12·0
Arable land	8·0	7·5	7·3	7·2
Grassland	5·7	5·5	4·8	4·6
Agricultural production[b] (DM billion at 1980 prices)				
Value of gross production	–	55·8	63·8	67·0
Intermediate inputs	–	27·2	33·4	32·4
Gross value added	25·0	28·6	30·4	34·6
Gross production				
(million t grain units)	42·0	54·1	61·1	66·9
(100 kg grain units per ha AA)	29·3	39·7	49·6	55·6
Apparent consumption of commercial fertilisers[c] (t million of nutrient)				
Nitrogenous fertilisers	0·6	1·1	1·5	1·5
Phosphate fertilisers	0·7	0·9	0·9	0·7
Potash	1·0	1·1	1·2	1·0
Pesticides (domestic sales in 1,000 t of active substance)				
Herbicides	–	14·9[d]	20·9	17·4
Fungicides	–	5·1[d]	6·5	8·5
Insecticides	–	2·1[d]	2·3	1·6
Others	–	2·3[d]	3·2	2·6
Feed use (DM million at 1980 prices)	–	8,204	11,489	10,900
Energy use (DM million at 1980 prices)	–	4,191	4,846	4,896
Number and size of farms				
Number (000)	1,385	983	797	721
Average farm size (ha per farm)	9·3	12·7	15·3	16·6
Number of livestock holders (000)				
Cattle	1,254	843	529	465[e]
Pigs	1,741	1,028	547	433[e]
Hens and poultry	2,808	1,305	425	353[e]
Livestock (million)				
Cattle	12·9	14·0	15·1	15·7
Pigs	15·8	21·0	22·6	23·6
Poultry	63·4	100·7	85·6	80·2
Fixed capital (DM billion at 1980 prices)	165·5	233·7	261·6	264·1
Mechanisation: tractors (number per 100 ha AA)	60·6	96·8	101·4[f]	171·2[e]
Labour use (labour units per 100 ha AA)	19·0	11·4	8·2	7·2

(a) Due to changes in the statistical concept data before and after 1979 are not exactly comparable.
(b) National accounts statistics: Agriculture, fisheries and forestry.
(c) Data refer to financial years and not to calendar years, e.g. 1970 = 1969/70.
(d) 1973.   (e) 1984.   (f) 1979.
(g) Food production less animal production on the base of imported feedstuffs.

Sources: Statistisches Jahrbuch für Ernährung, landwortschaft und Forsten, various issues; *Agrarbericht,* various issues.

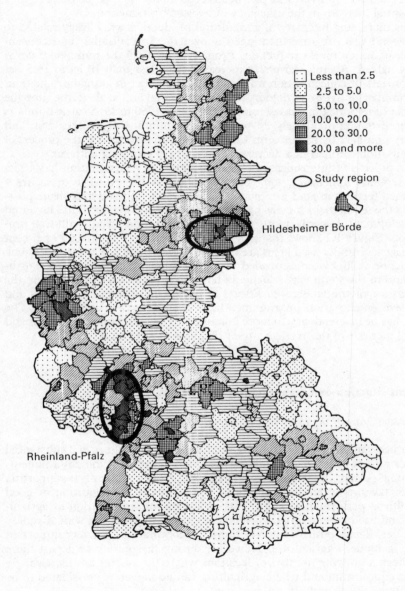

Less than 2.5
2.5 to 5.0
5.0 to 10.0
10.0 to 20.0
20.0 to 30.0
30.0 and more

◯ Study region

Hildesheimer Börde

Rheinland-Pfalz

Figure 1.1 Areas of intensive crop production, districts of the Federal Republic of Germany, 1983. 'Intensive' crops include sugar beet, rape, vegetables and horticultural products, fruit farming, tree nurseries and vineyards. *Source:* Schilling (1984)

higher density of livestock per hectare of about 40 per cent, and in a substantial increase in the efficiency of livestock production.

This increasing intensity of agricultural production was closely related to the second main feature mentioned above, the specialisation and concentration of agricultural production. From 1960 to 1985 the number of farms nearly halved. Average farm size increased from less than 10 ha to 16·5 ha. The process of concentration was much more distinct in animal production. The number of livestock holders decreased in total and at the same time the number of livestock increased remarkably: from 1960 to 1985 cattle numbers increased by approximately 20 per cent, pig numbers by 50 per cent and poultry numbers by 25 per cent. Concerning crop production, the process of specialisation resulted in a decreasing diversity of cropping patterns.

The process of intensification, specialisation and concentration has a distinct regional dimension, too. The relaxation of the pressure to secure a sustained fertility in land use by diversified production systems, made possible by the use of yield-increasing inputs from outside the farm, has favoured a development towards more regional specialisation in agricultural production. Figure 1.1 shows that in a number of regions more than 30 per cent of the agriculturally used areas are used for highly intensive crops. Livestock production is highly concentrated in a few regions, especially in the northwest and in the south-east. Changing farming systems led to a change in the appearance of the landscape. Efforts at improving farming conditions and extensive amelioration programmes have been a further component. The result has been a tendency to more homogeneity in cultivation conditions and a large increase in the average size of the plots of farm land.

Present situation and recent national initiatives

The current state of the environment in relation to agriculture

The process of intensification and specialisation in agriculture had substantial implications for the relationship between agriculture and the environment. For long periods this interdependence had mostly been of a supporting nature: favourable natural conditions were a necessary condition of good agricultural performance, and it was an important contribution of agricultural land use to maintain an open landscape and to support a wide diversity of species. These positive interrelationships continue to be a very important factor in modern agriculture. However, during the more recent past there have been a growing number of locations where the system has increasingly lost its equilibrium and where agriculture can no longer be considered to be in natural balance with the environment.

But non-agricultural factors tend to interfere with the equilibrium as well. Being located within a highly industrialised economy, rural areas in Germany went through considerable changes of physical infrastructure and non-agricultural resource demand which had various effects on agricultural land use systems. Moreover, extensive areas are affected by acid precipitation which has had detrimental effects on forest eco-systems and created problems for agriculture as well, at least in the form of additional costs of application of compensating lime fertiliser. Another external impact on agriculture is the contamination of soils by heavy metals due to the spread of sewage sludge on farm land in locations of high population density.

Concerning the impact of agricultural production on the environment, such interference is more or less directly related to the structural change in the production system and to the gradual substitution for land and labour of agro-chemicals and heavy machinery, respectively. The Council of Environmental Advisers (Rat von Sachverständigen für Umweltfragen, 1985) has recently presented a comprehensive assessment of environmental risks resulting from agricultural production. It distinguishes six domains of environmental stress which are summarized here in their perceived order of urgency:

1 Diminution and partition of natural biotopes.
2 Contamination of ground water by nitrate and pesticides.
3 Soil compaction and soil erosion.
4 Contamination and eutrophication of surface waters.
5 Reduction of food quality.
6 Air pollution through smell and dust.

The most important factors with more or less nationwide relevance are certainly the endangering of wild species, resulting from a destruction of biotopes, and the contamination of ground water. Of more regional relevance are soil compaction due to the use of heavy machinery and erosion, concentrated in locations with light soils and areas with the extensive row cropping of maize and sugar beet.

Eutrophication and contamination of surface waters are another severe environmental problem in Germany. According to a survey conducted in 1975, the annual run-off of phosphorus amounted to more than 100,000 tonnes. However, two-thirds of this came from household sewage, more than 10 per cent from industrial sources, and only 8 per cent was run-off from agricultural fields through drainage and erosion. Hence, on a national scale, agriculture seems to be only a minor contributor to eutrophication. But the problems vary between regions, and some rural surface waters are indeed significantly affected by phosphorus and nitrogen run-off as well as by contamination with residues of pesticides.

Although the use of agro-chemicals has intensified drastically during past decades, there seems to be no general indication of declining food quality which might have resulted from residues of fertiliser, pesticides, veterinary medication or feed additives. According to conventional methods of analysis, most chemicals found in foodstuffs remain clearly within the safety margins prescribed in the federal food law. Yet continued careful attentiveness is required here, especially since the test methods themselves need further refinement. Of particular concern are the nitrate content in some green vegetables and, in some more rare cases, residues of medical agents and feed additives in meat.

Air pollution by smell in the vicinity of piggery and poultry units is given a rather low rank in the list of the council.

The diminution and partition of natural or quasi-natural biotopes due to irrigation, land consolidation and intensification are the predominant reasons for the rapid decline of wild species in flora and fauna. According to estimates of the International Union for the Conservation of Nature (IUCN), the following percentages of species are more or less urgently endangered in the Federal Republic of Germany: vertebrate animals (44 per cent), dragonflies (51 per cent), butterflies (33 per cent), plants (26 per cent). Moreover the number of endangered species is gradually increasing. According to Sukopp (1981), major causes of the decline in species are such factors as the removal of small natural or quasi-natural sites, drainage, abandonment or change of land use, intensive application of agro-chemicals on crop land and intensive use of grassland (fig. 1.2). In total, agricultural land use was partly or fully responsible in 85 per cent of all cases of declining numbers of plant species. In the past many of those factors were closely related to regional programmes of land consolidation aiming at the establishment of more efficient farm organisation and rural infrastructure. Figure 1.3 illustrates for a region in northern Germany the steady removal of hedges which went with such programmes. Actually the share of natural or quasi-natural biotopes as part of the usable agricultural area (excluding forests including biotopes) is far below the recommendations of ecologists. Some regions with intensive agricultural land use barely reach a share of 2 to 3 per cent, and even this small area is partitioned into rudimentary sites, too small to serve as retreat or feed base for many of the endangered animals and too closely encircled by intensive crop land.

The second important environmental problem is the contamination of ground water by nitrate and residues of pesticides. Pumped water in quite a number of water catchment areas located within regions of intensive crop production, horticulture, vineyards or high livestock density on light soils is approaching or even exceeding the European Community limit of 50 mg NO_3/l.

According to recent observations, 5 to 8 per cent of drinking water exceeds this standard (Darimont et al., 1985). The regions with high nitrate concentration coincide in most cases with regions of intensive agricultural production.

The figures related to the nitrate problem are particularly alarming, because the maximum of nitrate leaching into ground water has certainly not yet been reached. This can be concluded from two observations. Water used today was contaminated ten to thirty years ago, at a time when fertilisation levels were significantly lower. Moreover, there are indications in some recent studies (Obermann, 1984) that the denitrification capacity of aquifers is steadily declining with time. As a result of such tendencies, various private pumping stations have been closed down already, and communal waterworks have had to deepen their wells or obtain water from distant catchment areas. Recently some waterworks have also reported that the concentration of certain residues of pesticides, particularly Attracine, used in maize production, in their water supply approaches or even exceeds the official tolerance levels as well.

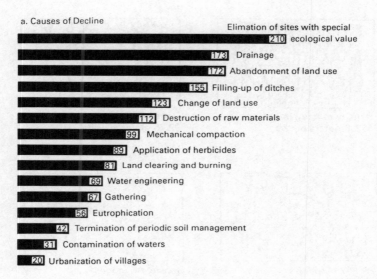

a. Causes of Decline

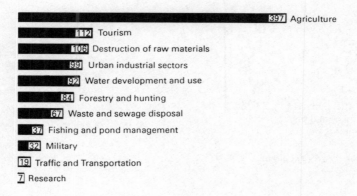

b. Sector of Origin for Decline of Species

Figure 1.2 Causes and sectors of origin of the decline of species. The figures show the number of species affected; multiple nomination of causes is possible. The total number of species which disappeared during the period of investigation was 581. *Source:* Sukopp (1981)

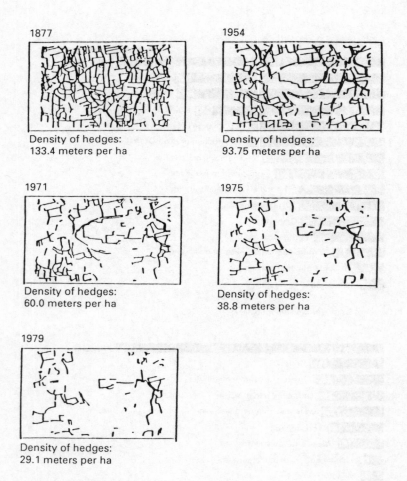

1877

Density of hedges:
133.4 meters per ha

1954

Density of hedges:
93.75 meters per ha

1971

Density of hedges:
60.0 meters per ha

1975

Density of hedges:
38.8 meters per ha

1979

Density of hedges:
29.1 meters per ha

Source: MARXEN (1979)

Figure 1.3   The destruction of hedgerows in an agricultural landscape during the course of a century. Density of hedges: 1877, 133·4 m/ha; 1954, 93·75 m/ha; 1971, 60·0 m/ha; 1975, 38·8 m/ha; 1979, 29·1 m/ha. *Source:* Marxen (1979)

Legislation and environmental policies

Environmental protection as an official policy goal has become an integral part of the overall policy strategy, in general policy as well as in agricultural policy. Officially the maintenance of natural resources and the environment has even gained an equal ranking with the traditional goals of agricultural policy. In practice, however, the farm income goal has clearly dominated decision-making in agricultural policy. The outstanding example which underlines this observation is the prevailing producer price support in pursuit of farm income objectives in spite of surplus production and environmental damage from highly intensive agricultural production.

Legislation for protection of natural resources and control of agricultural input use has been implemented at the federal as well as at the *Länder* level, the latter having a double competence through the Bundesrat (Senate), which participates in federal lawmaking, as well as through their own parliaments. The federal constitution authorises the federal parliament to enact framework legislation in such fields as nature conservation and water protection, which then may be specified by each *Land* in the form of region-specific laws and ordinances. There are other mostly input-related laws where the federal parliament is the only legislature.

Legislation related to agriculture and the environment may be grouped into three categories: resources/environment, input use and food. Table 1.2 gives an overview which underlines the rich potential of environment protection embedded in the current legislation. Of particular relevance for the two most urgent environmental problems, the diminution of biotopes and the decline of wild species on the one hand and the contamination of ground waters on the other, are the Federal Nature Conservation Act, the Federal Water Protection Act and the Waste Disposal Act.

The Federal Nature Conservation Act and the related *Länder* Acts aim at the maintenance of biological activity in nature, the preservation of landscape and the protection of habitats of wild plant and animal species. The law contains provision for various degrees of protection, ranging from landscape protection schemes with only a few restrictions to more rigidly controlled nature reserves and national parks. In so far as agricultural areas are located within nature reserves the authorities may restrict land use systems in various ways, but these areas only cover less than 1 per cent of the total area of the Federal Republic. Substantial income losses have to be compensated if they are comparable to a partial expropriation. Most critical from an environmental point of view is the general exemption of agricultural outside special protection areas from any control. The Act even explicitly states that 'orderly agriculture' serves the objectives of nature protection and is therefore not subject to any restriction. Yet the Act does not specify what is meant by orderly agriculture.

The protection of water resources against any kind of infiltration from external sources is the object of the Federal Water Protection Act and the respective *Länder* laws. Although the Act does not explicitly exempt 'orderly agriculture' outside specific areas from restrictions or sanctions, enforcable production restrictions are clearly defined for water protection areas only.

Table 1.2   Policies and legislation related to resources/environment, input, food and extensification/land diversion

RESOURCES/ENVIRONMENT

Nature Conservation Act
Spatial coverage: federal (framework) and *Länder* legislator
Goals: maintenance of biological activity in nature, preservation of landscape,
   protection of wild plant and animal species
Measures: delineation of nature reserves, levies on agricultural practices, optional
   region specific compensation
Issues: orderly agriculture exempted (but not defined)

Water Protection Act
Spatial coverage: federal (framework) and *Länder* legislation
Goals: protection of water resources against infiltration from external sources
Measures: delineation of water protection areas, levies on input use and land use,
   compensation assured by law
Issues: collective liability, compulsory permission of fertilisation, orderly agriculture
   not defined

Land Consolidation Act
Spatial coverage: federal legislation
Goals: improvement of efficiency in farm organisation and rural infrastructure
Measures: public organisation and co-ordination of inter-farm exchange of land,
   rural investment
Issues: more emphasis on ecology

Soil Conservation Concept
Spatial coverage: federal
Goals: research-oriented assessment of the state of the soil
Measures: finance for research and monitoring

INPUTS

Fertiliser Act
Spatial coverage: federal legislation
Goals: rules for production and trade
Measures: administrative admission of fertiliser
Issues: no relevance at farm level

Plant Protection Act
Spatial coverage: federal legislation
Goals: Protection of plants, food and feed against plant pests and diseases,
   elimination of damages caused by application of chemicals
Measures: testing, licensing, restrictions and prohibitions on application

Waste Disposal Act
Spatial coverage: federal legislation
Goals: preservation of soil and water
Measures: regulations with regard to storage and spread of sewage sludge and
   manure
Issues: several *Länder* have issued special ordinances which prescribe maximum
   areas per unit of livestock and which regulate the timing and level of the use of
   liquid manure

Table 1.2 – *continued*

FOOD

Food Act
Spatial coverage: federal legislation
Goals: protection of food against chemicals and residues
Measures: food control, production ceilings, limitation of food contamination
Issues: production, processing and trade in food

EXTENSIFICATION/LAND DIVERSION

Crop Edge Programme
Spatial coverage: *Länder* level, implemented in Schleswig-Holstein, Niedersachsen,
  Nordrhein-Westfälen, Hessen, Rheinland-Pfalz, Saarland, Bayern
Goals: preservation of wild plant and animal species
Measures: renunciation of plant protection
Issues: compensaiton ranges from 7 to 12·5 Pf/m^2

Grassland Extensification Scheme
Spatial coverage: *Länder* level, implemented in: Schleswig-Holstein, Niedersachsen,
  Nordrhein-Westfälen, Hessen, Rheinland-Pfalz, Bayern, Baden-Württemberg,
  Hamburg
Goals: preservation of wild plant and animal species
Measures: restrictions on fertiliser, plant protection, mowing, grazing, drainage,
  prohibition of changing grassland into crop-land
Issues: compensation varies considerably according to specific levy

Green Fallow Programme
Spatial coverage: Niedersachsen
Goals: reduction of surplus production
Measures: land set-aside
Issues: pilot project, compensation: DM 1,000 to DM 1,800 per hectare, according
  to valuation index of field

EC Structural Adjustment (scheme not yet implemented)
Spatial coverage: EC-wide framework, with nation-specific implementation
Goals: reduction of surplus production, with secondary emphasis on environment
Measures: subsidy on land diversion (entire farms, ten years) in combination with
  early retirement, subsidy on adjustment of production pattern in favour of non-
  surplus products, subsidy on reduced input application levels
Issues: political acceptability varies between member countries, norms for
  environmental protection not explicitly included

Two regulations in the law are of particular relevance to agriculture. One concerns collective liability, according to which a group of potential polluters, such as farmers in a catchment area, may be liable not just for stated damage but also for the endangering of ground water without a need to identify an individual originator. The other is a newly established provision of financial compensation for those farmers subject to certain restrictions on orderly agricultural land use where the resulting income loss does not imply some form of expropriation. Again, the Act does not define orderly agriculture.

Another law which, among other effects, aims at the protection of water systems is the federal waste disposal law. It contains regulations with regard to the spread of sewage sludge and excessive use of manure. On the basis of this law, several *Länder* have issued special ordinances which prescribe minimum disposal areas per unit of livestock and which regulate the timing and level of use of liquid manure.

In addition to the aforementioned legislation, various extensification policies and programmes have been implemented at the *Länder* level. Most of those programmes work on the basis of bilateral contracts and compensate farmers for the adjustment of certain land use practices aiming at the preservation of specific species or biotopes. Almost all *Länder* have so-called crop edge programmes which subsidise farmers who leave predetermined crop edges unsprayed. Eight out of eleven *Länder* offer grassland extensification schemes which subsidise farmers who refrain from spraying pesticides, reduce levels of fertiliser or leave meadows unused during main periods of hatching. Several aim at the expansion of nature reserves and compensate farmers for income losses due to restrictions on fertiliser and pesticide application.

These programmes fit the new aims of the European Community's Common Agricultural Policy, according to which subsidies may be paid for extensification and readjustment of production patterns. The primary aim, however, is the reduction of market surpluses rather than the protection of the environment.

In 1986 Lower Saxony launched its so-called green fallow programme, a subsidy scheme in which farmers receive a premium for leaving up to 20 per cent of their land fallow for one year. The premium was 1,000 to 1,200 DM/ha, depending on soil quality. Since this exceeds gross margins only on rather marginal soils, participation was below expectations throughout the country. For 1987 the programme has been adjusted and the premium now ranges from 1,000 to 1,800 DM/ha. There has not been any evaluation of the programme so far. However, it can already be said that ecological benefits, such as the support of endangered species, will be small as long as the programme works on an annual basis and allows rotation of fallow fields instead of stationary fallow over a number of years.

Finally, there are an increasing number of private organisations concerned with the protection of birds or other aspects of nature, fishing clubs, etc., which have concluded contracts with farmers related to certain extensification measures.

The institutional framework: some central issues

The apparent incongruity of a critical state of the environment on the one hand and a comprehensive set of environment-oriented laws, programmes and policies on the other needs some explanation. One important consideration is the dynamic relationship between pressure on the environment, social awareness and policy response. Concern with regard to many of the environmental problems is of only recent origin, either because pollution, particularly nitrate contamination, has been growing or because of the rapid disappearance of species, or because the priorities and the awareness of the urgency of environmental problems associated with aspects of landscape maintenance, preservation of species, etc., has grown. Moreover, legislation and policy implementation are retarded owing to the time-consuming process of public opinion formation and decision-making as well as the resistence of various interest groups.

Farmers form one of the most significant interest groups, and their representatives have repeatedly expressed concern about the income losses which would result from the various ecological restrictions. Hence they have been rejecting a strict application of the 'polluter pays' principle in so far as agricultural practices are concerned.

Another explanation of the apparent lack of environmental protection may have to do with the unsuitable distribution of political and bureaucratic competence. Until recently the Ministry of Agriculture had the political responsibility for both general agricultural policies and for policies related to the protection of nature and, partly, for water resources. Conflicts of interest between agricultural production and environment tended to be resolved in favour of producers' interests. In 1986 the responsibility for nature conservation and the environment was turned over to the newly established Ministry of the Environment, Nature Protection and Nuclear Safety. Experience with regard to the new mechanism of co-ordination between the two Ministries is too limited, however, for any conclusions to be drawn.

A third clue to the incongruity between the state of the environment and the availability of legislation are shortcomings in administrative execution of laws and policy programmes. Possible reasons for this cannot be discussed in detail here. One reason seems to be a lack of well defined environmental targets. Another is the lack of sufficient monitoring and evaluation of the state of the environment. Unless the regional state of soils and water resources, of species development, erosion and food quality is recorded on a regular basis, published and in so far as possible related to agricultural practices, the bureaucracy will hardly be in a position to identify the sources of environmental damage and apply relevant policies.

Another important issue is the impreciseness of the legal term 'orderly agriculture'. As was pointed out earlier, this term has not been officially defined. According to recent High Court decisions, 'orderly agriculture' might be defined in a strictly ecological sense, at least in so far as groundwater protection is concerned. This would imply that it applies only to those practices which meet location-specific ecological norms. Representatives of farmers as well as some policy-makers, however, tend to favour regulations and policies which define orderly agriculture in a much less rigid way which corresponds with real local practice. This would imply scope for more generous compensatory payments in cases of ecology-oriented levies. The debate is currently in full process.

Consequences of policy initiatives and options at a regional level

Selection of study regions

Two study regions with highly intensive crop production were selected in the Federal Republic: for the case of agricultural crop production the Hildesheimer Börde location, in the north-east, and for the case of horticultural and viticultural production the Rhein-Pfalz region in the south-west (fig. 1.1).

Table 1.3 Selected figures to characterise the Hildesheimer Börde and Rhein-Pfalz study regions, 1983

	Hildesheimer Börde	Rhein-Pfalz	Lower Saxony	Rhineland-Palatinate	Federal Republic[a]
Agriculturally used area (AA) of total area	62·2	48·6	58·6	37·3	47·5
Rural natural or quasi-natural biotopes % of AA[b]	0·85	0·41	4·79	0·70	2·36
		(per cent of agriculturally used area)			
Cereals	65	43	47	46	42
Root crops	25	11	10	7	6
Vineyards	0	38	0	9	1
Permanent grassland	8	4	39	31	39
Soil productivity index[c]	68	69	45	48	45
Farm size Average AA (ha)	35·2	7·2	22·3	10·8	15·4
% of farms above 30 ha	45	4	31	11	16
% of farms above DM 30,000 standard farm income[d,e]	147	137	100	104	100
Livestock units per farm	10	1	26	10	18

(a) Average of 237 rural districts.

(b) Moor, heath and fallow (without forest). The base for the percentage includes the agriculturally used area and the biotopes.

(c) Top-quality black soil: 100.

(d) Full-time.

(e) Federal Republic[a] = 100.

Source: Bundesforschungsanstalt für Landeskunde und Raumordnung, Data System Statistisches Bundesamt, Fachserie 3, Reihe 3.1.1 (Bodennutzung 1985).

The Hildesheimer Börde is characterised by the high share of agricultural land in the total area (table 1.3). Cereals and root crops dominate the use of arable land. The share of grassland as well as the livestock density of ten livestock units per farm are comparatively low. Excellent natural conditions and a favourable farm structure result in good production conditions for agriculture and a relatively high income potential.

The Rhein-Pfalz is characterised by intensive horticultural and viticultural production. Animal husbandry hardly plays a role. The agricultural production conditions and income potential are not as good as in the Hildesheimer Börde, owing to less favourable farm structure. Nevertheless the income potential is much higher than the Federal Republic average.

Method of analysis

The assessment of environmental problems and related policies in the two regions does not follow a single methodology, but uses a flexible approach, depending on the types of problems and the availability of data. Production systems are described on the basis of resource endownments and input–output relationships for a few selected crops on typical farms. Some quantitative measurements are presented both for the state of the environment as well as for the causal relationship between input use and environmental stress. Current policy efforts are described and analysed with regard to their effectiveness as well as their impact on farm incomes.

Recommendations for further action towards better integration of agricultural and environmental policies are then derived from this analysis. They are derived from quantitative projections of the impact of changes in input use and production systems on the environment, particularly nitrate leaching and the emission of residues from pesticides. A basic assumption throughout the analysis is that farmers try to respond to policy signals in such a way as to keep the resulting losses in income as small as possible. In the case of the northern region (Hildesheimer Börde) this assumption is explicitly embodied in a representative farm model, which is used to project the likely response to various variations of producer and input prices. The data base for the Rhein-Pfalz region did not allow the specification of a similar model.

The final policy strategy is based on the concept of economic efficiency, which assumes that policies which allow the meeting of environmental goals with lowest economic cost should be given priority. Economic costs, in this case, comprise forgone opportunities with regard to the achievement of other policy goals as well as the administrative costs of policy implementation.

The Hildesheimer Börde production system

The economic situation in the Hildesheimer Börde area indicates an unemployment rate of 12 per cent as against an average of 10·5 per cent for the rural districts of the Federal Republic and 14·6 per cent in Lower Saxony, the state to which the region belongs. While the rate of permanent and youth unemployment in Hildesheimer Börde as well as in Lower Saxony is clearly higher than the average for rural districts in the Federal Republic, earnings in industry and gross domestic product are both similar to the average of other regions.

The region is traditionally characterised by a high share of arable land, a low livestock density and a rather high land–labour ratio. In 1984/85 the average net profit per hectare was DM1,411 and per-family labour unit DM71,107.

Cereals and sugar beet are the main crops in the arable area, with winter wheat playing a predominant role and covering 44 per cent of total arable land. During the last ten years (1974/75 to 1984/85), application levels of fertilisers and pesticides have intensified considerably. The most important developments were as follows:

1   The volume of fertiliser inputs, measured as deflated cash expenditure, rose from 315 to 343 DM/ha and for pesticides from 136 to 291 DM/ha respectively.
2   Cereal yields rose by 30 per cent, sugar beet yields by 10·5 per cent.
3   Field sizes increased considerably.

Owing to rather high population density and intensive crop production there are few quasi-natural areas left in the Hildesheimer Börde landscape. Forests are located in only three small hilly areas. According to a recent survey, the share of areas classified as valuable for nature protection is only half as high in Hildesheimer Börde as in the state of Lower Saxony overall.

The most important agricultural practices contributing to environmental damage include chemical plant protection, over-fertilisation, soil compaction and erosion. High application rates of plant protectives, herbicides in particular, have reduced the diversity of species in flora and fauna below the levels of other regions (table 1.4). High fertiliser levels are another cause of environmental stress. According to a recent survey in the study region, the total input for mineral and organic fertilisers exceeds in many cases the crop uptake. The residue is either leaching into the ground water (nitrates in particular) or running off into surface waters (phosphates in particular). The quoted excess nitrogen application of 55 kg/ha is still significantly lower than in other studies, which have estimated excess nitrogen application in this region of above 100 kg/ha.

The quality of drinking water in the study region was investigated by Walther (1982). The nitrate concentration in ground water ranges from 20 to 90 mg NO_3/l. In 41 per cent of the investigated samples, most of them in arable areas, there has been a significant increase of nitrate contamination. Trend calculations for wells in arable areas indicate an increase of nitrate content of 0·44 to 1·86 mg NO_3/l per annum (Walther, 1982).

Evaluating these trends, it should be noted that nitrate contamination of ground-water resources in the Hildesheimer Börde region is not yet a severe environmental problem, at least as far as the tolerance level of 50 mg/l nitrate

Table 1.4 Comparison of pesticide use, yield and gross margin between conventional plant protection and integrated pest management, Hildesheimer Börde (3 years, average of 32 trials)

	Conventional plant protection system	Changes by integrated pest management (%)	
		Absolute	Relative
Costs of pesticides (DM/ha)			
Herbicides	141	−70	−50
Fungicides	150	−32	−21
Insecticides	40	−17	−42
Growth regulators	21	−6	−29
Total	352	−125	−36
Yield (dt/ha)			
Grain	83·7	−2·5	−3·0
Sugar beet	548·5	±0	±0
Adjusted gross margin[a] (DM/ha)	4,103	+35·9	+0·9

(a) Revenue less costs for plant protection, cleaning and drying.

Source: Wahmhoff and Heitefuss, unpublished, personal communications

in drinking water is concerned. Nevertheless, considering the time lag between the nitrate leaching and the pumping at the waterworks, there exists a clear danger of continuing increase of nitrates beyond tolerable levels of concentration.

Another form of environmental damage in the region, compaction, results from the regular use of heavy machinery. As a result there may be a decline in yields. Finally, there is also a problem of progressive erosion due to the expansion of sugar beet in the region. However, the resulting soil losses are of minor importance because slopes are not typical of this region.

The Rhein-Pfalz production system

Most of the indicators describing the economic structure and the labour market of the Rhein-Pfalz region are similar to those of other rural districts. Yet the overall income levels are below average, and the gross domestic product lies 18·2 per cent below the national average of rural areas.

The typical farm in Rhein-Pfalz is characterised by a high share of land under permanent crops, particularly wine and fruit, and small size.

Compared to the Hildesheimer Börde region, the net profit per hectare is quite high, at DM3,786 per ha, owing to the importance of highly profitable fruit, wine and vegetables. Yet because of an unfavourable farm structure, the net profit per family worker is less than half as high as in the Hildesheimer Börde and only DM33,684 per family labour unit.

The share of grassland as well as the livestock density is extremely low. As in the Hildesheimer Börde, the arable area of annual crops is dominated by cereals and sugar beet, which are produced with similar technologies. The input of fertilisers has declined over the past five years, whereas the intensity of pesticide use has increased.

Besides fruit, cereals and sugar beet the most important crops are wine and vegetables. On a typical viticultural farm in the Rhein-Pfalz 60 per cent of the wine is marketed in casks and 40 per cent in bottles. It should be pointed out that the wine production is very labour-intensive. Except for potassium, intensities of mineral fertiliser generally do not exceed those in cereal production. However, total nutrient inputs often exceed mineral fertiliser considerably, owing to the intensive use of organic fertilisers. Organic fertilisers in this case do not only comprise animal manure but also wine residues and garbage compost. Application rates of chemicals for plant protection are significantly higher than in cereal production.

In contrast, on a specialised horticultural farm in the region 90 per cent of the arable land is usually used for vegetables. Fertiliser application rates are extremely high, and for cauliflower are typically in the vicinity of 250 kg of N, 70 kg of P_2O_5, 350 kg of K_2O and 60 kg of MgO per hectare.

As in the Hildesheimer Börde the decline of species in the Rhein-Pfalz has been substantial. This is due not only to the intensive use of plant protectives but also to various landscape management activities, including drainage, land consolidation and increased nutrient content of soils. A decline of species occurs on agricultural land as well as on the small biotopes sprinkled around the landscape. Such biotopes along boundary strips, walls of vineyards and steps of terraces carry tree and shrub vegetation and are very important for the preservation of many endangered species. Their area has been steadily declining, especially owing to land consolidation measures, which have already covered nearly 50 per cent of the total vineyard area. Out of 583 endangered plants seventy-one depend on extensive agricultural land use (Korneck et al., 1986). Of course, the wildlife suffers as well, since it depends on some of the disappearing plant species.

A specific problem of the wine growing areas is the drastic increase in the copper content of the soils. The extremely high concentration of up to 2,500 mg/kg is nearly a hundred-fold higher than average soil conditions. Reasons for this contamination may be: intensive use of plant protectives with high copper content, use of copper sulphate for impregnation of stakes, as well as use of compost and manure with high copper content. While copper contamination seems not to affect the yields it has a detrimental impact on various other plant species.

The high intensity of nitrogen fertilisation leads to nitrate contamination of ground water and, for some nitrophile vegetables, to nitrate contamination of foodstuff as well.

Information about fertiliser levels and ground-water contamination is scant for the Rhein-Pfalz region. However, far more data are available for the neighbouring Mosel region, where the recommended fertiliser levels range from 80 to 240 kg N/ha, while actual fertiliser applications are usually in the vicinity of 250 to 300 kg N/ha, not taking into account the additional use of organic fertilisers produced from animal waste (Müller, 1982). Fertiliser levels are lower in the Rhein-Pfalz, yet one cannot exclude the possibility that relative levels in excess of norms may be similar to the Mosel region. Generally, overfertilisation in wine-growing areas leads to much higher leaching than to normal cropping areas owing to:

1   Short period of vegetation growth.
2   Small number of plants per unit area.
3   Low rate of nitrogen uptake associated with intensive nitrogen dynamics.
4   High rate of water leaching.
5   High potential for nitrogen mineralisation.

The nitrate leaching of vineyards in the Mosel area depends significantly on the type of soil. For instance, a nitrogen fertilisation of 225 kg N/ha was reported to result in nitrate concentration in leached water ranging from 200 mg to 700 mg NO_3/l (Müller, 1982). Investigations along the valley of the Mosel river found that 70 per cent of all analysed wells exceeded the nitrate limit of 50 mg NO_3/l, 40 per cent of the well water had a nitrate concentration of more than 100 mg/l, and 10 per cent reached a concentration of more than 200 mg NO_3/l.

The high intensity of nitrogen fertiliser in horticultural production endangers ground-water resources as well. Over-fertilisation is not unusual in horticultural production (Wehrmann, 1984). Other reasons for the high leaching potential are (Wehrmann and Scharpf, 1983, 1985b):

1   Considerable residual mineral nitrogen in soil at date of harvesting.
2   Voluminous, rapidly decomposable harvest residues containing nitrogen.
3   Cultivation of species with flat roots.

The high nitrate concentration in the pumped water of the Bruchsal waterworks, which is located within the study region, is mainly caused by the cultivation of vegetables, tobacco and asparagus, which require high fertiliser input on sandy soils (Rohmann and Sontheimer, 1985). In the case of vegetables (green and red cabbage and spinach) a nitrate concentration in leached water of 310 mg NO_3/l for a fertiliser level of 300 kg N/ha on sandy soils was measured.

The nitrate contamination of vegetables is another acuse of concern. It varies considerably according to the following factors (Wehrmann and Scharpf, 1985b):

1   Species, part of plant and variety.
2   Nitrate nutrition of the plant.
3   Exposure, location and season.

Impact of agricultural and environmental protection policies on production systems and the environment

It is impossible to provide a complete assessment of the environmental gains which can be expected from current policy efforts in the region as well as from potential policy changes. Yet there are various sources of information which provide an overview of activities and show the need for further action in the two study regions. The discussion will concentrate on current policy efforts first and then the estimated impact of various instruments within a hierarchy of policy measures. The Hildesheimer Börde region will be discussed in more detail, mainly because it clearly represents a larger area of intensive agriculture than the more specialised horticulture and viticulture of the Rhein-Pfalz.

Hildesheimer Börde
Current policy efforts may be subdivided into:

1   Research and extension to avoid over-use of fertiliser and plant protectives.
2   Subsidisation of unsprayed crop edges.
3   Measures to protect small biotopes and boundary strips of field.
4   Green fallow pilot programme.
5   Water protection.

Table 1.5   Effects of intensive and reduced pesticide use in sugar beet on the number of trapped arthropods in the soil surface (Hildesheimer Börde, 6 fields)

Treatment	Number of trapped arthropods per field[a]			
	Carabides		Spiders	
	1984	1985	1984	1985
1 Conventional pesticide use	1,517	2,026	3,982	2,922
2 Reduced pesticide use (−30%)	1,975	2,297	4,110	4,327
Differences:				
Absolute	+458	+271	+128	+1,405
Relative (%)	+30	+13	+3	+48

(a) Ten pit traps per field and treatment.
Source: Wahmhoff and Heitefuss, unpublished.

There has been a wide range of research efforts aimed at a reduction of fertiliser and chemical application. As regards the reduction of plant protectives, the most important recent improvement has certainly been the identification of economic thresholds for chemical treatments against weeds and aphids in grain production, and various pests in rape and sugar-beet production. Integrated pest management practices have been tested. They not only enable considerable savings in inputs and improvements of gross margins per hectare (table 1.4), but they also favour the survival of those species which are not the target of the respective treatment. An example for ground-dwelling arthropods under conventional and reduced pesticide use in surface beet is shown in table 1.5. The official extension service is gradually introducing integrated pest management practices to the farmers of the region. Unfortunately, these practices are not very successful against certain fungus diseases in wheat, one of the main cereals in the region.

There are several methods available to adjust fertilisation to crop demand and to avoid over-fertilisation (Buchner and Sturm, 1985; Wehrmann and Scharpf, 1985a). One example is the so-called Nmin method, which accounts for the reserve of mineral nitrogen in the soil profile as part of the first application of nitrogenous fertiliser at the beginning of the vegetation period. The method has been part of the extension package given to farmers in the study region. It is expected to appeal to farmers because it implies a reduction in the cost of fertiliser (Wehrmann and Scharpf, 1985a).

In practice the Nmin method seems not to be very attractive to cereal producers because it allows only slight nitrogen savings. Yet it leads to one to three dry tonnes per hectare higher sugar yields, and requires about 40 kg/ha less nitrogenous fertiliser than current practice in the case of sugar beet. Being concerned with good beet quality, the sugar-beet factories have been promoting the implementation of the Nmin method in agriculture for some time (Feyerabend, 1985). However, in the Hildesheimer Börde region only one-third of all farmers apply the Nmin method and nearly 50 per cent of these do not act in accordance with the recommendations for reduced fertilisation (Kuhlmann and Wehrmann, 1982; Finck, 1987).

The new unsprayed crop-edge programme by which the government subsidises those farmers who leave crop edges unsprayed is of minor importance, because most of the wild species which the programme intends to protect can no longer be found in the Hildesheimer Börde region. Therefore only some hilly areas are included in this programme.

A series of new laws and local ordinances to protect small biotopes and boundary strips have been implemented in recent years and are of particular importance for the region. These ordinances:

1   Prohibit the use of pesticides on non-agricultural lands.
2   Have amended the land consolidation law to include 'nature conservation' as a parallel goal of land consolidation programmes. (So far the impact of this amendment is not yet clearly visible and it needs further reinforcement).
3   Prohibit the conversion of forest land into agricultural land.

The green fallow programme has not been very attractive to most farmers in Lower Saxony, at least in its first year of implementation. While those farmers who participated registered for an average of five hectares per farm, corresponding to approximately 15 per cent of their land area, the overall coverage was only 1·2 per cent of the agriculturally used area. The percentage was only slightly lower in the Hildesheimer Börde (1·0 per cent). Thus, considering the short duration of a one-year fallow and the low area coverage, particularly within the intensive core areas of the region, the ecological effect of the green fallow pilot programme was certainly negligible.

As elsewhere in Germany, to protect water supplies the authorities have recently been pursuing a policy of expanding water catchment areas. By 1986 15·6 per cent of the overall area was officially delineated for water catchment (Lower Saxony: 9 per cent). Although the Water Protection Act provides for strict control of agricultural practices within these areas, only small zones close to the wells are subject to significant restrictions so far. Since the region has a comparatively low livestock density, the Liquid Manure Order in effect in Lower Saxony does not affect most farms, either. Nevertheless, some farms with a small land area and sizeable chicken or pig production are affected by the prohibition on the spraying of liquid manure during the winter season. Some of them have had to expand the storage capacity for liquid manure and, to this end, received a 20 per cent subsidy on their investment.

NEED FOR FURTHER ACTION

A tentative conclusion from the analysis of prevailing production systems and their environmental effects in the Hildesheimer Börde is that current efforts may not be sufficient to achieve a long-term stabilisation of the eco-systems. Further action may be needed. Such action has to be implemented in the most efficient way and therefore policy measures which combine improvement of the environment with the achievement of other policy goals at low economic cost should have priority. Top priority from this point of view should be a continuation of research and extension aiming at a decline of over-intensive, non-profitable use of fertilisers and pesticides. There is still scope for widespread improvement in on-farm research and extension services in this field. This should include better and more systematic information and the training of farmers in all matters of environmental protection.

Parallel to such extension campaigns, it will be necessary to enforce the existing legislation for more consistent preservation of the remaining small biotopes in the region.

The basic approach for a better integration of agricultural and environmental policies should be to reduce emphasis on those agricultural policies at the national and European Community level, price policies in particular, which stimulate market surpluses and at the same time tend to increase environmental stress through intensification and specialisation. If such policy reforms were to be implemented they would certainly apply to the country or even the Community as a whole, and not just to the study region. The most important policy change in this context would be to replace price support policies by other forms of support, such as direct income transfers in combination with payments for structural adjustment of farms and with various forms of environmental protection activities.

CONSEQUENCES OF PRICE POLICY MEASURES FOR FARM ORGANISATION AND THE ENVIRONMENT

While changes in output or input prices would affect agricultural production on a national scale (with only limited possibilities of adjusting input taxes according to regional differences in environmental priorities), there may be a specific need to apply additional levies or issue ordinances at the regional or local level. The most important example in this context is the expansion of water catchment areas and the stricter application of the federal water protection law, which allows the possibility of enforcing permanent green cover of arable land, of prohibiting the spraying of liquid manure, etc.

Table 1.6 Impact of reduced nitrogen fertiliser application on nitrate concentration in leaching water and related relative change in income[a]: short-run impact

	Average nitrogen fertiliser (kg/ha)			
	160[a]	140	120	100
Additional N supply from legumes (peas) (kg N/ha)	–	–	13	18
Nitrate concentration in leaching water (mg NO_3/l)	46·3	43·3	42·0	38·5
Relative net profit (DM/ha)	100	99·9	99·4	98·5

(a) Reference situation.
Source: Finck (1987).

To the extent that a liberalisation of agricultural price policies is politically unacceptable, or is inadequate to achieve overall environmental objectives, the next step would be to tax those inputs which cause particular environmental problems. It is for this reason that simulation runs are presented for a typical Hildesheimer Börde region in table 1.6 which assume a tax on nitrogen fertiliser and a tax on pesticides. The calculations are based on a farm programming model that provides for the possibility of adjusting both the land use pattern and the intensity of fertiliser and pesticide inputs to various policy measures. Yield response to inputs is derived from production function estimates which take into account the impact of the nutrient content in the soil on production. The relationship between fertiliser, yield and nitrate leaching into ground water is estimated from a comprehensive survey of related literature which is described above (Finck, 1987; Ohlhoff, 1987).

Table 1.7 Impact of reduced product prices on farm organisation, farm income and intensity of fertiliser and chemicals, Hildesheimer Börde

	Reference situation	Reduction in product price		
		10%	20%	30%
Cropping system (% of AA)				
Winter wheat	35·1	35·1	36·0	35·0
Winter barley	29·9	29·9	29·0	25·0
Winter rye				5·0
Sugar beet	35·0	35·0	35·0	35·0
Broad and field beans				
Cereals	65·0	65·0	65·0	65·0
Degree of specialisation[a]	1,117	1,117	1,121	775
Fertiliser (kg/ha)				
Winter wheat	165	165	165	165
Winter barley	165	165	165	165
Winter rye				125
Sugar beet	180	180	180	180
Broad and field beans				
Average of rotation (kg N/ha)	170	170	170	168
Rel.	100	100	100	99
Chemicals (DM/ha)				
Winter wheat	378	378	378	378
Winter barley	269	269	269	269
Winter rye				185
Sugar beet	523	523	523	523
Broad and field beans				
Average of rotation (DM/ha)				
Herbicides	233	223	223	223
Fungicides	110	110	110	106
Growth regulators	38	38	38	37
Insecticides	26	26	26	26
Plant protection (DM/ha)	397	397	397	392
Rel.	100	100	100	99
Yields (dt/ha)				
Winter wheat	64·9	64·8	64·8	64·8
Winter barley	59·6	59·6	59·6	59·6
Winter rye				52·0
Sugar beet	498	498	498	498
Broad and field beans				
Labour requirement MEU/farm	1,411	1,411	1,411	1,407
Expected net profit				
DM/farm	95,770	62,033	30,296	−1,505
DM/ha	1,197	775	379	−19
Rel.	100	66·2	32·2	–
Standard deviation net profit (DM/farm)	22,186	19,968	17,749	15,381

(a) The degree of specialisation, S, is defined:

$$S = \frac{1}{n} \sum f_i^2$$ where f = share of cultivated crops in arable land, and n = number of cultivated crops.

Source: Ohlhoff (1987).

The initial application of the model relates to the impact of the prevailing production system on the estimated nitrate concentration in ground water. It is based on short-run production functions. According to the results reported in table 1.6, the current system leads to an estimated nitrate concentration of 46·3 mg NO_3/l, which is below the tolerance levels but should cause concern, especially from a long-term perspective.

A reduction of the fertiliser level to 100 kg N/ha would reduce the estimated concentration to 38·5 mg NO_3/l. This reduction is comparatively small, because the model realistically assumes that farmers replace some mineral nitrogen by increasing nitrogen supplies from legume production. As mineral fertiliser is reduced to 120 kg N/ha some of the cereals are replaced by field peas, which also fix nitrogen. It is interesting to note that this relief would be accompanied by only a rather modest decline in income, which can be explained by the low marginal productivity of nitrogenous fertiliser, at least in the short run. Long-run income losses would certainly be higher, because the available stock of nitrogen in the soil would be gradually mineralised but not fully replaced at the lower level of fertiliser.

The estimated impact of a reduction of product prices on input levels and production patterns is shown in table 1.7. The simulated rates of price reduction are 10, 20 and 30 per cent respectively, whereby the latter rate would bring producer prices close to world market levels. Since fertiliser and pesticides are highly profitable at this location, a reducation of prices would not reduce optimal input levels very much, and hence one could not expect significant environmental relief in terms of reduced input levels. However, some relief could result from changes towards a more diversified production pattern. Except for the scenario with a 30 per cent price reduction, agriculture, as it is currently practised, would remain profitable under average conditions. Yet farm incomes would drop drastically. A 30 per cent price reduction would even result in slightly negative profits for the average farm. Hence one may conclude that price reductions by up to 30 per cent would probably have only minor effects on land use intensity but would result in a more or less significant decline of farm incomes. This would certainly imply further structural change and a more rapid growth of farm sizes.

The consequences of a tax on mineral nitrogen for farm organisation, nitrate leaching and farm income are shown in table 1.8. According to the calculations, increased prices for nitrogen lead to a significant reduction of nitrogen input from commercial fertilisers whereas nitrogen supplies from legume production and crop residues increase. At a price level of DM4·50 per kilogram of nitrogen the nitrogen fertiliser intensity is reduced to 48·2 kg/ha, compared to 151·8 kg/ha at the initial price of 1·50 DM/kg. This does not include the supply of an additional 25·6 kg/ha of nitrogen from farm residues and legume production. At this price level nitrate leaching would be diminished by 42 per cent as against the reference situation. Income losses would amount to 24·8 per cent and hence would be significantly lower than those resulting from a comparable rate of reduction in nitrate leaching achieved via reduced product prices.

Compared with a reduction of price support and a tax on nitrogen, the taxation of chemicals shows the largest environmental relief (tables 1.7

Table 1.8 Taxation of nitrogen and chemicals: impact on nitrate concentration, degree of specialisation, farm incomes, farm organisation and intensity of crop enterprises, Hildesheimer Börde

	Reference situation	Increase in price (%)		
		100	200	300
A. Nitrogen				
Price of Nitrogen (DM/kg)	1·50	3·00	4·50	
Gross margin (DM/ha)	2,225	2,061	1,950	
Rel.	100	92·5	88·9	
Net profit				
DM/farm	69,510	57,960	52,290	
DM/ha	993	828	747	
Rel.	100	83·4	75·2	
Cropping system (% of AA)				
Winter wheat	47	32	32	
Winter barley	32	21	21	
Sugar beet	21	22	22	
Field peas	–	25	25	
Fertiliser intensity (kg N/ha)				
Winter wheat	165	140	110	
Winter barley	140	110	100	
Sugar beet	140	80	80	
Average of fertilisation (kg N/ha)				
Commercial fertiliser	151·8	59·9	48·2	
Crop residues	–	25·6	25·6	
NO_3 concentration in leaching water (mg NO_3/l)	46·3	30·6	26·8	
B. Chemicals				
Cropping system (% of AA)				
Winter wheat	35·1	35·0	27·9	27·9
Winter barley	29·9	25·0	24·4	24·4
Winter rye		0·8		
Sugar beet	35·0	35·0	35·0	35·0
Broad and field beans		4·2	12·7	12·7
Cereals	65·0	60·8	52·3	52·3
Degree of specialisation[a]	1,117	619	690	690
Fertiliser[b] (kg N/ha)				
Winter wheat	165/0	165/80	105/80	105/80
Winter barley	165/0	135/80	135/80	135/80
Winter rye		58/0		
Sugar beet	180	176	167	167
Broad and field beans	–	–	–	–

Table 8 – *continued*

	Reference situation	Increase in price (%)		
		100	200	300
Average of rotation				
(kg N/ha)	170	153	120	120
Rel.	100	90	71	71
Chemicals (DM/ha)				
Winter wheat	378	283	80	80
Winter barley	269	131	131	131
Winter rye		91		
Sugar beet	523	53	53	53
Average of rotation (DM/ha)				
Herbicides	233	37	31	31
Fungicides	110	54	–	–
Growth regulators	38	35	17	17
Insecticides	26	26	24	24
Plant protection (DM/ha)	397	152	72	72
Rel.	100	38	18	18
Labour requirement MEU/farm	1,411	1,551	1,528	1,528
Expected value net profit				
DM/farm	95,770	74,765	70,762	64,902
DM/ha	1,197	935	885	811
Rel.	100	79·7	75·5	69·2
Standard deviation net profit				
DM/farm	22,186	24,823	29,381	29,381

(a) The degree of specialisation, S, is defined:

$$S = \frac{1}{n} \sum f_i^2$$

where f = share of cultivated crops in arable land, and n = number of cultivated crops.
(b) N fertiliser $CaCN_2$.
Source: A. Finck (1987). B. Ohlhoff (1987).

and 1.8). In the case of sugar beet a 50 per cent price increase for chemicals leads to the substitution of more labour-intensive mechanical weeding for herbicides and reduces the optimum level of nitrogen fertiliser. The optimal cropping system tends to become more diversified at chemical prices 100 per cent or more above initial levels.

Rhein-Pfalz
Many of the recent policy initiatives mentioned for the Hildesheimer Börde have also been taken in the Rhein-Pfalz region. Extensive areas of the region

are covered by grain and sugar-beet production systems with similar production conditions to those in the northern study region. Hence it would be
repetitive to discuss such on-going activities as the introduction of integrated
plant protection systems or efforts to reduce excessive fertiliser levels on
grains and sugar beet or to analyse the likely effects of product and input
price policies on these systems again. The following analysis will rather focus
on the particular problems of the vegetable and wine-producing areas of the
region.

CURRENT POLICY EFFORTS

Efforts to reduce an over-use of nitrogen fertiliser are not only aimed at a
reduction of nitrate leaching into the ground water. In the case of vegetables
it is also the nitrate content of green vegetables, green salad and spinach in
particular which is subject to growing concern. The nitrate content of
spinach increases significantly as total nitrogen supplies are raised above 200
kg/ha, whereas the yields increase only slightly at this level of fertiliser (fig.
1.4). Moreover, a high nitrate content has been observed during periods of
low light intensity and in plants with a high proportion of stems. Research
and extension have recognised these relationships and recommend that:

1   Farmers should never apply more fertiliser than is required for maximum
    yields.
2   Farmers should avoid production and harvesting during periods of low
    light intensity.
3   Plant scientists should breed for reduced shares of stems.
4   Attempts should be made to develop a preference for plant varieties with
    low nitrate content.

Recommendations for application rates of fertiliser on vineyards have also
been reduced recently, the argument being that the marginal productivity of
nitrogen with regard to quantities harvested is lower than previously
assumed. Yet some of the recommendations for wine growing are still above
the estimated take-up of nutrients. Moreover, application rates continue to
exceed the recommendations in many cases.

Various research projects in the region aim at a reduction of the volume of
chemical plant protectives used. First results of trials on integrated pest
management in apple growing show that many applications of pesticides can
be avoided without a significant loss of profit while resulting in a considerable
preservation of beneficial and non-production-damaging species of animals
and in a reduction of residues in harvested produce (table 1.9). As a result,
the extension service is deliberately trying to introduce integrated pest
management methods into practical farming.

Like other *Länder*, the government of Rheinland-Pfalz implemented a
crop edge programme in 1984. By 1986 the total length of crop edges on
cereal fields (width 3 m) left untreated with herbicides amounted to 50 km.
The result was that out of the thirty-two wild plant species which are
registered as endangered on cereal fields in the region, 30 per cent were found
on the untreated crop edges, whereas only 18 per cent were found on the

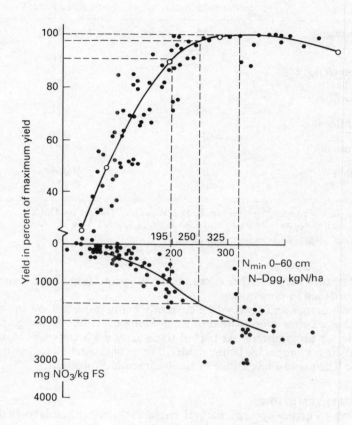

Figure 1.4 The impact of the supply of nitrogen on the yield and nitrate content of spinach, spring 1980 and 1981. *Source:* quoted in Wehrmann and Scharpf (1985b)

Table 1.9 Comparison between conventional and integrated plant protection in apple orchards, Rhain-Pfalz (1985)

Plant protection system	Integrated pest management	Conventional pest management
Number of treatments	15	24
Number of different pesticides	6	12
Sum of plant protection expense[a] (DM/ha)	1,562	1,826
Number of beneficial arthropods	348	104
Number of indifferent arthropods	2,071	1,061
Pesticide residues in apples[b] (mg/kg)		
Parathion-äthyl	n.a.	0·006
Captan	0·034	0·108

(a) Including the costs of planning ('conventional', 57 DM/ha; IPM system, 266 DM/ha).
(b) Maximum amount (food law): Parathion-äthyl, 0·5 mg/kg; Captan, 1·5 mg/kg.
Sources: Beicht *et al.* (1986); Krauthausen *et al.* (1986).

main fields (Oesau, 1986). Further measures for the protection of the special flora in vineyards are in preparation.

Various programmes and surveys are currently being implemented in the region, aimed at a better preservation of small biotopes within the on-going land consolidation programmes. As part of these activities a complete registration of existing biotopes is being made. The programme is under the control of the Rheinland-Pfalz Office of Environmental Protection.

NEED FOR FURTHER ACTION
Further efforts to attain optimal natural balances in the Rheinland-Pfalz region are urgent and should include the adjustment of macroeconomic conditions, i.e. the gradual elimination of price distortions along the lines analysed for the Hildesheimer Börde. Yet reforms of market policies or general input taxes may not be sufficient to solve some of the aforementioned specific local problems of the region.

Solutions to the problem of nitrate leaching into ground water will have to be found by strategies similar to those in other regions, i.e. expansion of water catchment areas and stricter application of measures prescribed in the Water Protection Act, especially where intensive vegetable and fruit production occurs in such areas.

Further action seems necessary to assure that nitrates in green vegetables stay within acceptable limits. As opposed to, for example, Switzerland and the Netherlands, legislation in Germany does not prescribe upper limits for nitrate content, except for baby and diet food. One option is to encourage the use of special contracts between producers and freezing companies with regard to the control of food quality through special practices and low input levels. There is a need also for consumer information campaigns aiming at more concern about quality rather than the mere exterior appearance of food.

The programmes for preservation of biotopes, which are currently being pursued, should be reinforced and continued with a view to establishing a network of interconnected biotopes, including crop edges, rather than the preservation of unconnected small sites. A reconsideration of on-going land consolidation programmes, at least to reflect the balanced recognition of the need for environmental restrictions, should have high priority. Integrated pest management should be the subject of further research, and the training of extension staff and farmers should be pursued.

Impact of policy options on other regions
The discussion in this chapter has, so far, concentrated on two regions with intensive crop production. For a better judgement of those policy measures the implementation of which would be on a national, rather than a strictly regional, scale, effects on other types of region are briefly discussed below. This applies to a reduction of product prices, taxation and quotas on inputs and nationwide set-aside programmes.

A reduction of product prices would lead to a drastic reduction of incomes in crop production systems. While the model results presented in table 1.7 indicate that this still might not lead to major changes of production patterns in high-fertility areas, changes might be more significant in less favoured areas. A trend towards more extensive land use systems and even abandonment of farms on marginal locations cannot be excluded. In any case, a more rapid reduction of product prices would accelerate the process of structural change. Income losses due to reduced product prices would be considerably smaller in regions with grain-based livestock production, because the prices of their major inputs would decline as well. Thus, for grain-based livestock regions, lower prices would most likely not be an appropriate measure to reinforce an improvement of the environment associated with reduced livestock numbers, diversification of land use and lowering of input intensities.

Taxation of nitrogen fertiliser would have to be drastic if it were to bring fertiliser levels in intensive crop production down significantly. However, there are remarkable variations between regions (Schulte, 1984). Unfortunately, there exists no feasible approach to taxing the nitrate emissions directly. The only practicable base of taxation may be the total nitrogen input. Thus the implementation of the 'polluter pays' principle would require an exemption of certain basic quantities from taxation. A related proposal was made by the Council of Environmental Advisers. Since the ratio of nitrogen use and emissions through leaching varies between locations, the exemption would even have to be region-specific. Moreover the taxation would have to include the use of organic fertilisers as well. Otherwise, crop production systems would be discriminated against through a tax on nitrogen, although the severest problems occur in regions with high livestock density. A tax based on excessive quantities of total nitrogen applied per hectare would have several effects: higher profitability of measures to improve the effectiveness of animal manure (e.g. storage), larger radius of profitable transport to areas with low livestock density, and reduced optimum levels of application.

A tax on plant protectives would result in a rather drastic reduction of input levels in the intensive crop production areas, assuming the same rates of taxation as calculated for nitrogen fertiliser. The response might be less elastic in other locations where the marginal productivity of pesticides is much higher owing to lower current levels of application and/or higher risks (Schulte, 1984; Ohlhoff, 1987).

There are only small local differentiations regarding the consequences of a quota on nitrogen fertilisers. As a result of the low input of commercial fertilisers, farms with intensive livestock production would be the least affected. They might even realise positive income effects due to a rising demand for organic fertilisers.

Calculations from the farm programming models as well as experience from the green fallow pilot programme show that there is little incentive for farmers to join land diversion or set-aside programmes in intensive crop production areas unless there is distinct regional differentiation in the premium system. Without such a differentiation it has to be expected that owners of marginal land would be the first to participate in land set-aside programmes, and this would lead to a concentration of fallow land in regions less favourable to crop production. Moreover the environmental benefits of land diversion and set-aside programmes, which are primarily designed to reduce agricultural surplus production, are doubtful unless they include provision for the pursuit of environmental targets. Short-term fallowing, for instance, yields no benefits for the achievement of nature protection goals and may even lead to greater nutrient leaching.

Finally, it should be pointed out that the selection of different policy instruments has to consider not only criteria like economic efficiency, ecological effectiveness, impact on farm incomes, political implementation or administrative applicability, but also the specific characteristics of the environmental problem. Impacts on other external effects have to be regarded, too.

Furthermore, the spatial extent of environmental nuisances is important. While price policies (e.g. input taxes) seen advantageous in those cases where the environmental problems go beyond regional levels, levies, ordinances or specific forms of collective bargaining are preferable to solve problems on a regional or local scale. The more it is necessary to regionalise the policy instruments the higher are the information and administrative requirements. These additional transaction costs, including the costs of monitoring, administration, bargaining and implementation (Whitby and Hanley, 1986) have to be balanced against the costs of misallocation which would result from the use of more global fiscal instruments.

Assessment of the potential for better integration of environmental and agricultural policies

The relationship between agriculture and the environment is currently at a rather critical stage. The population has become increasingly aware of ecological risks and of various negative impacts on the environment resulting from certain forms of agricultural production. This awareness coincides with increasing pressure on policy-makers for fundamental reforms of agricultural policies aiming at a long-term solution of the tremendous surpluses in various agricultural markets. This coincidence may imply a risk, but also a chance, for environmental policy. A risk is involved, because policy-makers may refrain from certain measures of environmental protection because they fear that the current unsatisfactory income situation on many farms rules out further pressure on agriculture on behalf of the environment. Since past experience has shown in many cases that agricultural interest groups are capable of influencing policies rather effectively, one may be sceptical with regard to the possibilities of implementing necessary environmental policies.

On the other hand, the current situation may also provide an opportunity for environmental policies, because it opens up potential to incorporate environmental goals into a new concept of agricultural policy.

Possibilities for a better integration of agricultural and environmental policies comprise the following strategic elements:

1 Cost reduction at farm level.
2 Abatement of price distortions and market intervention.
3 Agricultural contributions to cost effectiveness in pursuit of environmental goals.
4 Subsidised reduction of production capacity through diversion of land, readjustment of production patterns and extensification.
5 Orderly agricultural land use.

Policy measures which allow a reduction of production costs and hence raise farm incomes, in combination with environmental relief, have the best chance of acceptance by the farmers. Various current research and extension initiatives go in this direction. Promising approaches have been developed for the various components of production systems, including fertilisation according to plant up-take, integrated pest management in plant production, reduced or zero tillage practices, efficient feeding (e.g. avoiding over-supply of protein feed), etc. The available experience shows clearly that a better targeting of inputs, and of agro-chemicals in particular, enables considerable increases of income, reductions in contamination of ground water, the preservation of species and a reduced compaction of soils in many locations.

It will be the task of policy and extension services to improve the state of knowledge, to develop better methods of analysis and prognosis and to introduce related information into vocational training curricula. Many farmers still seem reluctant to apply methods of integrated pest management and better targeting of fertiliser levels because they are risk-averse and fear income loss in the event of unfavourable weather conditions or should unusual infections occur. Yet it has been shown in recent empirical studies that an excessive use of inputs is costly and inappropriate as part of a strategy of risk avoidance. To overcome hesitation or resistance to the more rational use of inputs consideration should be given to the introduction of insurance schemes for pest infections, which would cover such risks for those farmers who renounce the use of certain inputs or practices in a controllable way. In summary, the possibilities of combining a more effective use of inputs with environmental protection provide considerable scope for further applied research, and the potential for cost reduction is remarkable in this respect.

It was indicated earlier that there are currently good prospects for the simultaneous reform of agricultural policy and a better achievement of environmental goals. A gradual elimination of price distortions and a reduction of market intervention would improve the balance of supply and demand in agricultural markets. Moreover, lower product prices would reduce incentives for highly intensified and specialised production systems and thus contribute to environmental protection.

This study has shown that the environmental relief which would result from reduced prices may vary considerably between regions. Assuming typical production conditions and profit-maximising behaviour, a reduced product price results in a lower optimal level of agrochemicals. This applies to a given production function and even more in cases where the use of other inputs such as machinery and labour is decreased as well. Higher rates of price decline may even lead to extended fallow in locations where reduced gross margins no longer cover fixed costs. The use of agrochemicals would be reduced to zero in such cases. The response may be quite elastic in locations with low productivity. Farmers in areas of high soil fertility, on the other hand, including those in some of the study regions, may tend to maintain high input intensities in spite of product price reductions.

Empirical estimates tend to show that profit maximisation as a determinant of farmers' decision-making is not an unrealistic assumption for the majority of farms. Yet, parallel to the direct extensification effect of product price reductions, there are other factors which tend to induce continued intensification, in spite of reduced real product prices. One is further intensification on farms which so far have not reached their optimum level of agro-chemical use. The other is on-going technical progress tending to shift the production function upwards towards higher optimal input levels. In the past these two factors have even offset the extensification effect of reduced product prices. However, this may no longer be the case if future price reductions are to be larger than those observed in the past. In total, it is quite likely that reduced price support will lead to a reduced intensity in the use of agro-chemicals, not only due to a reduction of application rates for given crops but also due to changes in land use patterns. Farms on highly productive locations would hardly adjust their production systems, whereas those on less favoured locations can be expected to readjust and extensify their systems of production in order to keep the income losses resulting from lower prices as small as possible.

More extensive land use, including aforestation, will be generally welcomed from an ecological point of view in most locations. There are also locations, however, where an extensive form of land use such as aforestation and particularly the complete abandonment of farming would be undesirable. Examples of this latter undesirable trend are regions with a high share of forestry and a resulting high preference for open landscape. Such regional differences, however, cannot be used as an argument against the elimination of price distortions from an environmental point of view. They can only be used as an argument for additional location-specific measures which may have to go along with a liberalisation of agricultural markets.

A phasing out of market intervention would result in a more or less drastic reduction of farm incomes. Since the support of income has always been the predominant objective of agricultural policy, such negative prospects for farm income associated with a reduction of prices have been the major obstacle to market reforms in the past. So far, agricultural interest groups have successfully resisted alternative options of direct income transfers instead of high producer prices, although such transfers would certainly be more efficient from an economic point of view. More recently it has been pointed out that direct income transfers could even be combined with ecological necessity by the appropriate definition of the conditions under which farmers would be eligible for such transfers. If this is done, then it will be an important political task to assure the farm population that it can rely on direct transfers just as they have relied on transfers via protected prices, although fiscal transfers are more visible in government budgets.

A reduction of production intensity in agriculture is not only desirable because it contributes to a more efficient allocation of resources for market-oriented production. It also contributes to greater cost effectiveness in pursuit of environmental goals. A good example for this hypothesis is the case of nitrate contamination of ground waters. In principle various new technologies exist for eliminating excessive nitrate from drinking water, through diffusion with less contaminated water or by various biological or physical measures. Of course, application of such measures would assure that the European Community standard of 50 mg N/l in drinking water could be met. But the remaining ground water, which may be needed by later generations, would continue to be contaminated. This alone would represent a violation of the principle of precaution in environmental policies. Investment in technology for the elimination of nitrates would also be significantly more costly than the economic value of production forgone which would result from adjustments of agricultural production systems. Empirical estimates which underline the latter statement are presented in table 1.10. According to these results, the economic costs of reduced agricultural production, evaluated at shadow (world market) prices, are always lower than the costs of nitrate elimination, even in locations with light soils and high livestock density which are known to have a high leaching potential. Alternative calculations for different rates of denitrification within the aquifer do not change the basic result.

Since a rigorous liberalisation of agricultural markets would face severe political difficulties, the political process has recently concentrated on new programmes which would allow government subsidies for the reduction of agricultural production capacity, diversion of land in particular, and for the

Table 1.10 Costs of meeting the given standard of nitrate concentration in drinking water

Region	NO$_3$ content in leaching water (reference situation) (mg NO$_3$/l)	Cost of agricultural measures (DM/ha)		Cost of water treatment (economic cost) (DM/ha)[a]
		Income forgone, valued at market prices (financial cost)	Income forgone, valued at shadow prices (economic cost)	
Hildesheimer Börde				
$\alpha = 1.0$[b]	46.3	–	–	–
$\alpha = 1.0$	57.9[c]	4	−8	520
Südoldenburg-Münsterland (low livestock density):				
$\alpha = 0.66$	96·5	123	68	471
$\alpha = 1.0$		385	242	552
Südoldenburg-Münsterland (high livestock density)				
$\alpha = 0.66$	101.8	132	129	470
$\alpha = 1.0$		574	440	561

(a) Average size waterworks.
(b) $\alpha = $ NO$_3$ content in pumping water/NO$_3$ content in leaching water (rate of nitrate persistence).
(c) Figure derived by adding 25 per cent increment on average nitrate concentration in this location. Only in this case would the 50 mg/l limit be exceeded and costs of meeting the standard would arise.

Source: Finck (1987); Hasse (1987).

readjustment of production patterns. So far the main objective of these programmes has been the reduction of market surpluses. The approach is clearly supported by farmers' groups, although it represents a second best solution from an economic point of view, especially if the control of production capacity is combined with a continuation of price support. Politically these supply reduction programmes are appealing because they promise a rapid reduction of surplus production and because the envisaged compensatory payments promise acceptability on the farmers' side. The ecological value of such programmes depends on the specific arrangements. If the programmes go along with a continuation of price support or even with a rise in prices, they might result in further intensification on the remaining land, causing further environmental risks. On the other hand, the programmes provide fallow land which could be used for ecological purposes, especially if such land becomes part of an interconnected network of biotopes. Unfortunately, the current programmes have no provisions which would assure the establishment of such networks, because the choice of locations and fields is determined exclusively by farmers' decisions. A better integration of agricultural as well as environmental objectives could be achieved if the areas in which farmers may participate in such programmes were predetermined from the ecological point of view and if provision were made for medium and long-term fallow rather than for one year as in the green fallow programme of Lower Saxony.

A similar criticism relates to the subsidies for a readjustment of production patterns in favour of non-surplus products. The current concepts neither assure consideration of ecological criteria in the selection of participating areas nor do they make provision for reduced intensities in the production systems. These arguments also apply to the envisaged subsidy on the production of renewable raw materials.

Extensification programmes can be much better used for environmental purposes because they can be focused on location-specific ecological aspects. Experience with a multiplicity of regional programmes is quite promising. The provision of bilateral agreements with farmers allows a very close consideration of individual preferences and opportunities. A complete monitoring of the state of the environment and a definite specification of targets for the relevant components of the environment at a local level are a necessary condition for successful extensification. Otherwise a lot of money will be spent without a reasonable environmental return. Fulfilment of these conditions requires a clear statement of environmental goals and close co-ordination between public and private organisations in respect of targets and measures of environmental protection. Such monitoring and co-ordination is the major responsibility of the administration for environmental protection which exists at the *Kreis* (county) level.

Concerning the possibilities for a better integration of environmental policies into agricultural policies, there is finally a more fundamental issue which relates to the relevance of economic efficiency and social equity in defining the term 'orderly agriculture' for practical policies and legislation. In the long run a liberalisation of agricultural markets would certainly result in reduced incentives to intensive and specialised agricultural production and thus enable environmental relief. But liberalisation alone might not 'automatically' assure complete harmony of economic and ecological objectives, even if losses of farm income were compensated through direct income payments. It cannot be excluded that, even in a situation of reduced price protection and improved efficiency in resource allocation, some objectives of environmental policy would not be satisfactorily met. So far it has been the general policy direction under such conditions that the originators or intolerable pollution should be subjected to appropriate restrictions without monetary compensation. This would be implied in the 'polluter pays' principle. Although the 'polluter pays' principle is not usually enforced in the agriculture sector, it seems unlikely that such exemptions will be continued and generalised to all situations where agriculture and the environment are in conflict. Indeed, various regulations and ordinances are already in effect which farmers have to accept without compensation, e.g. the restrictions imposed by regional liquid manure orders.

The applicability and suitability of ordinances on production practices face two particular problems in the case of agriculture, however. One problem results from the fact that agricultural production conditions differ greatly between locations and that norms which seem appropriate at one location may be totally inappropriate at another. Hence ordinances have to be applied within a region-specific approach. The other problem of ordinances on agricultural production relates to the difficulties in administration and control.

From an economic point of view, ecological, objectives can be better and more efficiently achieved by taxes and licensing. Generally, the elasticities of input response with regard to changes in input prices tend to be similar to those in respect of opposite changes in product prices. Yet the resulting income losses are considerably smaller in the case of taxes. This implies that input taxes may be preferable if drastic changes in price ratios are required in order to achieve a reduction of input use. Taxes on certain inputs and practices which involve risks for the environment not only encourage substitution and reduction of intensities but initiate technical progress in favour of environmental objectives. Unfortunately, however, input taxes are confronted with the same problems as levies and restrictions on agriculture. The computations presented in this chapter show that taxes on nitrogen fertiliser as well as taxes on pesticides have to be quite high in order to achieve a sufficient reduction of input intensities from an ecological point of view. Moreover the results show that the likely response of input intensities and the related environmental relief would both vary considerably between regions. Since environmental damage does not result from the use of agro-chemicals *per se*, but rather from excessive levels of input use, taxation would have to be based on the excessive quantities only. Indeed, the German Council of Environmental Advisers has proposed a tax on mineral nitrogen fertiliser combined with a reimbursement of a basic fixed amount per hectare for this reason. Unfortunately, a tax on mineral fertiliser would exclude all organic fertilisers, although the nitrate problem is of particular importance in regions with high livestock density. The consequent conclusion would be to include the quantities of nitrogen applied through organic fertilisers in a tax system. However, difficulties of administration and control would have to be overcome in that case. A pragmatic solution will possibly have to be a combination of input levies or taxes with restrictions focusing on the local conditions.

In conclusion, politicians and legislatures are confronted with the difficult task of:

1   Making sure that 'orderly agriculture' as mentioned in the legislation is defined to be in agreement with local ecological needs.
2   Avoiding policies which pursue the support of agricultural incomes through incentives to intensive and specialised production.
3   Applying the 'polluter pays' principle to agriculture in the same way as it is applied to the rest of the economy.

Finally, it has to be pointed out here that agricultural production does not have only negative external effects on the environment. Agriculture also performs a multiplicity of positive functions for the environment which have tended to be neglected in the public discussions of the recent past, although the representatives of the farm sector have been emphasising them repeatedly. New initiatives for a reform of agricultural policies are under way that aim at recognising these positive 'services' to the environment. Although examples of new forms of financial support and bilateral agreements on the basis of extensification and readjustment of production have been presented, it would certainly be unrealistic to expect that any income loss resulting from environment restrictions will provide a new justification for a general transfer of resources for the support of agricultural incomes. The present federal constitution and the existing legislation do not justify financial compensation for the mere elimination or avoidance of actual environmental damage. Except for social support in cases where economic existence is endangered, compensation is conceivable only for positive contributions to environmental protection which go beyond what has to be considered as an obligatory ecological element of orderly agricultural practice.

References

Agrarbericht der Bundesregierung Deutschland (various years), Deutscher Bundestag, Bonn.

Bach, M. (1985), 'Stickstoff-Bilanzen der Dreise der Bundesrepublik Deutschland als Grundlage einer Abschätzung der möglichen Nitrat-Belastung des Grundwassers durch die Landwirtschaft', *Mitteilungen der Deutschen Bodenkundlichen Gesellschaft*, 43(II), 625–30.

Beicht, W., Günther, M., Harzer, U., and Weber, R. (1986), Vergleichende Untersuchungen zwischen integriertem und konventionellem Pflanzenschutz im Apfelanbau, I, Pflanzenschutz-Konzeptionen und bisherige Auswirkungén der Massnahmen sowie Förderung des integrierten Pflanzenschutzes', *Gesunde Pflanzen*, 38, 398–408.

Buchner, A., and Sturm, H. (1985), *Gezielter düngen—intensiv, wirtschaftlich, unweltgerecht*, Frankfurt.

Bundesminister für Ernährung, Landwirtschaft und Forsten (various years), *Statistisches Jahrbuch über Ernährung, Landwirthschaft und Forsten der Bundesrepublik Deutschland*.

Darimont, T., Lahl, U., and Zeschmar, B. (1985), 'Landwirtschaft und Grundwasserschutz—Massnahmenkatolog zur Verringerung der Nitratbelastung', *Wasserwirtschaft*, 75(3), 106–10.

Feyerabend, I. (1985), *Zuckerrübenproduktion und Stickstoffdüngung—Umweltschutz als Nebeneffekt*, Internationales Institut für Umwelt und Gesellschaft, IIUG-rep. 85–3.

Finck, H.-F. (1987), 'Anpassungsmöglichkeiten der Landwirtschaft zur Verminderung der Nitratbelastung des Grundwassers—Eine ökonomische Analyse für ausgewählte Standorte', dissertation, Göttingen.

Gaertel, W. (1985), 'Belastung von Weinbergsböden mit Kupfer', *Berichte über Landwirtschaft*, SH 198, 123-33.

Hasse, K. (1987), *Emissionsvermeidung oder Immissionsbeseitigung zur Einhaltung des Nitratgrenzwertes im Trinkwasser—Eine gesamtwirtschaft liche Kosten-Nutzen-Analyse. Abschlussbericht zur Nutzen-Kosten-Untersuchung 'Beitrag der Landwirtschaft zur Verminderung der Nitratbelastung des Grundwassers'*, Göttingen.

de Haen, H. (1985), 'Interdependence of prices, production intensity and environ-
    mental damage from agricultural production', *Zeitschrift für Umweltpolitik*, 3,
    199–219.
Korneck, D., Lang, D., and Reichert, H. (1986), *Rote Liste der in Rheinland-Pfalz
    ausgestorbenen, verschollenen und gefährdeten Farn- und Blütenpflanzen*, Ministerium
    für Umwelt und Gesundheit, Rheinland-Pfalz.
Krauthausen, H.-J., and Günther, M. (1986), 'Vergleichende Untersuchungen zwis-
    chen integriertem und konventionellem Pflanzenschutz im Apfelanbau, II,
    Aufwand für Planung und Durchführung des Pflanzenschutzes', *Gesunde
    Pflanzen*, 38, 409–16.
Kuhlmann, H., and Wehrmann, U. (1982), 'Landwirte düngen oft mehr als die
    LUFA's empfehlen', *top agrar*, 2, 54–7.
Landwirtschaftskammer Hannover (various years), *Betriebsstatistik*.
Landwirtschaftskammer Rheinland-Pfalz (various years), *Buchführungsergebnisse*.
Marxen, H. (1979), 'Entwicklungen einer Agrarlandschaft, dargestellt am Beispiel
    des Raumes Süderbrarup und Sterup', dissertation, Agrarwissenschaftliche
    Fakultät   der   Universtität   Kiel,   Institute   für   Wasserwirtschaft   und
    Landschaftsökologie der Universität Kiel.
Müller, W. (1982), 'Nährstoffaustrag aus Weinbergböden der Mittelmosel unter
    besonderer Berücksichtigung der Nitrat', dissertation, Bonn.
Obermann, P. (1984), *Nitratauswaschung und Nitratabbau im Bereich des Grundwassers
    agrarspektrum*,   Schriftenreihe   des   Dachverbandes   wissenschaftlicher
    Gesellschaften der Agrar-, Forst-, Ernährungs-, Veterinär- und Umweltforschung
    eV, 7, *Agrarstruktur im Wandel*, Munich, 341–51.
Oesau,   A.   (1986),   *Förderung   der   Artenvielfalt   von   Ackerwildkräutern*,
    Landespflanzenschutzamt Rheinland-Pfalz, Mainz.
Ohlhoff, J. (1987), *Spezialisierung im Ackerbau aus ökonomischer und ökologischer
    Sicht—Bestimmungsgründe und Reaktion auf veränderte Preis-Kosten-Verhältnisse*,
    Kiel.
Rat von Sachverständigen für Umweltfragen (1985), *Umweltprobleme der
    Landwirtschaft. Sondergutachten*, Stuttgart and Mainz.
Rohmann, U., and Sontheimer, H. (1985), *Nitrat im Grundwasser—Ursachen,
    Bedeutung, Lösungswege*, Karlsruhe.
von Schilling, H. (1984), *Räumliche Bedeutung der Konflikte zwischen
    Landbewirtschaftung und Umwelt*, Bundesforschungsanstalt für Landeskunde und
    Raumordnung, Informationen zur Raumentwicklung, 6, 525–38.
Schulte, J. (1984), *Begrenzter Einsatz von Handelsdünger und Pflanzenschutzmitteln*,
    Schriftenreihe des Bundesministers für Ernährung, Landwirtschaft und Forsten,
    Series A, Angewandte wissenscahft, 294, Münster—Hiltrup.
Schwille, F., (1969), 'Hohe Nitratgehalte in den Brunnenwässern der Moseltalaue
    zwischen Trier und Koblenz', *gwf-wasser/abwasser*, 110(2), 35–44.
Sukopp, H. (1981), *Veränderungen von Flora und Vegetation in Agrarlandschaften.
    Berichte über Landwirtschaft*, N.F., SH 187, 255–64.
Sunkel, R. (1983), 'Nitratauswaschung über die praktische Anwendung von
    Schadensschwellen für Unkräuter in Wintergerste', dissertation, Göttingen.
Walther, W. (1982), 'Veränderung der Beschaffenheit von Trinkwässen aus Äckern
    und Waldgebieten in Südost-Niedersachsen', *Veröffentlichungen des Instituts für
    Stadtbauwesen der TU Braunschweig*, 34, 215–39.
Walther, W. (1984), *Untersuchung der Veränderung von Wasserinhaltsstoffen in
    Trinkwassergewinnungsanlagen   im   Raum   Südost-Niedersachsen*,   Bericht
    Siedlungswasserwirtschaft, TU Braunschweig.
Walther, W., Scheffer, B. and Techgräber, B. (1985), *Ergebnisse langjähriger*

Lysimeter-, Drän- und Saugkerzenversuche zur Stickstoffauswaschung bei landbaulich genutzten Böden und Bedeutung für die Belastung des Grundwassers, Schriftenreihe des Instituts für Stadtbauwesen der TU Braunschweig, 40.

Wehrmann, J. (1984), 'Einfluss der, Düngung auf die Nitratauswaschung und den Nitratgehalt im Sickerwasser', *Landwirtschaftliche Forschung*, 37.

Wehrmann, J., and Scharpf, H.-C. (1983), 'Stickstoffaustrag in Abhängigkeit von Kulturart und Nutzungsintensität in Intensivkulturen', in Deutsche Landwirtschafts-Gesellschaft, *Nitrat—ein Problem für unsere Trinkwasserversorgung*, Arbeiten der DLG, 177, 95–113.

Wehrmann, J., and Scharpf, H.-C. (1985a), 'Sachgerechte Stickstoffdüngung-Schätzen, kalkulieren, messen, *AID*, 17.

Wehrmann, J., and Scharpf, H.-C. (1985b), 'Nitrat in Grundwasser und Nahrungspflanzen', *AID*, 136.

Whitby, M., and Hanley, N. (1986), 'Problems of agricultural externalities: conceptual model with implications for research', *Journal of Agricultural Economics*, 37(1), 1–11.

2. The effects of Swedish price support, fertiliser and pesticide policies on the environment

K.I. Kumm

Structural change in Swedish agriculture

Since the 1920s Sweden's agricultural area has rapidly decreased. Initially it was meadows and natural pastures which decreased, but during the 1950s and 1960s the area of cultivated land also decreased rapidly (fig. 2.1). These reductions resulted largely from more intensive cropping, with higher yields per hectare, fertilisers and pesticides having been of great importance for yield increases (fig. 2.2). A smaller agricultural area has sufficed for the domestic supply of foodstuffs, and agricultural production for export has not been profitable.

During the 1970s, however, the agricultural area stabilized, largely because subsidies on foodstuffs increased domestic consumption during the decade. Today most of the foodstuff subsidies have been removed, while the yields per hectare have continued to increase. It is expected that the area of agricultural land will continue to decline over the next few decades. Hitherto the reduction in the area of agricultural land has occurred mainly in forest areas and in northern Sweden.

Figure 2.1 also shows that the area devoted to different forms of grassland has decreased on arable land. On the other hand, the proportion of land devoted to grain crops has increased, particularly over the last thirty years. The shift from leys to grain depended to some extent on a shift from cattle to pigs (fig. 2.3). The replacement of horses by tractors also reduced the need for grassland, while grain now makes up a larger proportion of cattle feed than earlier.

Animal husbandry has tended to become concentrated in certain regions, involving more specialised enterprises, particularly expanding pig production. To a great extent this concentration has occurred in areas where the soil type and precipitation conditions are such that the abundant production of manure may cause considerable water pollution. Halland, on the west coast of Sweden, is one such area (fig. 2.4). In other regions the number of animals has decreased. Increasingly large numbers of farms no longer have

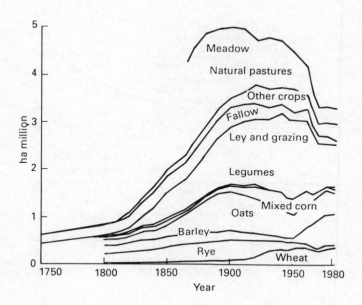

Figure 2.1 The extent and composition of agricultural land, 1750–1984. *Source:* Mattson (1985)

animals and depend completely on artificial fertiliser for their supply of plant nutrients.

Until 1950 both annual crops (mainly grain) and perennial leys were included in the rotations used on almost all farms. During later years, leys have sharply decreased and have been concentrated on certain farms with large herds of cattle. Today, an increasing proportion of farms are managed entirely without roughage-consuming animals and leys. On such farms the crop rotations are based entirely on annual crops, which strongly favours weeds.

These structural changes would have been impossible without chemical herbicides. On many farms the lack of diversity in crop rotation also leads to problems with plant diseases and pests, increasing the need for fungicides and insecticides. The cropping structure which has developed during the post-war period is thus highly dependent on pesticides.

Agricultural policy and structural change

During the post-war period Swedish agricultural policy has had three major goals. An income goal has aimed at providing people employed in agriculture

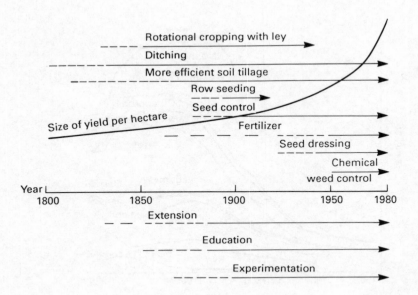

Figure 2.2 Time diagram showing when new aids and methods can be expected to contribute to yield increases in Swedish crop production. *Source:* Mattson (1985)

an economic situation equivalent to that enjoyed by those in other industries. A production goal, aimed at ensuring a certain level of domestic agricultural production, has been modified from time to time. In 1967 this policy goal aimed to reduce production, whereas in 1977 it was decided to allow considerable surplus production of grain. At the time of writing (1987), the aim is once again to restrict production. An efficiency goal has aimed at rational production, allowing good profitability in agriculture in conjunction with low prices of foodstuffs. In 1967 emphasis was placed on specialisation and shifts to large-scale production for improved efficiency. Ten years later, emphasis was placed on establishing and maintaining efficient family enterprises.

The means which have been, and still are, used to achieve these goals are price regulation, rationalisation, agricultural policy and research, experimentation and education. Agricultural price regulation aims at protecting domestic prices from the negative influence of price movements on the world market. Normally, domestic prices are higher than those on the world market. Domestic prices are supported through import duties and export subsidies (economic support) for domestic surplus production.

In the 1977 agricultural policy a goal adopting a fundamental ecological approach was introduced. It was emphasised that efforts should be made to develop production technology which associates demands for high efficiency

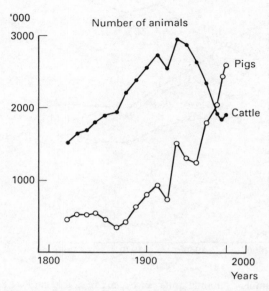

Figure 2.3 The development of cattle and pig rearing in Sweden. Grain now forms a larger proportion of cattle feed than it did. *Source:* official statistics compiled by Andersson (1986)

with demands for economical use of natural resources and ecological balance. The agricultural rationalisation goal set in 1977 was thus conditioned by ecological considerations. In the 1985 food policy, a goal dealing with environment and resources was also included for the first time. This goal implies that agriculture and food production must respect environmental quality and the requirement for long-term, rational, economical use of natural resources. To the greatest extent possible agriculture must use environmentally appropriate cropping methods which contribute to efficient use of soil, water and nutrients. To a reasonable extent, agriculture must be conducted so that it contributes to preserving genetic diversity and valuable varieties of flora and fauna in the countryside. Food quality is also emphasised.

Most of the structural alterations to agriculture described above occurred before ecological and environmental goals were included in agricultural policy or, at least, before environmental goals had become of general public interest. Structural adjustment has been influenced by measures chiefly directed at reaching income, production and efficiency goals. These have had unintended environmental effects. For example, price support has probably increased cropping intensity and thereby yields per hectare. With a production goal corresponding to domestic consumption (but not exports), this reduced the requirement for arable land. On the other hand, price support has resulted in continuing agricultural production in some regions where

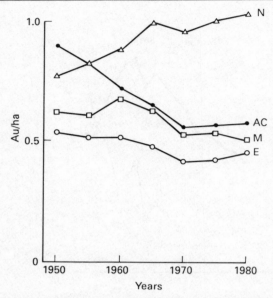

Figure 2.4 The number of animal units per hectare of arable land. *N* province of Halland in south-western Sweden, *AC* province of Västerbotten in northern Sweden, *M* province of Malmöhus in southern Sweden, *E* province of Östergötland in eastern central Sweden. *Source:* official statistics compiled by Andersson (1986)

agriculture would otherwise have disappeared. It is noteworthy that since environmental goals have been included in agricultural policy the means used to implement policy have not changed.

At least as much as agricultural policy, technical developments and changes in the price structure of agricultural products have influenced the technical and structural development of agriculture. The development of inexpensive and efficient pesticides, of inexpensive methods of industrial nitrogen fixation and of new high-yielding grain varieties with good straw strength to tolerate high nitrogen rates are examples of such technical developments.

Between 1950 and 1982 farm worker wages increased thirtyfold whereas machinery prices increased only fivefold and fertiliser prices only fourfold (Uhlin, 1980). This has speeded up the change to a labour-saving production structure. The use of leys has decreased in extent but still requires about twenty work hours per hectare whereas grain production, which has increased, requires only half this time. The production of pigs, which has increased in extent, requires little more than two work hours/100 kg/meat, whereas cattle production on grazing land requires at least twice that amount. The rationalisation policy and the extension service in agriculture have only speeded up changes which are basically caused by technical and economic development outside the framework of agricultural policy.

Environmental consequences of structural changes

Residual phosphates and nitrates in the aquatic environment
The structural changes in agriculture during the last thirty to forty years have influenced the run-off of plant nutrients to lakes, the sea and ground water in four ways:

1 A large proportion of the abandoned arable land has become forested either by planting or by self-seeding and as losses of plant nutrients from forest are lower than from arable land the aggregate loss of plant nutrients has decreased. However, the abandoned land has primarily been in forest areas with extensive plant production and consequently the reduction in leaching is probably limited. In addition, the negative effects of nutrient leaching on the environment are less significant in those parts of Sweden where the abandoned land is found.
2 The shift in plant husbandry from grassland (leys) to cereals has increased the leaching of plant nutrients, as leaching is greater from cereal crops than from grassland (Joelsson *et al.*, 1986).
3 The concentration of livestock in fewer but larger herds has increased the risk of leaching, since production and use of manure have been correspondingly concentrated. An increased number of farms have insufficient land area for spreading of manure. Swedish investigations also demonstrate that very excessive fertilising only occurs in connection with the use of manure (Andersson, 1986).
4 The increased use of fertilisers has increased leaching. The leaching of nitrogen, in particular, increases with the fertilisation level, particularly when the rates applied are in excess of the normal level (fig. 2.5).

Together, these effects show that artificial fertilisers cannot be considered in isolation from the environment.

The overall agricultural structure also has an influence on, for example, the extent of water pollution. Nutrient leaching into waterways, lakes and the sea may result in more intensive growth of aquatic vegetation. During the summer, phytoplankton may occur in such large amounts that the water becomes green-coloured and toxic. In a few Swedish lakes, after the algal bloom during the summer, dead and decaying algae drift towards the shores or accumulate in thick sediment layers on the bottom, with the heavily nutrient-loaded lake quickly becoming altered beyond recognition. The vegetation which earlier grew only along the shores starts to cover increasing areas of the lake and finally the lake rapidly becomes overgrown. Nutrient leaching also influences aquatic fauna. Zooplankton, fish and bottom animals are initially favoured by the increased availability of nutrients in the form of algae and other plants. Increased competition and other disturbances will, however, lead to numerous species being eliminated. For example, all salmonid fish gradually disappear, making room for an increased number of low-quality fish.

A particularly serious impact on the environment is the deficiency of oxygen which may occur in the deeper parts of nutrient-rich lakes and seas.

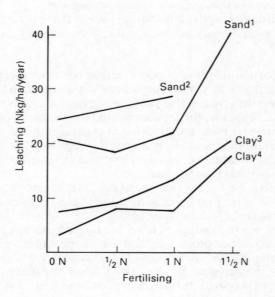

Figure 2.5 Research results showing the influence of fertilisation on leaching. N Normal recommended fertiliser level. *Source: 1* Joelsson *et al.* (1986), Kjellerup (1983), Kjellerup and Kofoed (1979); *2* Ivarsson and Brink (1985); *3* Kjellerup (1983), Kjellerup and Kofoed (1979); *4* Brink and Lindén (1980)

This is caused by the decomposition of micro-organisms and dead plants and animals which have sunk to the bottom. During decomposition the micro-organisms utilise the oxygen dissolved in the water.

During the 1970s most urban areas in Sweden built effective sewage treatment works so that today most of the phosphorus found in lakes and rivers originates from agriculture and forestry. Phosphorus is the chief eutrophicating element in most Swedish lakes. In the sea, on the other hand, it is frequently nitrogen which stimulates plant production most. For example, in Laholm Bay on the west coast of Sweden, which suffers from severe eutrophication, 60 per cent of the nitrogen originates from agriculture (National Environmental Protection Board, 1986).

Residues of pesticides in drinking water and food
High nitrate concentrations have been measured in the ground water, particularly in areas of intensive agriculture and sandy soils. The reasons for the leaching of nitrogen to the ground water are largely the same as for nutrient

leaching to rivers and lakes. Since the water moves at a maximum rate of a couple of metres per year in sandy soil and considerably more slowly in clay soil, it may take up to several decades before increased leaching is reflected in the nitrate concentrations in ground water at deeper levels. Against this background, and since fertilisation intensity increased during the 1960s and until the mid-1970s, it may be expected that the nitrate concentrations in ground water will most probably continue to increase in the near future, irrespective of changes in current agricultural policy. Nitrate in drinking water implies a health hazard, mainly for infants (methaemoglobinaemia). Intake of nitrate with drinking water or food may possibly also contribute to the formation of carcinogenic nitrosamines in the body. In Sweden, however, there have been no known cases of illness as a result of methaemoglobinaemia.

The heavy metal cadmium can be taken up by plants from arable land and thus transferred to humans via food. The supply of cadmium to arable land is considerably greater than the rate of removal, and, at present, the total concentration of cadmium in arable land is increasing at a rate of $0 \cdot 4 - 0 \cdot 5$ per cent per year. In the long term this represents a threat to humans. The total annual supply of cadmium to arable land in Sweden is, at present, 4 tonnes from phosphatic fertilisers, $2 \cdot 5$ tonnes from atmospheric deposition and $0 \cdot 5$ tonnes from municipal sewage sludge.

On the basis of available toxicologal information, the residue levels of chemical pesticides found at present in foodstuffs are not considered to imply any particular health hazard for the average consumer. However, the methods used to estimate risks are simplified models with clear limitations. Also, exposure to pesticide residues through consumption of vegetables may not only be difficult to calculate but may also vary strongly among different individuals. In addition, the analyses conducted at present provide an incomplete picture, since only part of the spectrum of pesticides used internationally can be routinely analysed.

Official control of pesticide residues is largely focused on fresh fruit and vegetables. Pesticide residues in grain products and drinking water, however, are not subjected to regular public control measures. Investigations of residues conducted at the Swedish University of Agricultural Sciences suggest that the occurrence of residues from agricultural pesticides may imply a health hazard (National Board of Agriculture, National Environment Protection Board and the National Chemical Inspectorate, 1986).

Spray drift from pesticides during application
As a result of wind drift, unintentional spraying of gardens and another areas adjacent to arable land occurs in some parts of Sweden. Studies conducted at the Swedish University of Agricultural Sciences provide evidence that unintentional spraying of land outside the intended area may be considerable. A study of pesticide use on 120 randomly selected farms in southern Sweden shows that filling of farm sprayers frequently occurs direct from rivers or wells, using the suction pipe on the sprayer. Carelessness and accidents during the filling and cleaning of farm sprayers have occasionally led to fish mortality in rivers.

Spray drift from aerial spraying has been studied very little. As the spraying height is naturally higher than when using a farm sprayer, a slightly larger spray drift can be expected from aerial spraying. According to the few details available it may be expected that at a distance of 50 m from the target there will be an application equal to about 10 per cent of that on the target itself. Thus, in Sweden, as a result of these concerns the aerial spraying of chemicals has been banned since 1986.

Long-term effects of chemicals on soil quality and productivity
An abundant supply of nutrients in the form of fertilisers and/or manure is essential if the yield capacity of arable land is to be maintained. A low supply of nutrients will decrease yield levels, particularly in plant rotation without leys and manure. Consequently, farms without animals are greatly dependent on commercial fertilisers.

These relationships were illustrated by the results of long-term fertiliser trials started during the 1950s in southern Sweden. In crop rotation without leys or animal manure the yield on unfertilised plots is only 40 per cent of that on plots with the highest levels of fertiliser use. In the crop rotation which includes leys and animal manure, however, the yields are 55 per cent of the maximum achievable through the liberal use of fertilisers. The latter rotation thus suffers less if the use of commercial fertilisers is reduced or completely ended. In crop rotation without leys and manure, drastic reductions in the use of fertilisers would lead to larger yield reductions in the long-term than in the short-term perspective (SOU, 1983: 10).

Unintended effects of pesticides and fertilisers on species other than those on which the pesticide/fertiliser is being applied

Flora
The number of vascular plant species in Sweden today amounts to about 1,700. Of these, 700–800 species occur in different agricultural environments. The older farming systems with grazing land, meadows and small fields resulted in a highly varied countryside with many different biotopes and a correspondingly rich and varied flora. Today about 300 vascular plants are considered to be acutely threatened, at risk or rare in Sweden and about 200 of these are found in agricultural areas. Approximately another seventy plants are considered to require more careful evaluation.

From the 1940s to the present, wooded pasture plants have suffered setbacks while the composition and properties of weeds have been influenced by changes in agricultural practices. Changes taking place today include the planting of forests on agricultural land in certain parts of the country; the removal of hindrances to cultivation practices; irrigation, reduced mowing and grazing management; and increased use of fertiliser and pesticides.

The fertilisation of arable land has led to an improvement in the ability of cereals to compete against individual weed species. This has implied a setback for certain weeds with weak competitive ability whereas others have been favoured. Fertilisation of formerly unfertilised natural grazing land results in the severe reduction of many species as 'nitrogen-favoured' grasses and herbs out-compete the original flora of the wooded pastures.

The use of herbicides in agriculture has partly influenced the composition of weed flora in the fields and also nearby plant communities. Moreover the proportion of weed species susceptible to herbicides has been reduced, whereas the density of hard-to-control species has increased. Through wind drift, spraying may also influence plant communities beyond the field. Examples of these affected communities include field borders, small areas of moraine or rock frequently found in Swedish fields, roadsides and forest edges. These plant communities are frequently complex and may be regarded as remnants of former, more widely spread eco-systems, which can be changed drastically if pesticides are applied to them.

Fauna

The use of chemicals in agriculture has obviously influenced Swedish fauna as well as flora. Possibly the most manifest example was observed during the 1960s when the dressing of seed with alkyl mercury was banned in 1966. In the following years there was a clear increase in several of the species which had suffered toxic effects.

The use of DDT as an insecticide is another example. The most noticeable effect is that birds which have this poison in their bodies lay eggs with thinner shells and poorer possibilities for the embryo to develop. DDT was banned in Sweden in 1970 but the problem remains to some extent, owing to atmospheric and waterborne transport from countries which still use these compounds.

The most important effect today of chemical pesticides, however, may be on the wild higher fauna, since important food sources like insects and weeds for certain animals have been eliminated by chemical spraying and weed control. Although fairly little is known about the amounts and extent to which pesticides occur in surface and ground water, in the investigations conducted so far not only have detectable quantities been found but also, in some cases, concentrations which may be hazardous to aquatic organisms. The use of herbicides, in particular, has resulted in the reduced availability of weeds and weed seeds which are important foods for many insects and birds. Chemical pesticides may also have very negative effects on batrachians (frogs) and reptiles. It is known that many plankton organisms are unusually susceptible and that they easily accumulate different substances. Small pools, etc., which are used as spawning places for such species, are particularly exposed to pollution in agricultural areas. The use of insecticides may locally eradicate entire populations. Similarly, the control of brush in the early summer along ditches between fields can have serious consequences, since lower vegetation covering the soil is essential as protection for these animals. When the brush disappears food availability in the form of insects, etc., is also reduced.

Spraying of insecticides generally results in a reduction of insects in the field, which in turn affects insectivorous birds, predatory insects and spiders, who are the natural enemies of pests. Some insecticides and fungicides also have negative effects on earthworms.

The present situation and recent national initiatives

Product price support

Product price support results in the prices received by farmers in normal circumstances being considerably above those on the world market. This also implies that exports of surplus production must be subsidised. This is illustrated for wheat in fig. 2.6. The situation is similar for most major crops and also for animal production. The reasons for this price support in Sweden are associated with a number of agricultural policy goals discussed earlier. In addition to these reasons, in northern Sweden a higher differential support price is offered for some agricultural products such as milk and meat. The reasons for this greater support involve national security, regional employment and development and the protection of the countryside.

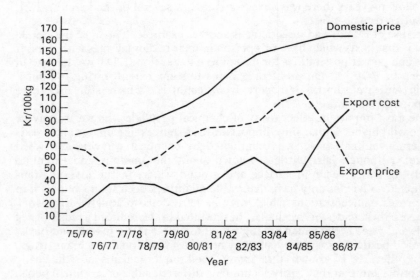

Figure 2.6 The domestic price, export price and export cost of wheat. The export cost is the difference between the domestic price and the export price. *Source:* Ds Jo (1986, p. 6); National Agricultural Marketing Board, pers. com.

From the environment viewpoint, product price support has at least two consequences. Firstly, it is commercially profitable to apply more than would be the case at the prevailing world market prices (fig. 2.9.). The same applies in principle to the input of pesticides. Secondly, product price support means that agriculture is economically feasible in areas which would probably be removed from agricultural production if the producers received only world market prices. At present low world market prices, most of Swedish cereal production would cease without price support. The reason is that the world market prices do not even cover the variable costs. A corresponding situation applies to most agricultural products.

Charges and taxes on fertilisers and pesticides

In order to reduce surplus production in agriculture and to finance the cost of exporting existing surpluses, a price regulation charge on fertilisers was introduced in the early 1980s. It has been gradually increased and now represents about 20 per cent of the price of fertilisers. In 1986 a price regulation charge was placed also on all agricultural pesticides, so that each application of a pesticide will cost an additional SKr 29 per hectare.

Table 2.1 Implementation of fertiliser charge and tax in Sweden from 1982 to 1987

Period	Charge (SKr/kg)			Tax (Skr/kg)			Charge + tax, % of fertiliser cost, excl. charge and tax		
	N	P	K	N	P	K	N	P	K
1982/83	0·30	0·58	0·18	–	–	–	7	7	7
1983/84	0·60	116	0·36	–	–	–	12	12	12
1984, second half	0·65	125	0·39	0·30	0·60	–	17	17	12
1985, first half	0·72	138	0·43	0·30	0·60	–	17	17	12
1985/86	0·93	179	0·56	0·30	0·60	–	20	20	15
1986/87	112	243	0·76	0·30	0·60	–	25	25	20

N nitrogen, P phosphorus, K potassium.
Source: National Board of Agriculture (1986).

In addition, in 1984 an input tax of 5 per cent of the price of fertilisers and pesticides used in agriculture was introduced. Funds from that tax are used for research on, for example, farm practices which reduce or avoid the use of agricultural chemicals. Some funds are also used for conservation measures and environmentally oriented extension. Table 2.1 illustrates how these charges and taxes on fertilisers have developed over the years.

It is difficult to evaluate the effects that the charge and tax have had on sales of fertilisers. At the same time that they were introduced and increased, fertiliser prices excluding tax and these surcharges also changed. In addition,

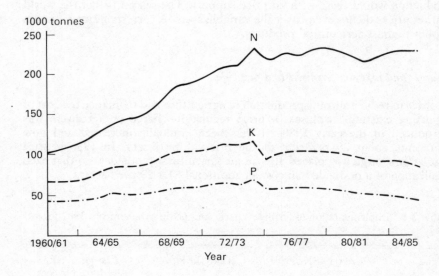

Figure 2.7 Sales of nitrogen (N), phosphorus (P) and potassium (K) to agriculture and horticulture, 1960–85 (000 tonnes). *Source: Jordbruksstatistik arsbok* (National Agricultural Statistics), 1986

the protein bonus payments on grain have increased and programmed cereal production has grown during this period in a manner which has tended to increase nitrogen consumption.

Figure 2.7 shows that there have not been any great changes in sales of fertiliser since the charge and tax were introduced. The largest change took place in the early 1970s, when the former upward trend in fertiliser use was arrested. Amongst other reasons this change in the 1970s can be explained by the fact that fertiliser prices were increasing at a faster rate than product prices and that animal production and consequently manure availability increased by 20 per cent from 1972 to 1980. The increased production of manure corresponds to about 15,000 tonnes of nitrate, 30,000 tonnes of phosphate and 15,000 tonnes of potassium. Prior to the 1970s the real price of fertilisers was falling and many farmers were applying it at sub-optimal rates.

The most profitable level of fertilisation for the farmer depends on the ratio between the fertiliser price and the product price. Table 2.2 illustrates how this ratio develops for nitrogen and barley. The price ratio of other cereals is similar to that of barley. The ratio between the prices of nitrogen and barley increased after the introduction of the charge and tax in the early 1980s but this increase is only partly explained by the charge and tax. The other

Table 2.2 Development of barley and nitrate fertiliser prices from 1971 to 1985 (the prices in brackets are the prices excluding all fertiliser charges and taxes)

Year	Nitrate fertiliser (SKr/kg N)[a]		Barley (SKr/kg)	Ratio N:barley (SKr/kg)	
1971[a]	1·50	(1·50)	0·45	3·3	(3·3)
1974[b]	2·10	(2·10)	0·54	3·9	(3·9)
1981[c]	3·90	(3·90)	0·96	4·1	(4·1)
1982[c]	4·60	(4·30)	1·09	4·2	(3·9)
1983[c]	5·40	(4·80)	1·17	4·6	(4·1)
1984[c]	6·50	(5·55)	1·22	5·3	(4·5)
1985[c]	7·50	(6·50)	1·25	6·0	(5·2)

(a) Calcium nitrate, including charges and taxes.
(b) Area calculations by the Swedish University of Agricultural Sciences.
(c) Agricultural statistics, 1986.

explanation is that nitrogen prices, excluding charges and taxes, have increased faster than product prices. The drastic increase in the ratio of about four in 1981 to about six in 1985 has not reduced the consumption of nitrogen. On the other hand, we do not know what would have happened without the increases in price.

Theoretical calculations using a linear programming model at farm level show that a 30 per cent increase in the nitrogen price should reduce nitrogen fertilisation by 10–15 per cent (Mattsson, 1986). This suggests that the charge and tax on nitrogen, together amounting to 25 per cent, have a considerable effect on nitrogen consumption. Mattsson's calculations suggest that a twofold increase in the price of nitrogen would reduce consumption by 30–40 per cent.

The tendency of the higher nitrogen prices to reduce the amount of nitrogen supplied are counteracted by increased protein payments for grain. During the 1980s protein payments were increased for wheat, with the result that the optimal application rate for wheat increased by slightly less than 10 per cent. Protein payments for barley were introduced in 1986. This will probably increase the economically optimal rate of nitrogen application to barley by 10 per cent on farms which produce feed grain for the market (Mattsson, 1986).

Legislation

The following legislation in Sweden is of importance for environmental protection in agriculture:

1 The Environment Protection Act, which covers air and water pollution and noise.
2 The Act on Products Hazardous to Health and to the Environment, which applies to the handling of poisons and pesticides.

3 The Nature Conservancy Act, which deals with conservation and management of natural assets.
4 The Water Act, which concerns drainage and the use of water supplies.
5 The Hunting Act, which is also intended to protect fauna.
6 An additional section recently added to the Act on Agricultural Land Management to facilitate the inclusion of nature conservation considerations in agriculture.

The first four above-mentioned Acts have regulations affecting agricultural activities. The Hunting Act has regulations against hunting. The last-mentioned Act is of another kind. Here environmental considerations are really seen as an integrated part of normal agricultural practice, which is of great importance for changing attitudes to environmental protection in agriculture.

The Nature Conservancy Act and the Act on Agricultural Land Management may to some extent influence the use of chemicals in agriculture. This particularly applies to applications outside arable land, where the flora and fauna may be affected along ditches, in wooded pastures, on the small 'islands' of rock or moraine frequently found in Swedish fields, etc. However, these restrictions have a negligible effect on the production and profitability of Swedish agriculture.

The Environment Protection Act has an important impact on the handling of animal manure in agriculture. The handling of fertilisers may also be affected. The law prescribes that permission must be sought to start new livestock production enterprises with more than 100 Animal Units (one Animal Unit = one cow, ten slaughter pigs or 100 poultry). The permission is issued by the provincial government to the Concessionary Board and conditions are placed on, for example, the location and design of livestock buildings as well as the storage and spreading of manure. Guidelines for environmental protection in livestock keeping have been issued by the Environmental Protection Board since 1970. These guidelines have also been used in monitoring the environment in cases where livestock keeping is on a similar scale.

The Environment Protection Act was extended in 1984 to include a provision making it possible for the government to declare water catchment areas which needed to be protected against pollution as 'especially sensitive to pollution'. In those areas the government can specify, for instance, how farms should be managed in order to reduce pollution, as regards the sorting and spreading of manure and cultivation practices, etc. This provision has been used for one area of Ringsjön in the province of Skäne and another is being prepared in the Laholm Bay region in the province of Halland, both provinces being in southern Sweden.

Advisory services

The importance of a good advisory service for farmers has been emphasized. As a result of the changes in food policy in 1985 it is important that the agricultural extension service should increasingly provide information on production methods reflecting increased consideration for the environment. To enable the extension service to adapt to these requirements the local Boards of Agriculture have been granted extra funds amounting to 5 million Skr per year. This intensified extension service covers, for example, problems involving plant nutrients and plant protection as well as alternative cropping methods. The financing comes from revenue obtained from the environmental tax imposed on pesticides and fertiliser.

As will be discussed below, it is frequently economically profitable to reduce the utilisation of fertiliser and pesticides. In such cases the extension service can enhance the environment while improving on-farm profitability.

The programme to halve the use of pesticides in Sweden

The Swedish government aims to reduce the use of pesticides by 50 per cent in five years. To this end, the three authorities most affected—the Board of Agriculture, the Environmental Protection Board and the Inspectorate of Chemicals (1986)—have prepared, at the request of the government, a proposal for a programme to reduce pesticide risks in agriculture. The proposal includes the following measures:

1 Changing to using substances with fewer health and environmental risks.
2 Reduction in the use of pesticides.
3 Special measures to protect health and the environment.

The proposals have been prepared in co-operation with the agricultural organisations, which support them.

Using substances with fewer health and environmental risks
Before a pesticide may be used in agriculture it must be approved by the Inspectorate of Chemicals. The tests preceding approval include an evaluation of the risks and the advantages of using the substance. In 1985 new legislation on the control of chemicals was introduced whereby, for example, certificates of approval will be restricted to five years and the actual need for the pesticide must be examined in connection with the approval. Subsequently there will also be a tightening up of demands for documentation of the effects of the pesticide on the external environment, such as persistence and leaching.

Today it is generally considered that herbicides are used excessively. In response the three authorities have proposed that herbicides must be tested at at least three application rates instead of only one as at present. Two of these application rates should be below the lowest rate that gives full control. In this way, basic information for registration will be improved and farmers will be able to make more informed decisions about how much chemical to apply.

Reduction in the use of pesticides
This measure groups various kinds, including research, extension inputs, approval of sprayer equipment and efficiency testing of farm sprayers. The

proposal aims at preventing chemicals being used if other economically similar control measures are available. It should also ensure that the chemical used corresponds to an actual requirement, and that the rate of application is based on the principle of minimum use needed to obtain an acceptable effect.

Special measures to protect health and the environment
Special legislation deals with the application of pesticides. Spraying from the air, for example, is banned. It is proposed to revise and strengthen this legislation, for example by banning the filling of farm sprayers direct from lakes and waterways and preventing filling or cleaning where spillage of pesticides can easily pollute surface water or ground water.

Today training in the use of pesticides is necessary, and permission to use the most toxic pesticides is required. It is proposed that this policy be extended to cover all substances used in agriculture. Finally, increased control of pesticide residues in foodstuffs is proposed.

Effects of the proposal
Various authorities expect that the amount of active pesticide ingredients used in agriculture will be halved within the next five years. At least half this reduction will be achieved through the proposed measures, assuming that they are introduced. The rest of the reduction is expected to be achieved through a transition to low-dose compounds, etc. In the long term it is expected that research into new technology and non-chemical methods should yield further reductions. From the risk angle, the proposals should reduce the negative effects on the external environment. A transition to compounds with lower health risks and better protective equipment will also reduce health risks for the people handling pesticides.

Costs and financing
At present the annual costs of the new programme amount to about SKr 27 million. About 75 per cent of the costs go to research, experiments and developmental work. It is suggested that the new proposals be mainly financed with funds from the tax on fertiliser and pesticides. Provided that manufacturers and the trade do not compensate for the smaller market, it is estimated that the value of farmers' purchases of pesticides will decrease by SKr 150 million per year within five years. The cost of these proposals to the State, together with the costs to agriculture, amount to about SKr 50 million per year. Since the modified and reduced use of fertilisers should not reduce yields, increased revenue should amount to SKr 150 million–50 million = 100 million per year. In addition, there will be environmental improvements.

Agricultural research policies

In June 1984 the Swedish government requested the Research Council for Forestry and Agriculture to prepare a research programme dealing with alternative forms of production in agriculture. In discussing this request the Minister of Agriculture commented that the form of agriculture that had developed during post-war years was high-yielding and efficient, which was an advantage to both consumers and farmers. But, at the same time, he realised that this form of agriculture led to certain undesirable consequences, for example negative environmental effects and risks associated with the utilisation of fertilisers and pesticides. The problems associated with present forms of production were considered by the Minister of Agriculture to be such that alternative forms of production must be found which retain high efficiency while counteracting environmental risks. Alternatives to present forms of production should be characterised by their ability to preserve the productive capacity of the soil and to minimise the use of chemicals. In parallel with the programme on alternative forms of production in agriculture, the Research Council was requested to prepare a long-term programme of forest research. Increased timber production and improved possibilities of utilising the forest for other important purposes, such as recreation and open-air activities, were emphasised in the request.

The Research Council for Forestry and Agriculture was very optimistic in its inquiry (1985b) with regard to the potential for research to develop new technology linking high yields and efficiency with environmental benefits. With new technology, it is estimated that in the future Sweden will require only 2 million ha of arable land as opposed to 3 million at present, and 400,000 dairy cows as opposed to 600,000 at present to cover national food requirements. New agricultural technology is also considered to be capable of fulfilling highly valued environmental demands.

In conjunction with crop production and the utilisation of chemicals, it is proposed that research be conducted on:

1 Cropping systems which reduce nutrient leaching through utilisation of minimum tillage and direct seeding practices and, also, catch crops.
2 The better utilisation of manure in crop production by developing technology for uniform spreading without evaporation losses, etc.
3 Ways of reducing the need to use pesticides, including the development of healthy seed, genetic resistance, cultivation technology, biological control methods, better application techniques, etc.
4 Cropping systems that do not require utilisation of water-soluble salts for plant nutrient supply and chemicals, such as synthetically manufactured pesticides.
5 The increased use of leys in crop rotations, which would reduce nutrient leaching and the requirement for chemical control inputs but accounts for the need for a larger proportion of forage in animal production.
6 Alternative uses of land which is no longer required for food production, such as energy crops and forests.

The Research Council proposed a budget of SKr 22·5 million per year for research into alternative forms of agricultural production. For the current budget year (1986/87) a total of SKr 17 million was funded, using the revenue from taxes placed on pesticides and fertilisers.

Measures to decrease overproduction in agriculture

During the 1970s consumption of domestically produced foodstuffs increased strongly, partly as a result of the introduction of price support for agricultural products. The increased consumption and occasionally strong world market prices for grain resulted in favourable price trends and improved profitability for the farmers. This stimulated considerable new investment and production increases in agriculture. At the same time, an increasingly important goal of Swedish agricultural policy during the 1970s was to exploit all available agricultural land and, as a result, considerable extra costs were accepted for the construction of dwellings and roads in order to maintain agricultural land which could be tilled.

In the early 1980s the consumption of domestically produced foodstuffs stagnated or decreased, partly as a result of the reduction in price support. But at the same time production continued to increase and an increasingly large proportion of the production had to be exported. Particularly in the case of animal products, world market prices were far below prices paid to domestic producers. As this resulted in higher 'export costs', reduced producer prices and lower profitability in agriculture, it became an important goal to reduce the surplus animal production. Consequently, the following measures were introduced to reduce the production of surplus cattle meat, pig meat and milk:

1   A ban on new investment in livestock buildings.
2   The withdrawal of bonus payments for older milk producers.
3   Retirement support for older milk, pig and specialised cattle producers.
4   A voluntary two-level price system for milk.
5   A charge per hen for egg producers with more than 1,000 layers.

At present the animal production surplus is disappearing and the above production-restricting measures appear to be fulfilling their goal. It would have been possible, however, to link reduced animal production with environmental improvements by targeting these measures on farms and in areas with surplus manure and soils susceptible to leaching. In such cases the reduced number of livestock would result in lower application rates of manure and reduced water pollution.

Until 1985 surplus production of grain was not considered a problem and it was assumed that arable land not required for domestic food supply should be used to produce grain for export. Increasing surpluses and decreasing export prices resulted, however, in recent attempts to reduce the surplus. A working group has been appointed by the government to analyse the prospects of reducing the grain surplus. For 1987 it has been proposed that fallows (land set aside) should be subsidised. This proposal is to be introduced for about 100,000 ha. Fallow as a means of restricting production has been criticised from the environmental standpoint, since fallowing leads to greater nutrient leaching than land with vegetation cover.

The group found that the present surplus of grain corresponds to 0·4 million ha of the 2·9 million ha of arable land in Sweden. Assessments prepared by the Swedish University of Agricultural Sciences suggest that the surplus area will be 0·8 million–0·9 million hectares at the turn of the century and will increase as a result of increased productivity. The group considers that international developments in the grain sector indicate a long-term trend towards low and decreasing grain prices. This assessment is partly based on OECD's draft report on agricultural trade.

The Working Group on Cereals has prepared estimates of the socio-economic effects of future surplus production both in the short term for the rest of the 1980s and for the long term. In the short-term scenario, the costs of machinery, buildings and land improvement, etc., are fixed whereas in the long term they will be fluid. Table 2.3 illustrates that the socio-economic deficit, that is, the cost to the nation, using current export prices, is almost 2,000 SKr/ha/year in the short term and about 3,000 SKr/ha/year in the long term. For the individual farmer, however, production is frequently profitable, since the producer price is two or three times higher than the export price. It should be emphasised that these calculations concern the southern agricultural areas of Götaland, which is the highest-yielding agricultural area in Sweden. In areas with lower yield levels the socio-economic profitability of surplus production for export is even lower.

Table 2.3 Socio-economic consequences of continued surplus production of grain in the southern farming areas of Götaland in the short term until the end of the 1980s and in the long term, at 1985 prices

Item	Short-term 40 SKr/dt[a]		Long term 40–65 SKr/dt[a]	
Gross from barley		1,750	1,900	to 3,100
Less Expenses				
Essentials[b]	2,100		2,250	
Machinery	700		1,850	
Land improvement and buildings			350	
Labour	770		770	
Interest	30		30	
Other items			300	
Total expenses		3,600		5,550
Socio-economic effect of continued grain production (SKr/h)		−1,850	−2,450 to 3,650	

(a) Free at central silo.
(b) Seed, commercial fertiliser, pesticides, etc.

Source: Ds Jo (1986, p.6).

The working group has prepared economic calculations for alternative use of land which is at present under grain. In doing this they assumed an oil price of SKr 120 per m^3 and a yield of 12–14 tonnes of dry matter per hectare per year and found that energy forestry is economical from the socio-economic perspective, contrasting with the large deficit on grain production for export according to the calculations mentioned earlier. Price support for grain and available machinery capacity for grain production mean, however, that in the absence of profitable alternatives grain is more profitable than energy forestry on most farms, at least in the short term. In addition, energy forestry requires capital and does not yield income for three or four years, which could cause liquidity problems for some farmers. Consequently it is considered that energy forestry would require economic support during the start-up phase in order to become established.

Calculations show that energy grass or ethanol and engine oil produced by grain and oilseed is a marginally profitable socio-economic alternative to the continued exportation of grain, as the production cost per energy unit for energy grass is 50 per cent higher than that for energy forestry. Neither do agricultural products used as industrial raw materials outside the foodstuffs sector appear to be profitable. It is proposed that surplus areas not suited to energy forestry be planted with conventional forests (spruce or birch). This may concern, for example, drought-susceptible land, small and irregular fields surrounded by forest, etc. However, planting of forest is of doubtful profitability. If the planned rate of interest is 3 per cent the anticipated net present value will be about zero. At higher interest rates there would be an eonomic deficit. However, this is considerably smaller than the socio-economic deficit in grain production for export. Price support for grain will improve the economic profitability of future agricultural production, however, and consequently the planting of forest will require subsidies to become profitable for the individual farmer.

Table 2.4 Comparison of the area of land to be diverted to other forms of production with area of grain land which is surplus to Swedish requirements (ha)

	Until 1990	Until 2000
Peas	10,000–20,000	20,000–40,000
Energy forestry	8,000–10,000	100,000–200,000
Planting forest	2,000–10,000[a]	50,000–100,000[a]
'Niche' R&D inputs		30,000–60,000
Total proposed	20,000–40,000	200,000–400,000
Surplus area	400,000	500,000–900,000

(a) Availability of high-yielding cuttings and machinery will limiit the rate of introduction.
Source: Working Group on Grain (1986).

Apart from energy forestry and the planting of forests, the Working Group on Grain proposed an increase in the production of peas and certain 'niche' products to reduce the surplus. The effects of the group's proposals on the area of grain cultivation can be seen in table 2.4, which shows that the measures discussed eliminate only a small part of the surplus area.

Planting of forest will result in the sharp reduction or cessation of fertiliser and pesticide use, and leaching of nutrients would decrease considerably in comparison with continued grain production. In energy forestry the use of chemicals will not decrease as much but the negative environmental effects caused by nutrient leaching will probably be smaller than if grain production were to continue. The working group proposed no measures to encourage the establishment of forest plantations or energy forestry in areas where the reduced use of chemicals and/or reduced leaching is particularly important from the environmental viewpoint. Although it is obvious that large environmental advantages would be achieved by such measures.

The working group also analysed the potential to limit production through the reduced use of nitrogen and growth-regulating substances. On the basis of available material dealing with the input intensity problem, the group found that it is difficult to analyse the volumetric effect on grain production of reducing the intensity of input use, partly owing to the difficulty in predicting how farmers will adapt to a new situation. To decrease the use of nitrogen by 40 per cent, however, calculations made for the entire agricultural industry suggest that the price of nitrogen would have to increase by about 300 per cent. Other calculations made for different types of individual farms show a considerably smaller need for price increases in order to reduce the use of nitrogen by 30–40 per cent.

Bans or limitations placed on the use of growth-regulating substances are also discussed as a means of reducing surplus production. Growth-regulating substances make it possible to use higher rates of nitrogen without the risk of crop lodging. Growth-regulating substances are used on 150,000–170,000 hectares, mainly on rye and wheat. The working group considered that a ban on the use of growth-regulating substances would probably not lead to any noticeable yield reduction of grain, since the yield-promoting effect of nitrogen decreases before its protein-promoting effect. On the other hand, there is a risk that protein content would decrease, undermining quality.

The great uncertainty which exists about the effect on the grain surplus of a massive increase in the price of nitrogen, together with the difficulties involved in designing a system which makes it possible to return levied taxes and charges to the farmers, meant that the working group considered the available data insufficient as a basis for recommending the introduction of such economic measures to reduce agricultural production and improve the environment. The available data, in addition, are not unequivocal. The group considered that it was not possible to reduce grain production by means of measures to reduce intensive cultivation. This does not exclude the possibility of introducing measures on a minor scale to reduce the use of fertilisers and pesticides. The group considered that an expert commission should be set up with the task of evaluating research and experimental activities in this sector, to propose any further measures required, to design a concrete system whereby measures could be introduced to reduce grain production, and to analyse the consequences.

Consequences of policy initiatives and options at a regional level

Lake Ringsjön region

In April 1985 the government declared Lake Ringsjön and its catchment area a particularly pollution-susceptible area. The availability of plant nutrients in Lake Ringsjön has resulted in plankton algae developing in large quantities. This makes the lake less attractive for swimming and boating and has harmful effects on fish. In the long term, it is anticipated that the high production of plankton algae will cause increased sedimentation on the lake bottom, with decreases in depth and clogging by vegetation. In Lake Ringsjön it is mainly phosphorus that determines the plankton production level. Agriculture contributes more than 40 per cent of the phosphorus load in Lake Ringsjön and slightly less than 80 per cent of the nitrogen load (table 2.5). Before sewage treatment works were improved during the 1970s, municipal waste made up a large proportion of the load. If the supply of nutrients is to be reduced further, then measures must be introduced to treat spot discharges from rural dwellings and weekend cottages, etc., as well as introducing measures in agriculture.

Table 2.5 Total nutrient load in Lake Ringsjön

Source	Nitrogen		Phosphorus	
	Tonnes/year	%	Tonnes/year	%
Atmospheric deposition	30	3	0·2	1
Sewage works	43	5	0·2	1
Rural dwellings	14	2	5·0	37
Weekend cottages	3	–	0·9	7
Livestock	5	1	1·0	7
Arable land	650	75	5·0	37
Forest	80	9	1·0	7
Other land	40	5	0·4	3
Total	865	100	13·7	100

The catchment area of Lake Ringsjön covers 13,000 ha of arable land and 22,000 ha of forest and other land. There are about 14,000 cattle and 30,000 pigs in the area, corresponding to about 12,000 animal units. The number of animal units per hectare is thus slightly less than 1·0. Leys cover 40 per cent of the arable land, whereas annual crops (mainly grain) comprise 60 per cent. Only 10 per cent of the area is tile-drained.

The topography and soil conditions in the Lake Ringsjön area favour surface run-off of precipitation and snow melt water, contributing to a strong flush of nutrients in waterways leading into Lake Ringsjön. Certain areas are particularly susceptible to erosion. Of a total of 13,000 ha of arable land, about 1,000 ha are considered sensitive to erosion. The risk of particularly large flows of nutrients out of these areas is therefore significant. It has not been possible, however, to assess the magnitude of the flow of nutrients into Lake Ringsjön from these areas.

The county administration has prepared regulations and guidelines for the pollution-sensitive areas. These guidelines were based in part on a programme prepared in co-operation among the county administration, the local Board of Agriculture and the Federation of Swedish Farmers. The five most important regulations and guidelines and their economic consequences are as follows:

1. Adjustment of nutrient supply. Supply of fertilisers and manure must be based on analyses of the soil's nutrient content and must not exceed the recommendations of the Board of Agriculture.

   Economic analysis demonstrates that the phosphorus supply would decrease by 30 per cent and profitability would improve by SKr 150 per hectare if all farmers followed the recommendations of the Board of Agriculture. It is mainly farms with a high stocking density and thus large amounts of manure which should reduce the complementary use of fertilisers.

2. Ploughing-in of manure. Manure which is spread on unplanted land must be ploughed in within thirty-six hours. During the period 1 December – 28 February manure may be spread only if it is ploughed in on the same day as it is spread and may not be spread closer than 10 m to open ditches.

   These restrictions lead to few if any negative economic consequences.

3. Storage space for manure. Storage space for manure must correspond to at least eight months of production. Manure should be spread mainly in the spring on crops with a long growing period.

   Increased storage capacity permits increased spreading of manure during the spring. Changing from autumn spreading to spring spreading improves the utilisation of nitrogen, which implies that fertiliser costs can be reduced. Spreading of manure in the spring may, however, result in yield reductions and loss of income as a result of delayed seeding and soil compaction. It is doubtful whether increased spring spreading is economically profitable even when sufficient storage capacity is available. Farms which already have large storage capacity do not always utilise it for maximum spreading in the spring. Consequently it is not certain that increased storage capacity always leads to increased spreading in the spring. The regulations do not, in fact, require—only recommend— spreading in the spring.

   The annual cost of expanding storage capacity will be large, particularly on farms planning to cease livestock production within a few years. In such cases the increased storage capacity will be very unprofitable for the farm. On larger farms with constant livestock levels the loss will be smaller. The losses vary between Skr 5 and Skr 100 per tonne of manure per year and depend on the plans for the future development of the farm.

4. Treatment of waste water. Waste water from a farm milking shed for more than eight cows must be filtered through the soil or collected in the manure tank. This may result in costs of SKr 1,000–3,000 per farm and year.

5. Soil tillage for reduced erosion. When ploughing, a safety zone of at least 1 m must be left along open ditches. On steeply sloping arable land ploughing should be done so that erosion is prevented in cases of surface run-off. Autumn harrowing should be avoided in areas susceptible to erosion. These measures have negligible economic consequences for the farm.

It will be observed from the five key recommendations listed above that no demands are placed or proposed for changes in the choice of crops or cropping system. Neither are there demands for reduced fertilization levels below the economic optimum. The recommendations, however, involving considerable costs, relate to investment in storage capacity for manure and treatment of waste water. As the regulations were prepared in late 1985, it is too early to distinguish any resulting improvement in water quality. However, the quantity of phosphorus in Lake Ringsjön has decreased considerably during the last five years. The improvement may depend on natural variations or may be a result of all the water management measures taken in the catchment area prior to 1985. Improved manure handling after comprehensive survey and extension activities may be mentioned as an example. A comprehensive study of agriculture and water management has also been conducted among farmers in the area. These study circles are being conducted on the initiative of the Federation of Swedish Farmers.

The Laholm Bay region

During recent years serious changes in the ecological balance in Laholm Bay have been observed. These changes are considered to derive mainly from large quantities of nitrogen entering the bay. Nitrogen stimulates algal growth, and large quantities of organic matter are formed. Degradation of the organic matter utilises oxygen, and oxygen deficiency may occur, causing, for example, the death of fish and bottom animals. Phosphorus is regarded as of minor importance for algal growth, since this element is present in surplus quantities and thus does not limit production in Laholm Bay.

Agriculture in the catchment area is mainly found in southern Halland. Agriculture here chiefly involves intensive livestock keeping. A large proportion of the livestock consists of pigs, and the area of grassland is thus relatively small. In combination with permeable soils and high precipitation this implies a great risk of nitrogen leaching, evidenced by the nitrate-polluted ground water in the countryside and high concentrations of nitrogen in the surface water.

Laholm Bay
region

Lake Ringsjön
region

Figure 2.8 The Lake Ringsjön district and the Laholm Bay region

In May 1986 the government designated Laholm Bay and parts of its catchment area as a district particularly sensitive to pollution (fig. 2.8). At the same time the government requested the county administrations concerned to prepare proposals for regulations on protection measures, restrictions and other preventive measures in the area. Proposals were prepared by a working group consisting of representatives of the county administrations, the Boards of Agriculture, municipal authorities and the Federation of Swedish Farmers.

Table 2.6 illustrates the supply of nitrogen and phosphorus to Laholm Bay. The designated pollution-sensitive area comprises about 60,000 ha of arable land divided among 2,500 farms. The stocking density is 1·0 animal unit per hectare. The catchment area of Laholm Bay also includes arable land outside the designated area, where agriculture is located in forested areas and is less intensive.

In some parts of the designated pollution-sensitive area the number of pigs increased sharply during the 1970s, owing to the establishment of a small number of large-scale pig production enterprises.

According to guidelines drawn up by the Swedish Environment Protection Board the number of animal units per hectare should not exceed 2·0. Larger densities produce manure containing considerably more nutrients than needed for normal crop production. More than 30 per cent of the number of

Table 2.6 Calculated and measured supply of nitrogen and phosporus with rivers emptying into Laholm Bay (the areas which are not drained by the large rivers comprise 0·6 per cent of the catchment area and have been included in the calculations)

Origin	Nitrogen		Phosphorus	
	Tonnes	%	Tonnes	%
Arable land	3,000	61	40	27
Forest and other land	1,245	26	50	35
Industries	55	1	10	7
Sewage works	500	10	25	17
Rural areas	100	2	20	14
Total	4,900	100	145	100
Measured in the rivers, mean, 1972–81	4,270	–	120	–

animal units in the area are, however, on farms with more than 2·0 animal units per hectare. As much as 16 per cent are on farms with more than three animal units per hectare. If the guideline of 2·0 animal units per hectare is to be respected the number of animal units in the pollution-sensitive area should be reduced by 7,500. At present the problem of surplus manure is partially solved by marketing the manure.

On the basis of statements by ecological experts the working group has suggested a target reduction of the nitrogen load from the designated pollution-sensitive area of 25 per cent (or 700–800 tonnes per year). In addition, the phosphorus load should also be decreased.

Among possible measures to be taken the following are most likely to influence the nutrient load:

1 Adjustment of nutrient applications to crop requirements, which frequently implies a reduction in the application rates;
2 Increased spreading of manure in the spring; and
3 Increased establishment of grasslands and other crops with long growing periods and the cultivation of catch crops.

Adjustment of nutrient application

Experiments have demonstrated that leaching of nitrogen increases as the fertilisation rate is increased and that a reduction in the fertilisation rate from the level which is optimal for efficient production by 50 to 70 per cent would reduce nitrogen leaching by only a few kilogram per hectare (fig. 2.5). A reduction in the nitrogen rate to 70 per cent of the optimal level reduces farm profitability by about Skr 50 per hectare for barley production and by about

SKr 150 per hectare for winter wheat production. A socio-economic calculation using export prices for grain and with charges and taxes removed from the fertiliser price shows, however, that a 30 per cent reduction in fertiliser rates is profitable from a social viewpoint (fig. 2.11). Consequently price support makes it more difficult to adjust nutrient applications to environmental requirements.

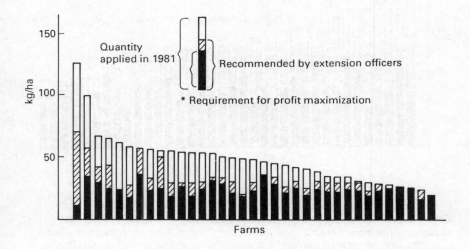

Figure 2.9   The average supply of phosphorus on thirty-six farms in 1981: the phosphorus requirement calculated from soil analysis and the crops grown together with the proposed fertiliser programme for 1983. Each column represents one farm

Nonetheless, even at present-day prices it is possible to reduce nutrient application and thus reduce the nutrient load in Laholm Bay. A study of thirty-six representative farms in the area found that actual nutrient application on many farms was considerably above the economically most profitable level and that only a few farms had a sub-optimal application rate (fig. 2.9 and 2.10). By adjusting nutrient application to the economically optimal level, it is estimated, annual net farm income could be improved by an average of Skr 100–150 per hectare on the farms studied and at the same time, by eliminating the overuse of nitrogen, leaching would be reduced considerably (fig. 2.5). In addition, by marketing surplus manure from intensive livestock farms to farms without their own manure it is possible to obtain further improvements in profitability and reductions in leaching.

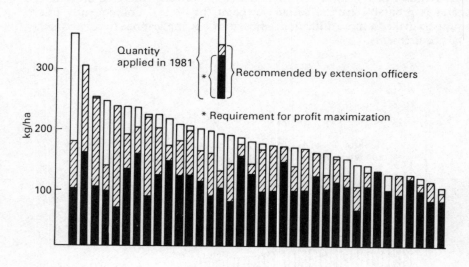

Figure 2.10   The average supply of nitrogen on thirty-six farms in 1981: the
nitrogen requirements of the crops grown and the proposed fertiliser programme,
using the best possible technology and prevailing planning conditions but not
corrected for the residual effect of manure. Each column represents one farm.
In all cases the levels refer to the total nitrogen content of the manure, which is
considerably higher than the level of plant-available nitrogen

The above study and the resultant extension activity on the thirty-six farms
took place in 1982/83 and were evaluated in 1984. The evaluation found that
a number of changes in agricultural management had occurred in order to
adapt the supply of nutrients to the requirements of the plants. The nitrogen
and phosphorus use, thus, had been reduced by 14 per cent and 38 per cent
respectively in comparison with the situation in 1981. In the province of
Halland as a whole, however, nitrogen utilisation increased slightly from
1981 to 1984. Phosphorus utilisation decreased slightly in Halland as a whole
but not as much as on farms which had received special extension advice.
Consequently, it has been concluded that specific extension inputs are one
means of improving the utilisation of plant nutrients and reducing the load of
nutrients emanating from agriculture.

Increased spreading of manure in the spring
About half the manure in the area is at present spread in the autumn.
Autumn spreading may cause serious nitrogen leaching, particularly if the
manure is spread on bare fields. A shift to spreading in the spring would

reduce the leaching of nutrients, while more nitrogen from the manure could be taken up by the crops. A shift to spreading only in spring requires, however, increased storage capacity. In addition, spring spreading may lead to soil compaction and delayed sowing, with yield reductions as a result.

The economic consequences of expanding manure storage capacity and increasing spring spreading vary strongly among different farms. The change in profitability may vary depending on animal species, type of manure, depreciation period and soil type, etc. The highest profitability is found in large expansions of slurry tanks for poultry manure on farms with large areas available for spreading and soils which are relatively tolerant to compaction. In most cases, however, it is unprofitable to expand storage capacity and to increase spring spreading. The economic cost of this extra storage in some cases amounts to more than SKr 50 per kilogram of reduced nitrogen leaching, particularly when only a small expansion of storage capacity is required and the farmer plans to cease keeping livestock within a ten-year period.

Increased grassland and catch crops
It has been estimated that nitrogen leaching from annual crops such as grain, potatoes, sugar beet, etc., is about 40 kg/ha/year in the Laholm Bay area and only 10 kg/ha from grassland in this area. Leaching from forest land is estimated at 2 kg/ha/year. As a result, it has been calculated by the working group that an increase in the proportion of grassland from 33 to 38 per cent and a corresponding reduction in the area of spring cereals would reduce leaching by 100 tonnes of nitrogen per year in this area.

Cultivation of catch crops may lengthen the growing period and thereby reduce leaching in, for example, grain production. The catch crop can be sown in the main crop. After harvest of the main crop the catch crop continues to grow and takes up part of the nitrogen which otherwise would have been lost during the autumn and winter. In experiments catch crops have reduced the nitrogen leaching by about 10 kg of nitrogen per hectare.

A reduction in leaching of 10 kg of nitrogen per hectare would imply a reduction in the load over the entire area of about 200 tonnes of nitrogen. On farms which are able to utilise the catch crops as feed, this may be a profitable measure. Otherwise there will be an economic deficit of about Skr 200 per hectare and year.

The most important measures considered by the working group both within agriculture and in other sectors with regard to costs and reduced nutrient leaching are summarised in table 2.7. In this context the figures referred to are farm-economic costs. The socio-economic costs probably coincide with the farm-economic costs except for 'reduced N-fertilising by 30 per cent of the optimal level for grain production'. A reduction of this order in the intensity of fertiliser use will not lead to socio-economic net costs at the prevailing low level of world market prices for grain. Instead the reduction is socio-economically profitable even when the positive environmental effects are disregarded. The problem is to find means of achieving the reduction.

In order to introduce measures to reduce leaching in agriculture the working group proposed, among others, the following measures:

Table 2.7 Estimates of costs and benefits of measures for reducing the nitrogen load

Measure	Investment cost (SKr million)	Annual cost (SKr million)	Reduced load (N tonnes/year)	Cost (SKr/kg N)
Sewage Works				
80% N reduction at the five largest sewage works	25	3–5	200	15–25
Rural dwellings				
More efficient treatment than sedimentation	50	5–7	9	500–800
Agriculture				
Extension of storage capacity for manure	84	5–10	300	18–35
Cultivation of catch crop		4	200	20
Adjustment of agriculture within economic possibilities		0[a]	100?	0
Reduced N application by 30% of optimal level in grain production[b]		2–5?	50–100	20–100?

(a) Requires farm water coonservation planning corresponding to an annual cost of about SKr 1 million.
(b) The basis of the calculations is relatively uncertain.

1 Information and farm-directed water conservation planning in order to improve or at least not undermine economic profitability by requiring all farms with more than 10 ha of arable land or 30 animal units to draw up a nutrient balance sheet. If manure production at best utilisation level exceeds the nutrient requirement for the farm the surplus manure should be spread on other farms.
2 The introduction and enforcement of regulations in accordance with the Environment Protection Act for eight to ten months storage space for manure.
3 Increased control to ensure that the regulations are observed.
4 Area subsidies of Skr 200–300 per hectare for catch crops and investment subsidies amounting to 50 per cent when existing manure storage facilities are expanded.

5 Access to the area and investment subsidies as a form of compensation when farmers decide to reduce livestock numbers rather than up grade their manure storage facilities.

The costs of extension inputs, monitoring and contributions could be covered within the framework of the environment tax placed on fertiliser and pesticides. No measures to reduce nitrogen fertilisation below the economic optimum for the farm are proposed.

The working group stated that a reduction in the area of grain cultivated in the Laholm Bay region is desirable with regard both to nutrient losses and to the grain surplus problem. Greater attention to environmental effects in the context of reducing surplus production is necessary. The proposals of the working group have not yet led to any decisions (March 1987).

Impact of policy options on other regions

The measures introduced or proposed for Lake Ringsjön and the Laholm Bay have had negligible influence on the total production volume of agriculture. Consequently they do not influence the country's surplus situation, or agriculture in other parts of Sweden. Within the prevailing institutional structure, however, the environmental authorities and individual farmers have not yet had sufficient incentive to undertake environmental measures which are unprofitable to the firm because of surplus production but are profitable from a socio-economic perspective. It should be emphasised also that in the Laholm inquiry there were discussions on the reduction of stocking density as well as of cereal growing in order to reduce nitrogen leaching. But policy decisions at a local level still have to be made.

Alternatively, measures could have been chosen which both reduce nutrient leaching and reduce the production of surplus products. Examples of such measures are:

1 A drastic reduction in the application of fertilisers.
2 Reduced livestock-keeping and thus reduced manure production.
3 Alternative land use (energy forests, conventional forests, energy grass, extensive grazing, etc.).

The latter two measures have been introduced and proposed, respectively, in order to limit production at the national level, but as yet there has been no attempt to focus these measures on the particularly pollution-sensitive areas of southern Sweden.

If one or more of these measures are chosen for the designated pollution-sensitive areas, the need for production-limiting measures would decrease in other parts of Sweden. As discussed above, in the year 2000 it is estimated that Sweden will have a surplus of between 500,000 ha and 900,000 ha of arable land. From the viewpoints of nature conservation and regional economics, it is important that as little as possible of the reduction in the area cropped takes place in the forested regions with little agricultural activity. In such areas it is important that agriculture continues in order to maintain biological variety and quality. In many forest areas alternative job opportunities are also lacking and consequently agriculture is important for employment. In the particularly pollution-sensitive areas in the agricultural regions of southern Sweden, more grassland or broadleaved woodlands would enrich the countryside. At the same time, the labour market in these areas is better than in many forest areas.

On the other hand, traditional efficiency targets must also be considered in reaching a decision on when to change, for example, from grain production to alternative land use.

Assessment of the potential for better integration of environmental and agricultural policies

There appear to be good possibilities of combining the objective of reducing fertiliser and pesticide use and their negative environmental effects with the objective of improving farm income. A national programme to reduce pesticide applications by half could simultaneously improve profitability in agriculture by SKr 100 million a year. Investigations at the farm firm level in the Lake Ringsjön area and in southern Halland demonstrate that a reduction in the application of phosphate fertilisers by 30 per cent together with a reduction in nitrogen application would improve profitability by about SKr 150 per hectare. In Sweden, increasing resources are being made available for extension advice in support of such profitable and environmentally favourable adjustments. In addition, a long-term research programme into alternative forms of agriculture is being started. This programme is based on the assessment that new technology can be developed which is both environmentally beneficial and efficient in the traditional context.

Perhaps the most important agricultural policy objective at present in Sweden is to reduce surplus production. Environmental considerations have been considered only to a limited extent in inquiries and decisions on measures to limit production. The only control measures which have both production-limiting and environmental objectives are the charges and taxes on fertiliser and pesticides. It is still too early to determine the impact of these charges and taxes on the application of chemicals, on the environment and on production volume. The greatest effect has likely been the generation of revenue used to pay for export subsidies, research and extension inputs, but the overall impact of these measures has probably been limited.

During recent years a number of measures have been taken to reduce livestock output. If the reductions had been concentrated on firms and regions with high stocking levels producing excessive amounts of manure, then significant environmental improvement might have been achieved, since large application rates of manure pose greater risks of nutrient leaching than fertiliser. However, no specific measures have been taken to encourage reductions among such firms or regions.

A proposal has been submitted with regard to reducing grain surpluses. The main emphasis in this proposal is on energy forestry and conventional forestry up to year 2000. If this proposal is implemented it should reduce the leaching of nutrients. Naturally, the environmental advantage will be largest if the shift from grain to forests occurs in particularly pollution-sensitive areas.

The working group which proposed water conservation measures in southern Halland estimated that nitrate leaching would decline from 40 to 10 kg/N/ha/year as a result of a change from grain to grassland production. In forestry areas the leaching would probably be less than 10 kg/N. If grain production ceases there could also be socio-economic savings of from SKr 1,850 to Skr 3,650 per hectare and year (table 2.3), since world market prices are now so low that production for export would not cover the cost of production.

Table 2.8 Comparison between reduced grain production and enlarged manure storage capacity as water conservation inputs in southern Halland (both measures reduce the nitrogen load by 300 tonnes per year)

Alternative measure	Socio-economic cost or saving (SKr million/year)	Reduced load (tonnes N/year)
Larger storage space for manure	5–10 (cost)	300
Transfer of 10,000 h of grain to grassland or forest[a]	18·5–36·5 (saving)	300
Gain obtained by reducing grain production instead of enlarging the manure storage capacity	23·5–46·5[b]	0

(a) Grassland and forests are assumed to give a socio-economic net yield of 0 SKr/h. Grassland can be utilised for energy or grazing. If it is utilised for grazing, then production must be very extensive, with low inputs of (e.g.) work, essentials and building investment. More intensive production with larger costs leads to large socio-econoomic deficits at the prevailing low levels of world market prices of meat.

(b) This gain is achieved at an export price for grain of 40–65 SKr/dt. At 115 SKr/dt (the domestic producer price) the two alternatives will be almost the same (−9 to +10 million SKr/year). If consideration is given to the fact that continued grain production gives more labour income on the farm and opportunities to utilise existing machinery, buildings and land improvements, continued grain production will be the best business-economic solution for most farmers.

Source: calculations by the author on the basis of material discussed in table 2.3.

In addition, the shift from grain production to grassland or forestry would reduce the need for further increasingly expensive water conservation measures. An example illustrates that large cost savings could be made if reduced grain production were utilised as a means of reducing leaching. By transferring one sixth of Sweden's cropped area (10,000 ha) to forest, instead of extending manure storage capacity, it would be possible to save SKr 23·5 million to SKr 46·5 million a year. Both measures would reduce nitrogen leaching by 300 tonnes/year. The world market price of grain would have to rise to the same level as the Swedish domestic price (a two- or threefold increase) before expansion of manure storage capacity became economically equivalent to replacing grain production with grassland or forests.

Corresponding socio-economic profits should be possible in the Lake Ringsjön region by growing grass or forest in particularly erosion-sensitive areas.

These socio-economic measures, attractive as they may be, are difficult to introduce at the farm level, given current product price support. Given the high prices received by farmers for their products, it would be expensive to shift from grain to grassland production or to forest. The commission developing measures to reduce the grain surplus proposes, however, that energy forestry and planting of conventional forests should be encouraged through subsidies. Such subsidies would make forestry economical, even in areas with agricultural potential. Opportunities for better integration of environmental and agricultural policies by targeting the subsidies to areas where reduced grain production would be environmentally desirable, however, have not been considered by the commission.

Table 2.9 Socio-economic and environmental implications of barley production and alternative land uses such as forestry or permanent grassland

	World barley price (SKr/kg)		N leaching (kg N/ha)
	0·40	0·80	
Barley production with 90 kg N/h	−1,600	−200	40
Barley production with 20 kg N/h	−1,500	−300	30–35
Alternative land use[a]	0	0	10

(a) Grassland with extensive grazing, enery forestry or conventional forestry.

Source: Author's own calculations, based on fig. 2·9.

Given the current low world market prices for grain it also appears socio-economically profitable to reduce nitrogen fertilisation rates. This has been illustrated by an economic assessment of fertilisation trials on barley, conducted in southern Sweden. The evaluation demonstrates that about 100 kg/N/ha optimises profits to the farm at the present level of domestic product price support but that at world market prices the optimal amount of fertiliser to apply would be much lower (table 2.9). Product quality and the residual effect of nitrogen fertilisation in future years were not considered in the calculations. Consequently it is not likely that the optimum fertilisation rate

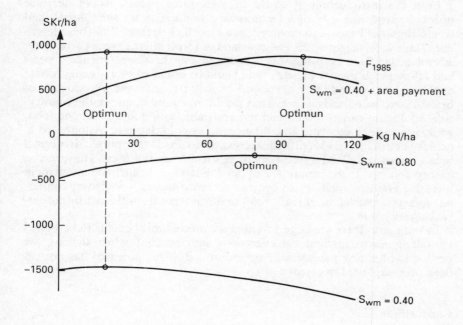

Figure 2.11 The relationship between nitrogenous fertilisation and total revenue less total costs (exclusive of land and basic machinery) for barley. F_{1985} = farm firm economics at 1985 prices. $S_{wm} = 0.40$: socio-economics, world market prices for barley, according to table 2.3, of 0.40 SKr/kg; fertiliser prices exclusive of charges and taxes. $S_{wm} = 0.80$: socio-economics, world market price for barley of 0.80 SKr/kg; fertiliser prices exclusive of charges and taxes. Area payment 2,400 SKr/ha. *Source:* author's calculations, based on data from the inquiry into means of reducing the supply of nitrogen and phosphate to Laholm Bay

would decrease as much in reality as the figure suggests. In addition, the introduction of world market prices would have different effects for different crops.

One of the problems connected with removing price support is that agricultural production would yield large economic deficits, and while world prices remain low, production would probably cease in most of the country. The removal of product price support would thus jeopardise the income and production objectives of agricultural policy. There is, however, one way of solving these problems. The introduction of world market prices could be supplemented with an area payment. In fig. 2.11 an example has been included showing world market prices plus an area payment of SKr 2,400 per hectare. Compared to the present economic situation the result would be lower optimal nitrogen rates but approximately similar profitability.

Both the introduction of world market prices, which would decrease nitrogen rates, and subsidies for grassland or forestry on agricultural land would thus yield socio-economic savings as well as decreased nutrient leaching. Table 2.10 suggests that the grassland or forest alternative is particularly advantageous from both the socio-economic and the environmental aspects and offers much greater benefits than could be obtained by reducing nitrate application rates. In highly developed agricultural areas the introduction of broad-leaved woodlands or grassland would, in addition, enrich the countryside. Biological variety would increase and game would find cover and food, while exposure to the wind and wind erosion would decrease. As opposed to fields of grain, forests also provide recreational areas for the public. In normal broad-leaved woodlands, fertilisers or pesticides are not used. However, in energy forestry these inputs are applied but to a smaller extent than in intensive grain production. In contrast to broad-leaved woodlands, coniferous forests on agricultural land entail disadvantages from the nature conservation viewpoint.

In summary, there are large potential socio-economic benefits to be gained by taking environmental considerations into account when looking for production-limiting measures in agriculture. But this potential has not yet been fully exploited in Sweden.

Conclusions

During the post-war period, agriculture has undergone rapid structural change, contributing to the fulfilment of the efficiency, productivity and income goals of Swedish agricultural policy. Structural changes have also had environmental effects. Crop production has become more specialised, with less grassland and more grain increasing the need for pesticides and nutrient leaching. Animal husbandry has been concentrated in certain regions and farms, while many other farms have ceased keeping animals. This has hindered efficient utilisation of manure and farms lacking livestock have become completely dependent on artificial fertiliser for their supply of plant nutrients. On other farms which specialise in livestock production high manure application rates have led to high levels of nutrient leaching.

The fundamental reasons for structural change in agriculture have been technical development and altered price relationships. Wage levels have increased considerably faster, for example, than the prices of chemicals used to improve production. Agricultural policy has facilitated and speeded up structural change. The environmental impact of this change started to attract attention in agricultural policy during the late 1970s. In agricultural policy decisions made in 1985, achieving environmental goals became one of the major objectives.

An important part of agricultural policy involves price support, so that domestic producer prices are considerably higher than world market prices. The aim of price support is to fulfil the income and production goals of agricultural policy. However, price support also has environmental consequences. Firstly, farmers use more fertiliser and pesticides than is socio-economically and environmentally desirable. Secondly, it means that agriculture can be sustained in many places where the land would be taken out of production or planted with forest trees if price support was not available.

In Sweden price support in combination with technical development has led to considerable surplus production of agricultural products and large related 'export costs'. A central objective of agricultural policy during recent years has therefore been to reduce surplus production. Attempts have been made to solve this problem without abandoning price support. The measures taken or proposed in official inquiries into the limitation of production generally also have environmental consequences but attempts to combine reduced production with improving the environment have not been fully exploited. This is illustrated by the following examples:

1 Measures have been taken to reduce animal production, which leads in turn to reduced manure production, but these measures have not focused on farms or regions with surplus manure production and soils sensitive to leaching, nor have these measures been directed at highly intensive livestock firms and regions.
2 In 1987 farmers who increased fallow areas received compensation, although the increased area of fallow increased nutrient leaching.

The only measure consciously designed both to reduce the surplus and to improve the environment is the charge and tax on fertiliser and pesticides. These charges and taxes, used to finance export costs and pay for research and extension inputs for environmental protection, have probably led to a slight reduction in the use of chemicals, but this has not been quantified. Very high charges on fertilisers would be required to reduce their use to the point where surplus production would be significantly reduced. Such high charges would create great problems in other sectors where the reduced use of fertilisers is not so desirable from the production viewpoint, such as forestry, horticulture and energy forestry.

Attempts to solve environmental problems associated with intensive crop production, fertilisers and pesticides in Sweden largely involve measures outside the price structure of agricultural policy. Extension inputs, research and development and also legislation are the most common approaches which have been tried. Available data suggest that there is good potential for using extension inputs and research and development to achieve simultaneously the reduced use of chemicals, environmental improvement and better profitability. The following examples may be cited:

1 Government authorities in the agricultural and environmental sectors have presented a programme designed to reduce the use of pesticides by half within five years to reduce related environmental risks without reducing yields and are expecting potential improvements in farm income of approximately SKr 100 million per year.
2 In the Lake Ringsjön area it is possible to reduce phosphorus fertilisation by 30 per cent and simultaneously improve farm profitability by SKr 150 per hectare per year.

3 On about forty representative farms in the Laholm Bay region it would be possible to reduce phosphate fertilisation by 30 per cent and nitrate fertilisation by 20 per cent, while simultaneously improving profitability by SKr 150 per hectare per year.

Legislation has been introduced to ban particularly hazardous individual pesticides. In addition, there are new regulations on pesticide use and spreading. For example, aerial spreading is now banned. Neither pesticide nor fertilisers, however, have quotas on their use. Requirements and restrictions are more comprehensive as regards storage and spreading of manure than for fertilisers.

The restrictions in force in designated pollution-sensitive areas described above apply mainly to the storage and disposal of manure and entail considerable costs for many farmers. Restrictions which lead to loss of profitability, however, are not applied to fertilisers.

Measures introduced or proposed aimed at reduction of nutrient leaching in pollution-sensitive areas have not been designed to simultaneously reduce surplus production. The following three measures, among others, would reduce both water pollution and surplus production:

1 First, a shift from grain production to grassland for extensive grazing, energy forestry or conventional forests, rather than increasing manure storage capacity, as both measures reduce leaching by approximately similar amounts and the savings from extensification are greater.
2 Second, reducing livestock production on farms where manure production exceeds the nutrient requirements of crops.
3 Third, reducing the use of fertiliser and pesticides to a level which is socio-economically profitable in relation to current low world market prices for agricultural products.

The problem with these measures is that notwithstanding their positive impact on the socio-economic level they are unprofitable at the farm level. This problem could be solved by making the proposed incentives for forestry and grassland higher in the designated pollution-sensitive areas. Correspondingly, measures to reduce livestock production could be targeted on particularly pollution-sensitive areas. In order to accomplish this third measure and simultaneously maintain agricultural income levels it would be necessary for area-related payments per hectare to be introduced as present price support policy is phased out. If water pollution originating from agriculture is primarily a local problem, then measures 1 and 2 are preferable to measure 3. From the socio-economic and environmental aspects, measure 1 should be preferable to measure 3.

In conclusion, the above examples demonstrate that:

1 There is good potential for using extension inputs and increased research and development to combine reduced application of chemicals with environmental improvement and better profitability in agriculture.

2   There is considerable scope for improving socio-economic returns and
    creating a better environment by paying greater attention to environmen-
    tal aspects when designing production-limiting measures in agriculture.

References

Most of the references are written in Swedish. In some cases there are summaries in
English.

Andersson, A.-K. (1985), *Water Management Measures in the Lake Ringsjön Area—an
    Analysis of the Economic Impact at Farm Level*, Report 257, Department of
    Economics and Statistics, Swedish University of Agricultural Sciences (English
    summary).
Andersson, R. (1986), *Losses of Nitrogen and Phosphorus for Arable Land in Sweden.
    Magnitude, Regulating Factors and Measures Proposed*, Swedish University of
    Agricultural Sciences (English summary).
Bergström, L. (1986), 'Nitrate leaching and drainage from annual and perennial crops
    in tile drainage plots and lysimeters', *J. Environ. Qual.*, 15.
Brink, N., and Lindén, B. (1980), 'What happens to the nitrogen in commercial
    fertilisers?' *Ekohydrologi*, 7, Swedish University of Agricultural Sciences (English
    summary).
County Administration of Malmöhus Province (1985), *Proclamation on the Regulations
    in the Catchment Area of Lake Ringsjön*.
County Administration of Malmöhus Province (1986), *Lake Ringsjön—a Lake in the
    Process of Transformation*.
Ds Jo (1986), *Measures to reduce Short-term Grain Production*, Ministry of Agriculture.
Ds Jo (1986), *Measures to reduce the Grain Surplus*, Ministry of Agriculture.
Gummesson, G. (1987), 'Can chemical control of weeds be halved with retained
    profitability?' *Weed and Plant Protection Conference*, Swedish University of
    Agricultural Sciences (English summary).
Ivarsson, K. and Brink, N. (1985), *Leaching from a Fine Sand Soil in Holland*,
    Swedish University of Agricultural Sciences.
Joelsson, A., *et al.* (1986), *Measures to reduce Nitrogen and Phosphorus Leaching into
    Laholm Bay*, Halmstad.
Joelsson, A., and Kumm, K.-I. (1984), *Advisory Work concerning the Application of
    Crop Fertilisers*, SNV pm 1852, National Environment Protection Board (English
    summary).
Kjellerup, V., and Kofoed, A. Dam (1983), 'Influence of nitrogeneous fertilisers on
    leaching of plant nutrients', *Tidsskrift Planteval*, 83, 330–48.
Mattsson, C. (1986), *An Economic Analysis of the Application of Nitrogen Fertilizer and
    Manure Handling*, Report 265, Department of Economics and Statistics, Swedish
    University of Agricultural Sciences (English summary).
Mattsson, R. (1985), *Agricultural Development in Sweden*, 'Aktuellt' series of reports
    from the Swedish University of Agricultural Sciences, 344, Uppsala.
National Board of Agriculture, National Environmental Protection Board and
    National Chemical Inspectorate (1986), *Proposal on measures to reduce health and
    environmental risks when using pesticides*.
National Environment Protection Board (1986), *Agriculture and the environment: a
    programme of action*.
Prop 1983/84: 107, Appendix 6. The section of the research policy proposal dealing
    with the Ministry of Agriculture, Ministry of Agriculture.

Research Council for Forestry and Agriculture (1985a), *The Research Council for Forestry and Agriculture takes stock*.

Research Council for Forestry and Agriculture (1985b), *The Research Council proposes . . .*

SOU (1983), *Use of plant nutrients. Report of Inquiry into the use of Chemicals in Agriculture and Forestry*, Ministry of Agriculture.

SOU (1983), *Control of Plant Pests and Weeds. Report of Inquiry into the use of Chemicals in Agriculture and Forestry*, Ministry of Agriculture.

SOU (1984), *Agriculture and Food Policy. Report of Food Committee*, Ministry of Agriculture.

Uhlin (1982), in Research Council for Agricultural Forestry (1985a), p. 11.

3. The effects of US agricultural policies on water quality and human health: opportunities to improve the targeting of current policies

R.M. Wolcott, S.M. Johnson
and C.M. Long

Introduction

The problem

When chemical fertilisers and pesticides were first used on a large-scale basis, farmers and consumers praised the benefits of increased productivity and relatively low prices. However, in recent years, evidence that these chemicals may pose substantial threats to human health and the environment has led to questioning of the net benefits of chemically intensive agriculture. The increased concern over agricultural chemicals in combination with the financial instability of the agricultural sector, and escalating federal farm programme costs, indicates that improved integration of agricultural and environmental policy could lead to improvements in farm income and environmental quality.

Most fertilisers and pesticides in the United States are used to ensure high crop yields. If high crop yields were associated with high net farm income, then negative externalities from agricultural chemical use could represent an acceptable social trade-off. However, the American agricultural sector has suffered from the most severe economic downturn since the Great Depression owing to overproduction and unfavourable foreign market conditions. Federal farm programme participants producing rice, corn, wheat, cotton, and soya beans at supported prices and under conditions of supply control through acreage reduction applied 90 per cent of all herbicides and insecticides used domestically (Conservation Foundation, 1986).

The present condition of US agriculture provides a unique opportunity to integrate environmental and agricultural policies that address farm income and commodity supply. Historically, federal and state agencies and interest groups concerned with these goals have tended to act in relative isolation. The farm community has focused generally on policies designed to improve crop yields, technology, markets and incomes. Meanwhile the environmental community has supported policies to reduce human health risk and environmental degradation.

Recently many agricultural and environmental interests have been addressed in one piece of legislation, the Food Security Act of 1985. This Act limits conversion of highly erodible uncultivated land into crop production through the 'sodbuster' provision. The 'swampbuster' provision limits the conversion of wetlands. It also requires conservation compliance to address soil erosion concern. In addition, the Conservation Reserve Programme idles crop land that is highly erodible or located adjacent to streams, lakes and estuaries from crop production for a period of ten years.

Intensive agriculture in the United States

Agricultural production in the United States today is quite unlike that of fifty years ago, when farms were diversified and agricultural chemical use was virtually non-existent. At that time, farmers tended to mix crop and livestock production and grow a variety of crops in rotation. This form of agricultural production was labour- and land-intensive. Between 1945 and 1985 agricultural productivity doubled as off-farm inputs were introduced to the production process (Fedkiw, 1986). Several institutional, economic and technological forces acted together to transform agriculture into its present highly specialised, capital- and purchase-input-intensive state.

The primary institutional forces stimulating the transformation in agriculture were agricultural research, education and extension services. These programmes were initiated in the late 1960s in response to farmer concern over low productivity and farm income. The resulting US Department of Agriculture and land-grant college systems were instrumental in advancing plant breeding, plant and animal disease control, and farm equipment technology. By 1920 these systems had developed hybrid seed corn, drought- and disease-resistant varieties of wheat, and mechnical harvesters. Most of these improvements, however, were not widely adopted until after World War II. Adverse economic conditions associated with the Great Depression precluded the necessary investment (Cochrane, 1979).

The outbreak of World War II and the subsequent high demand for agricultural products that lasted until the 1960s was the major economic force that enabled US farmers to invest in agricultural technology. Between 1933 and 1970 farmers increased machinery use by 212 per cent and increased purchases of feed, seed and livestock by 270 per cent (Cochrane, 1979).

The greatest change, however, was in the use of agricultural chemicals, which increased by 1,800 per cent from virtually no use in the 1930s. Nitrogen use alone increased by 1,600 per cent (see fig. 3.1). Meanwhile,

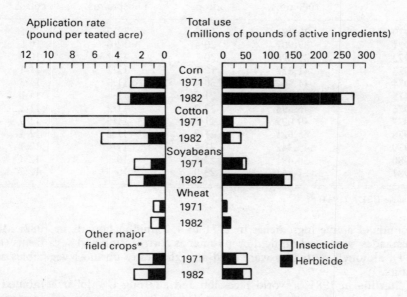

Figure 3.1 The rate of herbicide and insecticide use in the United States, 1971 and 1982. Other major field crops include sorghum, rice, peanuts, tobacco, rye (1971 only), barley, oats, hay and alfalfa. *Source:* US Department of Agriculture, cited in Conservation Foundation (1986)

land under cultivation remained essentially constant and labour inputs decreased by 7 per cent. While all these advances in farm technology contributed to decreased labour requirements and increased yields per acre throughout the period, the greatest source of increased productivity was from commercial nitrogen, potash and phosphorous fertilisers (Cochrane, 1979).

By 1970 a second major economic force began to further transform farming into a more specialised and purchased-input enterprise. The combination of a declining US dollar, world food shortages and improved economic conditions in foreign countries greatly increased export demand for major crops (Conservation Foundation, 1986). By 1976, United States farmers supplied 40 per cent of all wheat, 65 per cent of all coarse grains and 64 per cent of all soya beans in international trade (Conservation Foundation, 1986). While farm output increased by 51 per cent during the 1970s, most of the increase was due to cropland expansion of 55 million acres (Cochrane, 1979).

Between 1970 and 1981 total fertiliser inputs increased from 38 million tons to 52 million tons (table 3.1). During this period, primary nutrient use (nitrogen, phosphorus and potash) increased from 21 million tons to 23 million tons. Meanwhile, total pesticide use increased from 405 million

Table 3.1   Use of ferilisers in the United States, 1970–81

Year	Quantity (000 tons)	% nitrogen	% phosphorus	% potash
1970	38,292	19·5	12·0	10·5
1971	39,902	20·4	12·0	10·6
1972	39,896	20·1	12·2	10·9
1973	41,822	19·8	12·2	11·1
1974	44,964	20·4	11·3	11·3
1975	40,606	21·2	11·1	11·0
1976	46,893	22·2	11·2	11·1
1977	49,099	21·7	11·5	11·9
1978	45,621	21·8	11·2	12·1
1979	49,342	21·7	11·4	12·7
1980	50,491	22·6	10·8	12·4
1981	51,761	23·0	10·5	12·2

Source: USDA (1984)

pounds of active ingredients in 1971 to 452 million pounds in 1982. Most pesticides were used on field crops such as corn, cotton and soya beans (fig. 3.1), although pesticides were used at higher rates on most vegetables and fruits.

Starting in 1981, a world recession and a strong US dollar combined to reduce United States export demand drastically (Conservation Foundation, 1986; Fedkiw, 1986). Between 1981 and 1986 the value of US farm exports declined from $44 billion to $26 billion. Corn, wheat and soya bean crops were most affected, experiencing an average 40 per cent decrease in market price. Market prices for rice and cotton fell more substantially but accounted for smaller shares of total exports. For federal farm programme participants, incomes were protected by the Food Security Act of 1985, which made payments to farmers who idled a portion of their acreage. Approximately 30 per cent of the cropland base in 1986 was idled owing to reduced acreage and paid diversion programmes. The Conservation Reserve became a more important source of cropland decreases after 1986.

In spite of the downward trend in market prices, agricultural production remained at high levels, owing to government price support programme incentives and unusually favourable weather. Between 1981 and 1986 federal expenditures increased from $4 billion to over $25·5 billion. As a by-product of federal intervention to support prices and maintain farm income, massive surpluses mounted. By 1985 approximately 43 per cent of maize production, 25 per cent of soya bean production, 78 per cent of wheat production and 73 per cent of cotton production were surplus stocks (Jones, 1987). According to research conducted by the Food and Agricultural Policy Research Institute (FAPRI), market prices will remain below target price levels through to 1996 (fig. 3.2).

During the period between 1981 and 1986 total commercial fertiliser use decreased from 54 million tons to 44 million tons. In terms of primary

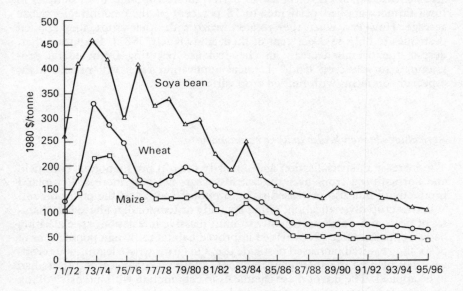

Figure 3.2 The real US Gulf port prices of maize, wheat and soya beans, 1971/72 to 1985/86: crop years and projections to 1995/96. *Source:* US Food and Agricultural Policy Institute, No. 3–80

nutrient use, consumption fell from 24 million tons to 20 million tons (USDA, 1987). Meanwhile, pesticide use increased from 452 million pounds in 1982 to 507 million pounds in 1984 and fell to 476 million pounds by 1986. During the 1960 to 1986 period, consumption of nitrogen fertilisers experienced a nearly fourfold increase (Taylor, 1988). Future consumption of both fertilisers and pesticides is projected to decline. The major reason for the expected decrease in agricultural chemical use is that crop land will continue to be taken out of production through acreage reduction restrictions governing participation in farm programmes and the Conservation Reserve Programme (USDA, 1987).

The combination of high export demand, which started in the 1940s and greatly increased in the 1970s, and federal price support payments induced farmers to specialise in growing fewer crops and intensifying capital input use. Moreover, as the technology improved many farmers such as livestock producers switched to field crop production to reduce their need for labour. Meanwhile, production of horticultural crops (fruit and vegetables) became increasingly specialised on farms and concentrated in areas such as the Pacific and Gulf coasts.

Each of these forces has worked to create a production system dependent on agro-chemicals. Farmers not only substituted chemicals for labour, they also increased their chemical use to permit monocropping. For example, in Iowa farmers applied pesticides to 78 per cent of their continuous maize acreage. However, when they rotated maize with other crops, they applied pesticides to only 15·3 per cent of their land (Kuch, 1987). It appears that, despite the overall decrease in chemical use resulting from federal programmes to idle crop land, chemical application rates will remain high, especially on farms with limited crop variety.

Agriculture, human health and the environment

The increase in specialisation and intensity of crop production of both field and horticultural crops over the last fifty years not only increased cropland productivity but also increased risks to public health and the environment. Whereas crop diversification, either spatially or temporally, altered the habitat of pests and weeds sufficiently to limit massive infestation, specialisation had an opposite effect. It provided improved habitat to sustain populations of pests and required increased pesticide use. On farms where leguminous crops were not planted in subsequent rotations, additional chemical fertilisers had to be applied. The use of these chemicals in conjunction with intensive tilling practices led to erosion of sediment and movement of pesticides and nutrients into surface waters and subsequently ground water, causing considerable concern over adverse effects on human health and the environment.

Public health effects
Pesticides are found in all environmental media (food, air and water) and public exposure to them occurs through ingestion, dermal contact and inhalation. People are exposed to residues of pesticides in and on food. Residues of pesticides and fertilisers are found in ground and surface waters. During pesticide application, farm labourers and neighbouring populations may be exposed through dermal contact or inhalation. While some pesticides have been found to cause a wide spectrum of cancer and non-cancer health effects, many chemicals remain untested and their health effects are unknown. Some chemicals do not appear likely to cause problems at the expected levels of exposure.

Farm workers generally receive the highest total exposure to pesticides and, as a result, incur the highest individual health risks. Pesticide exposure in food, although generally much lower than worker exposure, results in the most significant national aggregate mortality and morbidity risk because of the larger exposed population. In 1987 EPA ranked pesticide-related cancer and non-cancer risks, primarily related to food residues, highest among all forms of public health risk. In spite of this consensus, little is known of the absolute aggregate public health risk of consuming food containing pesticide residues. Likewise, little is known of the absolute size of the aggregate risk associated with the other exposure paths.

Only one study has attempted to provide estimates of the magnitude of dietary cancer risk from consumption of pesticide-contaminated food. In 1987 the National Academy of Sciences Board of Agriculture examined the cancer risk from dietary exposure to twenty-eight pesticides applied in fruit and vegetable production. It concluded that the mean risk associated with continuous consumption of these substances at the tolerance levels ranged from one in a thousand to one in ten thousand. However, it is noted that not all foods contain these pesticides and, if they do, they will usually be well below the tolerance level.

In the last decade, pesticide ground-water contamination has become a growing concern. A continual stream of monitoring data is providing details on the magnitude of potential exposure through drinking water supplies. In 1985 EPA identified seventeen pesticides in the ground water of twenty-three states (USDA, 1985). Later, it found several additional pesticides in ground water throughout the United States. EPA concluded that this contamination resulted from normal, recommended pesticide use. In addition to finding pesticides, in 1984 the United States Geological Survey (USGS) found ground-water contamination from fertiliser nitrates in over 20 per cent of the 124,000 wells sampled.

Environmental effects
In addition to potential adverse human health effects from intensive agricultural practices, evidence of pervasive damage to terrestrial and aquatic ecosystems from both intensive and extensive practices continues to mount. In 1979 EPA found that cropland run-off was the primary cause of water quality problems in 30 per cent of US stream miles (Judy *et al.*, 1984). In 1986 cropland soil run-off accounted for more than 30 per cent of the sediment and 40 per cent of the nitrogen entering US waterways (Gianessi *et al.*, 1986). Loadings of soil, nutrients and pesticides cause substantial damage to lakes, streams, estuaries and wetlands by blocking spawning grounds, depleting oxygen and damaging aquatic vegetation. The economic cost of cropland run-off to the United States in 1986 alone was estimated to be $2·2 billion (table 3.2) (Conservation Foundation, 1986).

Wetlands, in addition to other forms of wildlife habitat, are also threatened by conversion to agricultural use. In recent times, nearly 400,000 acres of wetlands have been drained or otherwise destroyed each year. Over 80 per cent of these losses have been attributed to agricultural production practices and expansion of the cropland base (Heimlich and Langer, 1986).

Integrated policy to protect human health and environmental quality

Evidence of adverse effects from agricultural chemical use continues to increase. As pesticide use in agriculture grows either as a result of market price or farm programme incentives, concern over the negative impact will intensify. Enlightened policy development to replace current incentives that encourage high chemical use and surplus production will require comprehensive analysis. Pesticide use is closely tied to policies for agriculture, environment and public health. The US Environmental Protection Agency, US

Table 3.2 Cost estimates of damages from erosion-related pollution in the United States ($ million at 1980 values)

Type of impact	Range of estimates	Single estimate	Crop land's share
In-stream effects			
Biological	No estimate	No estimate	
Recreational	950–5,600	2,000	830
Water storage	310–1,600	690	220
Navigation	420–800	560	180
Other	460–2,500	900	320
Sub-total in-stream	2,100–10,000	4,200	1,600
Off-stream effects			
Flood damage	440–1,300	770	250
Water conveyance	140–300	200	100
Water treatment	50–500	100	30
Other	400–920	800	280
Sub-total off-stream	1,100–3,100	1,900	660
Total effect	3,200–13,000	6,100	2,200

Source: Conservation Foundation (1986).

Department of Agriculture and the Food and Drug Administration share the responsiblity of implementing more efficient policies to achieve higher social benefits.

The Comprehensive Economic Pesticide Policy Evaluation System

Given the diversity of policies and their sometimes irreversible impact on agriculture and the environment, correctly anticipating policy outcomes and their interrelationships is essential. To examine the outcomes of these policy alternatives in this broadened context, the Comprehensive Economic Pesticide Policy Evaluation System (CEPPES) is being developed. This project was initiated in 1986 through a co-operative agreement between EPA and Iowa State University.

CEPPES is being designed as a comprehensive modelling system that will integrate economic, biological and geophysical systems to determine the impact of agricultural and environmental policies on pesticide use, agricultural sector performance, human health and the environment. The system should also identify key information gaps in assessing the comprehensive implications of pesticide and related policies that are designed and administered by different government agencies.

Organisation of CEPPES

CEPPES has four components for tracking the impacts of pesticide use—the policy component, the agricultural decision component, the biogeophysical component, and the health risk component. These four components will be able to integrate many data bases and models that yield pesticide fate and the costs and benefits of various pesticide application regimes (fig. 3.3).

The policy component will examine the effects of alternative agricultural and environmental policies on pesticide use, pesticide fate, environmental quality and the agricultural sector. Direct policies such as drinking water standards, pesticide use restrictions and taxes as well as indirect policies, including commodity programmes, conservation programmes and crop insurance or risk reduction programmes, can be examined.

The agricultural decision component will model the effects of alternative pesticide regulations and agricultural policies to determine the nature, extent and location of production and cultural practices. It will also evaluate changes in government programme costs, pesticide application, farm income, crop prices, surplus stocks and rents to fixed resources.

The biogeophysical component will incorporate climatic, biological and geophysical process models to determine the yield levels, yield variability and pesticide fate associated with alternative policy options. Data on the crop canopy, the crop residue, the soil condition and the crop in the previous year will be utilised to determine the extent to which rainfall will run off or be absorbed into the soil. In addition, this component will examine exchanges in the root zone, the unsaturated zone and the aquifers. To measure the availability of ground water and accompanying nutrients, data on previous crop rotations, cultivation practices and soil types will be utilised to determine the effects of crop production and cultivation choices on the condition of the soil surface.

The health risk component will utilise data on the population distribution of the United States, water supply sources and water supply treatment to estimate the health risks from consumption of contaminated water. It also will estimate the health risk associated with eating food containing pesticides. By integrating aggregate food supply statistics gathered from the USDA and pesticide data from EPA, implied consumption levels for pesticide residues will be estimated. Data on the application method, the climatic conditions and the location of pesticide application will be used to examine the inhalation risk. Agricultural data on patterns and frequency of pesticide applications will be used to determine applicator risk. Data on the pesticide application, method and climatic conditions will be used to determine the volatilisation of pesticides in the air.

To determine the health risks from the four types of pesticide exposure, a risk reference dose will be calculated. Accumulations of these doses incorporating time and population densities are then used to generate estimates of health risks or summary measures of health damage.

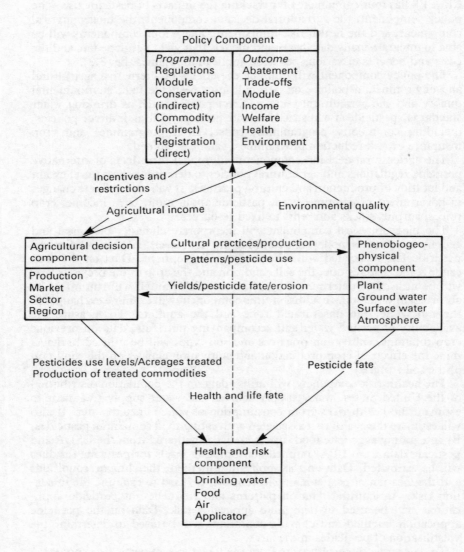

Figure 3.3 The organisation of the Comprehensive Economic Pesticide Policy Evaluation System (CEPPES)

Status of CEPPES

Although CEPPES is designed to be national in scale, not all the four components are ready for national-level analysis. Currently, only the agricultural decision component and the policy component are ready for this level of analysis. However, pesticide data which are used in both these components and the biogeophysical component are still limited to one pilot region located in the upper Mid-west. Consequently, CEPPES can at present provide two types of analyses—an economic analysis of the national agricultural sector and a partial agricultural/environmental analysis of the pilot region.

Currently, fertiliser data have not been included, although it will be possible to include them later. Therefore the health risks from fertilisers and combinations of fertilisers and pesticides cannot be assessed.

CEPPES will be able to model the economic, environmental and societal implications of specific pesticides for selected crops. Currently, CEPPES can only assess the impact from pesticides that are used on the major cash crops—maize, wheat, soya beans, cotton and rice—for the pilot region. While this modelling effort is important from the societal, economic and political perspectives, it does not address concern about the risks from chemical application in fruit and vegetable production. Because fruit and vegetables are produced on a limited scale, however, it may be more appropriate to use regional models for these crops.

The case study using CEPPES

Background

The Food Security Act of 1985 (FSA 85) introduced the Conservation Reserve Programme (CRP) to meet the dual goals of protecting environmental quality and limiting surplus crop production. The objective of the programme was to idle a total of 45 million acres of crop land for ten years by the 1991/92 crop year.

To be included in the programme, farmers had to calculate and propose a monetary bid (i.e. the rent they required) for crop land that met the Conservation Reserve Programme criterion. For the first five sign-up periods the criterion was that crop land had to be highly erodible land (HEL) according to a US government erosion definition.

Once all the bids were made for each sign-up period, the US government accepted only those bids under a specified bid level. This maximum bid level was based on the productivity of crop land and government budget concern. However, if the government wished to include additional land, it would increase the maximum bid level. The total bids were then constrained so that no more than 25 per cent of all crop land in each county could be enrolled.

As an additional incentive to participation, farmers in the Conservation Reserve Programme were offered a subsidy equivalent to half the cost of establishing plant cover. Additional incentives were provided for planting trees.

By the fifth sign-up only 23·1 million acres of crop land had been idled, most of which was located west of the Mississippi river, where wind erosion poses a more serious threat to agricultural resources. The relatively low participation implied that crop production would remain at a high level and its associated programme costs would be relatively high. In addition, concern began to be expressed about the lack of eligibility criteria specifically addressing surface water degradation, as well as loss of wetlands and wildlife habitat from agricultural production.

Therefore, in January 1988, new criteria were established that would allow strips of crop land adjacent to rivers, streams, lakes and wetlands, known as filter strips, to be eligible for the Conservation Reserve Programme. Filter strips, defined as 20 m to 30 m wide, were used as an additional criterion because they limit the movement of agricultural chemicals and sediment into surface waters. Their vegetative cover may also prevent the infiltration of chemicals into ground water. To some extent, they also provide habitat for wildlife.

In addition to the newly established filter strip criterion, another agricultural Bill was recently introduced in Congress known as the Nunn–Cochrane Bill. If passed, it would expand the Conservation Reserve Programme to 65 million acres and allow up to 35 per cent of the crop land in a county to be enrolled.

An additional Bill, Senate 2045, would create an Environmental Conservation Acreage Reserve Programme (ECARP) to idle 5 million to 20 million acres and expand the Conservation Reserve Programme by 5 million acres. For this Bill, however, the enrolment criteria are centred on water quality, soil damage, soil salinity, siltation and other problems stemming from pesticide use and from cultural practices.

Selection of a case study

To select an appropriate case study that would be representative of policies instrumental in maintaining agricultural production and limiting government programme costs, while minimising environmental risk, several criteria need to be met. These include ensuring that adequate data are available, that the policy option is relevant to the problem, and that the model results are generalisable.

Using these three criteria, it was apparent that a national case study evaluating both the targeting of filter strips within the current Conservation Reserve Programme and expanding it to 65 million acres would be appropriate. However, because of inadequate data, changes in farm income, cropping practices and pesticide use will be supplied only for the upper Mid-west pilot region.

Targeting Conservation Reserve Programme acreage using the filter strip criteria poses the possibility of both improving surface water quality and reducing government costs by lowering deficiency and other payments to farmers and controlling excess production capacity. Expanding the Conservation Reserve Programme to 65 million acres would likely result in a higher level of environmental protection, with cost savings to the government.

By selecting this case study, several goals can be met. First, the impact of targeting and expanding the Conservation Reserve Programme on the agricultural sector could be measured by determining changes in the level of crop production, the nature of cropping practices, farm income and acreage idled. In addition, the location of future Conservation Reserve Programme enrolment can be determined. Second, the impact on US Treasury outlays can be measured in terms of cost savings from altering the Conservation Reserve Programme.

Within the case study, four alternatives involving the two policy options were evaluated:

1 Targeting 5 million acres in filter strips out of the original 45 million acres in the Conservation Reserve Programme, hereafter referred to as 45/5.
2 Targeting 20 million acres in filter strips out of the original 45 million acres in the Conservation Reserve Programme (45/20).
3 Targeting 25 million acres in filter strips in the 65 million acres in the Conservation Reserve Programme as proposed by the Nunn–Cochrane Bill, using the current maximum bid level (65/25 low).
4 Targeting 25 million acres in filter strips in the 65 million acres in the Conservation Reserve Programme as proposed by the Nunn–Cochrane Bill, using a 40 per cent higher maximum bid level (65/25 high).

Option 3 was included to provide a scenario in which the government continued to accept bids from farmers up to the current maximum level established by the government. However, another very likely scenario is that the government's maximum bid level would need to be increased in order to obtain enough land to meet the goal of 65 million acres. Therefore an additional scenario is examined, known as 65/25 high. Under this 65/25 high option, bid prices were arbitrarily and uniformly increased by 40 per cent.

Methodology

The first step in developing the case study was to estimate the number of acres of filter strips by land class and county. Land class data were necessary to determine the number of acres that were highly erodible and the number of acres that were considered Class I, that is, land having low erosion potential but located in areas characteristic of filter strips. Data for making statistically acceptable county-level estimates do not exist. Therefore data from the 1982 and 1977 US Department of Agriculture National Resources Inventory were used to determine available filter strip land and the share of this land privately held on a county basis (Holt *et al.*, 1988). According to this method, there were 28 million acres of eligible filter strip land privately held. Most of this land is located in the upper Mid-west and in the other wetland areas of the United States.

A second step was to estimate the eligible filter strip acreage that was already included in the Conservation Reserve Programme because it met the highly erodible land criterion. As with the other filter strip data, records of this overlap are not available. Therefore the eligible filter strip acreage already included in the previous Conservation Reserve Programme sign-ups was assumed to be proportional to the eligible filter strip acreage found in all crop land. Using this method, the overlap estimated already to be included in the Conservation Reserve Programme was 3 million acres. Therefore 25 million acres were eligible for future Conservation Reserve Programme enrolment. For the 65/25 options, the width of the filter strips had to be increased to 300 ft to make the eligible acreage meet the targeting requirement of 25 million acres. For this option, the total acreage could be enrolled up to 35 per cent of the total cropland base in a county.

To model these options, a baseline for 1991/92 was developed (45/0), based on the first four sign-ups (17·7 million acres), past actions of the USDA concerning FSA 85, international agricultural policies and macro-economic conditions. International agricultural policies were assumed to remain similar to the current policies throughout the evaluation period. The macro-economic conditions were assumed to be similar to those currently prevailing in the United States and in the major international markets for programme commodities.

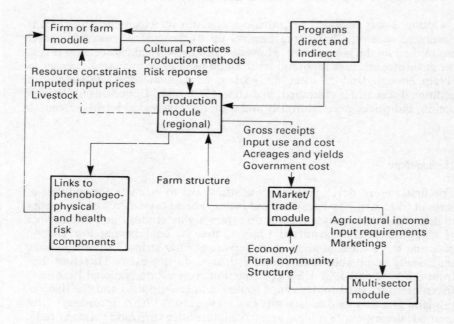

Figure 3.4   The interaction between the models used in the case study

The total quantity of land eligible for the Conservation Reserve Programme using both erosion and filter strip criteria was allocated to counties based on the distribution of land in the previous sign-up periods, using the quadratic programming model depicted in fig. 3.4. Cropland base acreage, which is used to compute commodity programme benefits, was projected similarly. It was necessary to include data on the base acreage because it could represent a source of savings to the US government through reducing the land base eligible for commodity programme benefits (Westoff and Meyers, 1988).

The analytical approach for evaluating the national case study was to use three modules of the Agricultural Decision Component of CEPPES (fig 3.5). These modules, known as the market, regional and national modules, were selected on the basis of the state of CEPPES development and the requirements for policy information.

The market module was used to determine where Conservation Reserve Programme enrolments would occur and to calculate changes in government cost savings, market prices, planted acres and base acreage associated with the commodity programmes. The regional module, representing an area of high filter strip eligibility, was used to estimate Conservation Reserve Programme costs of the targeting options and to indicate the impact of the Conservation Reserve Programme options on farming practices, e.g. changes in pesticide use, farm income and regional bid prices.

The national module, which comprises 105 linear programming models associated wth the 105 production areas in the contiguous United States, was used to estimate changes in net farm income, crop production, planted acres, base acres associated with the commodity programmes, and commodity programme participation. It also evaluates changes in tillage methods and aggregate pesticide use.

The selection of the representative pesticides examined was based upon the frequency of use in the pilot region, as estimated by a national chemical use data base and from discussions with scientists. For maize the pesticides used in conventional and conservation tillage were 'Alachlor' and 'Atrazine'. Pesticides used for no-till maize production were 'Dual' and 'Atrazine'. With regard to soya beans, 'Treflan' and 'Sencor' were used in conventional and conservation tillage. Pesticides used for no-till soya bean production were 'Dual' and 'Sencor'.

Results

The results of the Conservation Reserve Programme case study are presented for both the national level and the pilot region level. These results are summarized in three preliminary reports (Holt *et al.*, 1988; Holtkamp *et al.*, 1988; and Westhoff and Meyers, 1988).

National-level results

Most of the filter strips included in the Conservation Reserve Programme were located in the Mid-west because this area contained the largest quantity of available filter strips. This meant that priority was given to filter strip criteria over the original soil erosion criteria.

Government costs associated with the analysis include both the costs of the commodity programmes (i.e. support prices for maize, etc.) and the cost of paying annual rents to Conservation Reserve Programme participants. The costs associated with the alternatives are expressed relative to the base case of no targeting (45/0).

The costs of targeting 5 million acres to filter strips in the national Conservation Reserve Programme (45/5) were roughly equal to those projected using the highly erodible land criterion (45/0). When the targeted acreage increased to 20 million acres, government costs increased to approximately $0.6 billion over the period extending to fiscal year 1992. However, under the 65/25 alternative, government costs were transformed into savings, as the Conservation Reserve Programme took an additional 20 million acres out of production. The cost savings associated with the 65/25 low option, in which bids were maintained at their current level, was $3 billion over the projection period. For the 65/25 high option the cost savings were less, at about $1 billion over the projection period.

Pilot region results
Farm income results differed, depending on the alternatives. For the 45/5 alternative, farm income changed relatively little. When 20 million acres were targeted under the 45/20 option, farm income increased an average of 10 per cent. Targeting the 20 million acres may have resulted in taking out the more highly productive land. Therefore, when the supply decreased, market prices increased and farm income increased.

The greatest change occurred under the 65/25 option, in which farm income increased by 17 per cent. The primary reason for this increase was that restricting the crop land in production resulted in decreased supply and higher market prices. Therefore participation in government programmes other than the Conservation Reserve Programme and deficiency payments was lower under the 65/25 alternative.

Within the pilot region, targeting filter strips and expanding the Conservation Reserve Programme acreage changed cropping patterns. For the 45/5 alternative, the crop mix and cropping practices did not change significantly.

When 20 million acres were targeted to filter strips, acres planted to wheat declined by 20 per cent in Iowa and southern Illinois. However, it should be noted that relatively little wheat is grown in these areas. In northern Illinois, the only area in which 'no-till' practices were used, no-till agriculture declined by 65 per cent. The reason for this large decrease is that this acreage was targeted as filter strips in the Conservation Reserve Programme.

The most significant changes in cropping practices occurred in northern Illinois under the 65/25 alternative. In this area, small grain and hay production decreased by 100 per cent. However, this area experienced a 113 per cent increase in Conservation Reserve Programme participation. In northern Wisconsin soya bean production decreased by nearly 40 per cent and decreased the area under production by 50 per cent. However, like northern

Illinois, this area experienced a relatively large increase in Conservation Reserve Programme enrolment, amounting to 50 per cent.

The only major decreases in pesticide use occurred in areas where Conservation Reserve Programme enrolment replaced no-till agriculture. 'Metalachlor', a no-till pesticide used in maize and soya bean production, decreased in each of the alternatives. Despite the decrease in acres planted under the 65/25 option, pesticide use did not change significantly.

The results derived from this modelling system were dependent upon the assumption that agricultural programmes, macro-economic policy and trade policies would remain similar to those of 1987. Radical changes in any of these factors could alter the results substantially. In addition, the analysis was conducted assuming that crop yields in the United States, and in all countries participating in world agricultural commodity markets, would remain at average levels. Changes in yields due to weather could alter the results as well.

Finally, the estimate of eligible filter strip acreage is the first available one for the United States. The location of eligible filter strip acreage significantly affects bid prices. When the results of the current Conservation Reserve Programme sign-up, which includes filter strips, are available, it will be possible to verify these estimates to some extent.

Future developments in CEPPES

The preliminary case study shows that the CEPPES can be used to analyse current environmental and agricultural issues and that integrating policies for both agriculture and the environment can be politically and economically attractive. The case study, however, also demonstrated that CEPPES needs additional work to provide an in-depth analysis of environmental and health concern. In particular, problems must be addressed that relate to data aggregation, time–space scales, technical information, health risks and economic decision responses.

Within CEPPES, aggregation of data is a major problem in the biogeophysical component. To determine the level of aggregation, either representative watersheds, a sampling of watersheds or an aggregate model approximating all watersheds must be selected. Aggregation problems in respect of economic data must be addressed in the agricultural decision component.

Currently the pilot project is plagued with different time–space scales among its components. These differences have contributed to the difficulty of developing the pilot version of the model. Reconciliation of these time-scale differences will require additional technical information and analysis.

A major problem for the pilot project has been modelling with incomplete information on relationships between crop yields and use of specific pesticides. Only experimental data are available. For the pilot project, these yield–pesticide use relationships have been simulated.

Another important information gap concerns the unsaturated zone adjacent to ground water as noted in the biogeophysical component. To provide the linkage between pesticide use and pesticide fate, information about the biological and chemical processes in both the root zone and the unsaturated zone is required (Pimentel and Levitan, 1986). More research and development will be required to measure and model pesticide fate accurately.

Within the health risk component, the carcinogenic impact of pesticide residuals is estimated. However, other health risks from pesticide residuals are likely to exist. This component of the modelling system may require extension to encompass other possible consequences of exposure to pesticides and pesticide residuals. Also several approximations for health risk, such as the acres treated with pesticides, that are used to estimate applicator risk will have to be refined and verified.

In developing the necessary data for the agricultural decision component, farm decisions and regional use patterns were estimated, using a non-linear programming framework. On the other hand, the demand and supply models were estimated econometrically. Therefore the resulting correspondence between these two sets of data is weak. In addition, limited empirical data exist for use in anticipating producer responses to policies designed to affect pesticide use rates. Each of these data elements must be improved.

A final concern is adequately assessing risk from pesticides (Miranowski et al., 1974). The information available for estimating the coefficients of the response function in the agricultural decision component is historical and experimental, and often does not reflect producer behaviour in relation to the regulations or policies to be explored (Pope, 1982).

Once these improvements have been made, CEPPES will be more useful in conducting other analyses. Other policy options that could be considered for analysis include controlling production either through decoupling existing commodity price supports from programme commodities or employing acreage restrictions and target prices. Another option would be to expand Conservation Reserve Programme eligibility criteria to include endangered species habitat. In addition, a variety of incentives or sanctions could be examined that would reduce the intensity of cropping practices. For example, a tax could be applied to more dangerous pesticides. While a tax may not fully capture the total external effects of pesticide use, the resulting revenue could be targeted to manage conservation reserves, monitor pesticide loading in ground water, or to establish an information clearing house on integrated pest management.

Conclusion

As noted at the outset, agricultural and environmental policy has been developed to meet often narrow objectives, such as increasing crop productivity, supporting farmer income and restricting individual pesticides. The resulting increases in crop-land productivity have come at the cost of over-production of crops, low farm incomes and adverse environmental conditions. A potential catalyst for transcending these problems is the development of a comprehensive assessment system that portrays the multiple policy implications of farm and pesticide policies.

The Comprehensive Economic Pesticide Policy Evaluation System provides a diverse and comprehensive data and modelling system useful for analysing alternative agricultural and environmental policies affecting pesticide use. The intention of CEPPES is to narrow the range for judgement in the policy arena. A major contribution of CEPPES in policy analysis is to focus the policy debate on the issues about which there is true uncertainty. The result will be more enlightened and socially desirable policies for all sectors of the economy affected by agricultural and pesticide regulations.

References

Abt Associates Inc. (1986), 'Pesticide Drinking Water Exposure/Risk Methodology', Washington, D.C.

Alder, Kenneth. (1986), United States Environmental Protection Agency, personal communication.

Batie, Sandra S. (1987), 'Institutions and Ground Water Quality', paper presented at the fifth National Symposium on Ground Water Pollution, Agricultural Chemicals and Ground Water Pollution Control, 26–7 March, Kansas City, Missouri.

Blackmer, A.M. (1984), 'Losses of fertilizer-N from soils', in *Proceedings from Iowa's 37th Annual Fertilizer and Ag-chemical Dealers' Conference*, Iowa State University, Co-operative Extension Service, CE-2081.

Blackmer, A.M. (1985), 'Integrated studies of N transformation in soils and corn responses to fertilizer-N', in *Proceedings from Iowa's 38th Annual Fertilizer and Ag-chemical Dealers' Conference*, Iowa State University, Co-operative Extension Service, CE-2158.

Carson, Rachel. (1979), *Silent Spring*, Boston, Mass., Houghton Mifflin.

Cochrane, W.W. (1979), *The Development of American Agriculture: a Historical Analysis*, Minneapolis, Minn., University of Minnesota Press.

Cohen, S.Z., Creeger, S.M., Carsel, R.F., and Enfield, C.G. (1984), 'Potential pesticide contamination of groundwater from agricultural uses', in R.F. Kruger and J.N. Seiber, eds., *Treatment and Disposal of Pesticide Wastes*, American Chemical Society, Washington, D.C.

Conservation Foundation (1986), *Agriculture in a Changing World Economy*, Washington, D.C., Conservation Foundation.

Eriksen, Milton H. (1976), *Use of Land Reserves to Control Agricultural Production*, Economic Research Service ERS-635, Washington, D.C., United States Department of Agriculture.

Fedkiw, John (1986), *The Evolving Use and Management of Our Forests, Grassland and Croplands*, Washington, D.C., United States Department of Agriculture, Office of Budget and Programme Analysis.

Gianessi, L.P., Peskin, H.M., Crosson, P., and Puffer, C. (1986), 'Nonpoint Source Pollution: are Cropland Controls the Answer? A report prepared for the United States Environmental Protection Agency by Resources for the Future', unpublished.

Headley, J.C., and Lewis, J.N. (1967), *The Pesticide Problem: an Economic Approach to Public Policy*, Baltimore, Johns Hopkins University Press.

Heimlich, Ralph, and Langer, Linda (1986), *Swampbusting: Wetlands Conversion and Farm Programmes*, Washington, D.C., United States Department of Agriculture, Economic Research Service.

Holt, Matthew T., Haney, Douglas A., and Frohberg, Klaus K. (1988), 'Buffer Strip Area and Future Conservation Reserve Programme Allocation', preliminary report, EPA contract CR-813498-01-2, Center for Agricultural and Rural Development, Iowa State University.

Holtkamp, Derald J., Traxler, Greg T., and Frohberg, Klaus K. (1988), 'Impacts of Conservation Reserve Programme Options on Farm Income, Production, Land Use and Rental Values in five Midwest Producing Areas', preliminary report, EPA contract, CR-813498-01-2, Center for Agricultural and Rural Development, Iowa State University.

Hoyer, Bernard E., *et al.* (1987), *Iowa Groundwater Protection Strategy*, Environmental Protection Commission, Iowa Department of Natural Resources.

Jones, J.J. (1987), 'The Agricultural Sector', unpublished paper, United States Environmental Protection Agency.

Judy, Robert D., *et al.* (1984), *National Fisheries Survey, 1, Technical Report: Initial Findings*, prepared for United States Department of the Interior, Fish and Wildlife Service and United States Environmental Protection Agency, Office of Water, FWS/OBS-84/06, Washington, D.C., United States Department of the Interior, p. 28.

Kelley, R.D., Hallberg, G.R., Johnson, L.G., Libra, R.D., Thompson, C.A., Splinter, R.C., and DeTroy, M.G. (1986), 'Pesticides and ground water in Iowa', National Water Works Association, *Agricultural Impacts on Ground Water Conference*, Omaha, Nebraska.

Kuch, Peter J. (1987), 'Agricultural Programmes and the Groundwater Contamination Problem in Iowa', paper presented at the Iowa Public Health Association and Iowa Environmental Health Association Annual Conference.

Miranowski, John A., Ernst, Ulrich F.W., and Cummings, Francis H. (1974), 'Crop Insurance and Information Services to Control Use of Pesticides', paper prepared for the United States Environmental Protection Agency, EPA-600/5-74-018.

Nielsen, Elizabeth G., and Lee, Linda K. (1986), 'The Magnitude and Costs of Groundwater Contamination from Agricultural Chemicals: a National Perspective', paper presented at the American Agricultural Economics Association meeting, Reno, Nevada.

O'Hare, M., Curry, D., Atkinson, S., Lee, S., and Canter, L. (1985), *Contamination of Ground Water in the Contiguous United States from Usage of Agricultural Chemicals*, Environmental and Ground Water Institute, Norman, Oklahoma.

Pimentel, David, and Levitan, Lois (1986), 'Pesticides: amounts applied and amounts reaching pests', *Bioscience*, 36(2), 86–91.

Pope, Rulon (1982), 'Empirical estimation and the use of risk preferences: an appraisal of estimation methods that use actual economic decisions', *American Journal of Agricultural Economics*, 64(2), 376–83.

Shortle, J., and Dunn, J. (1986), 'The relative efficiency of agricultural source water pollution control policies', *American Journal of Agricultural Economics*, 68(3).

Taylor, Harold (1988), United States Department of Agriculture, personal communication.

United States Department of Agriculture (1984), *Agricultural Statistics 1984*, Washington, D.C., United States Government Printing Office.

United States Department of Agriculture (1985), 'Inputs, Situation and Outlook Report', Economic Research Service IOS-7, Washington, D.C.

United States Department of Agriculture (1987), 'Inputs, Situation and Outlook Report', Economic Research Service AR-5, Washington, D.C.

United States Department of the Interior (1985), *National Water Summary, 1984*, United States Geological Survey Water Supply Paper 2275, Washington, D.C.

United States Environmental Protection Agency (1987), *Unfinished Business: a Comparative Assessment of Environmental Problems*, Appendix 1, Report of the

Cancer Risk Work Group, Office of Policy Analysis.

Westhoff, Patrick C., and Meyers, William H. (1988), 'Impacts of Conservation Reserve Programme Options on Agricultural Markets and Government Costs', preliminary report, EPA contract CR-813498-01-2, Center for Agricultural and Rural Development, Iowa State University.

Wintersteen, W., and Haztler, R. (1986), '1985 Iowa Pesticide Use Survey', Co-operative Extension Service, Iowa State University.

Part II
Intensive animal production and the management of animal manure

Part II

4. Intensive livestock production in France and its effects on water quality in Brittany

P. Rainelli

Location and development of intensive livestock rearing

The general position in France

Poultry raising in its modern form emerged during the 1950s in France. Intensive pig farming developed considerably in the following decade, but the spatial distribution and, especially, development were different. We look first at the picture in the 1950s and then at today's.

The early 1950s
The scanty statistics available from the 1955 General Agricultural Census suggest that poultry production was common on farms throughout the country except for the mountain regions. But poultry was especially conspicuous in the regions of Lyons, the north, the west and the foothills of the Pyrenees (fig. 4.1a). Although intensive poultry raising was emerging, farmyard production, highly dependent on the traditional family farm structure, and largely self-sufficient both in stock and in cereal feeds, still predominated. There was very little commercial marketing except for the Bresse region, which has possessed a trademark of origin since 1936.

The geographical location of pig raising enterprises in 1950 is shown in fig. 4.1b and is similar to the poultry pattern. The main differences are the presence of the Massif Central and the greater influence of western France among the production regions. Pigs were reared on farms producing potatoes and other field crops suitable for feed and also by dairy farms which could use the buttermilk left over from butter production. Cereals were also grown and feed was almost entirely self-supplied.

In contrast to poultry, though, there was already a considerable degree of commercial production of pigs in western France, although production in the Massif Central area was mostly for local consumption. There were also some

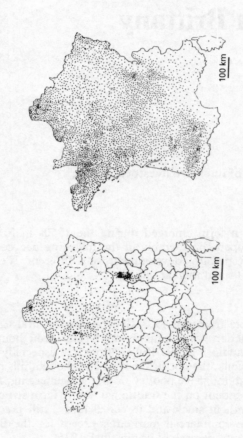

Figure 4.1a Intensive breeding in the early 1950s: poultry. *Left*, farms that sold over 100 poultry in 1955; each dot represents fifty operations. *Right*, farms breeding hens in 1955; each dot represents 1,000 birds. *Source*: RGA (1955), following Diry (1984, pp. 35, 37)

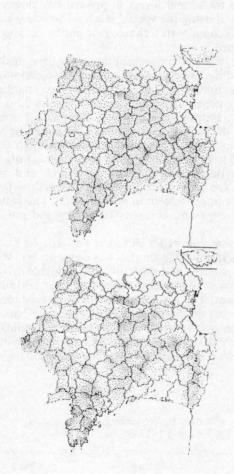

Figure 4.1b Intensive breeding in the early 1950s: pigs. *Left*, sows: each dot represents 1,000 animals. *Right*, pigs: each dot represents 200 animals. *Source:* Klatzmann (1955, pp. 73, 74)

large 'feeder' pig farms using dairy industry by-products and, near the larger cities, food and industrial wastes such as milling and oil by-products in the Marseilles area.

The present situation

The pattern has now clearly changed from farm-scale to industrial-scale production. The shift from traditional forms to present-day forms of production can best be illustrated using the special study of Brittany presented below, but the explanatory factors, both technological and in the way society has developed, can be outlined here.

On the technological side, a set of convergent advances have made intensive poultry and pig breeding possible. More productive, homogeneous breeds were developed as understanding of animal feed requirements improved. The considerable genetic potential could be exploited only with the right kind of nutrition, bearing in mind a minimum cost constraint. Hence the development of the animal feed industry. At the same time, in view of the disease risk inherent in large-scale husbandry, a high degree of veterinary expertise also had to be developed. This involved not only veterinary treatments but also stringent prophylactic measures and properly designed buildings. These factors, better mastered for poultry breeding, brought about a considerable improvement in performance. The pattern was successfully extended from eggs and chickens to turkey and guinea fowl farming in the late 1960s.

The evolution of pig raising followed a different pattern. At first, small farm units developed, where feed for the animals was based on beet with high dry matter content, barley, potatoes and skim milk, all available on the farm itself. Selective pig breeds were not highly developed, the buildings cost little, and labour was abundant. The use of industrial/commercial feed began to grow after the poultry crisis of 1964–65. At that time true battery production of pigs appeared, associated with milk and fodder feeding. Later farms which were exclusively devoted to pig production, either with cereal crops for feed, or else with no cropping at all, were developed.

Table 4.1 Share of intensive husbandry in five leading and ten leading *départements* in France overall, 1959–61 and 1983–85 (%)

	Poultry		Pork	
	1959–61	1983–85	1959–61	1983–85
Five leading *départements*	21·6	39·2	25·7	50·2
Ten leading *départements*	34·8	54·7	39·4	62·0

The percentages were calculated using series based on 1970 prices. An estimate of these series based on current prices yields a lower percentage at the end of the period but a higher one at the beginning, given the differing development of prices of the main factors in final production.

But intensive breeding could not have come about without the profound change in French society which occurred over the period (Saunier and Schaller, 1978). Rapid industrialisation fuelled by a large-scale flight from the land led to sweeping changes in eating habits. Poultry consumption, with its strong farm tradition, evolved accordingly, with a sharp increase of industrial poultry consumption among clerical and non-active groups while professional, managerial and blue-collar groups tended to increase their consumption of pork. France, too, represented a special case because 'the almost simultaneous emergence and development of true "industrial" meat production disrupted the "normal" growth of the industrial chicken market' (Saunier and Schaller, 1978). No such competition occurred elsewhere. Meanwhile, persistent food habits account for the fact that alongside standardised production of poultry for domestic consumption and export, quality labelled brands were also developed.

As table 4.1 shows, geographical concentration is more significant for pork than for poultry. From 1959–61 to 1983–85 the share of the five leading *départements* in the country as a whole increased, reaching 50 per cent. For poultry, the share of the five leading *départements* did not quite double and remained at a lower level.

Concentration of farms
Figure 4.2, drawn up on the basis of the 1979–80 General Agricultural Census, contrasts with fig. 4.1. Intensive rearing is markedly concentrated in western France, despite some significant concentrations in the south-west, the greater Lyons area, the Valence plain and Flanders. Today the high-output poultry and pig areas coincide to a remarkable extent, especially with the increasing weight of the western regions.

The situation in Brittany

With nearly 39 per cent of French intensive rearing output by value in 1983–85 (32·0 per cent for poultry, 46·3 per cent for pork), Brittany occupies a special place among the regions. Some historical background is followed by a more detailed review of the present situation.

The development of intensive rearing in Brittany
By the late 1950s, while poultry breeding was already definitely expanding, overall intensive production represented less than one-third of final agricultural output by region (slightly over half nowadays). Intensive production originated with the battery egg farm in 1929 near Rostrenen, Côtes-du-Nord. In 1931 the first incubator was built.

In 1961 Le Bihan, looking at the origins of this process, commented that intensive egg farming began to spread first in Côtes-du-Nord towards Guingamp, then from 1942 onwards into Finistère (Carhaix, Châteauneuf-du-Faou, Bannalec). It became widespread after 1948–49 because of the combination of factors already mentioned. In 1948 the Services Vétérinaires established an avian pathology laboratory in Côtes-du-Nord.

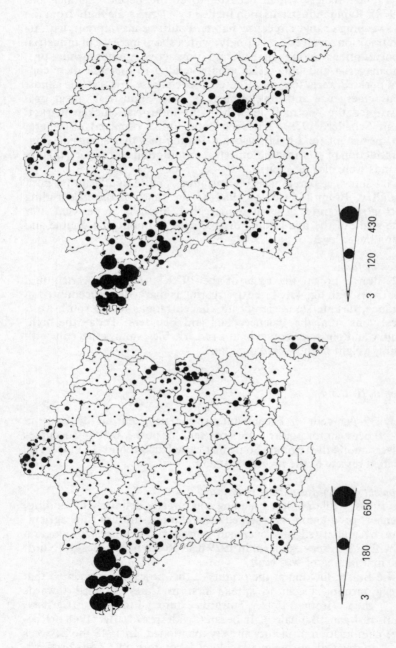

Figure 4.2 The geographical distribution of intensive rearing, 1959–85, showing each *département* as a percentage of total French production. *Left*, pig production. *Right*, poultry.

This coincided with the emergence of the livestock feed industry, providing not only technical assistance but also supplier credit.

At the same time, incubation centres proliferated and new strains were introduced. Hitherto reserved for an elite, owing to the difficulties of appropriate nutrition in the absence of compound feeds and disease control problems, modern poultry breeding became more widely accessible.

While intensive production of eggs for consumption centred in somewhat richer areas like Trégorrois and the Châteaulin basin, production of table chickens, which got under way in the early 1950s, tended to locate in the less prosperous cantons (Le Bihan, 1961, p. 44). As with egg production, this spread from the Côtes-du-Nord towards Finistère but also towards Morbihan (Gourin), Locminé, Saint-Jean-Brévelay, Rochefort-en-Terre. The development of table chicken rearing came about partly under pressure from livestock feed firms, but especially under (downstream) pressure from the processing industry, particularly after the difficulties of 1957. The creation of a powerful processing industry focusing on slaughterhouses played a key role.

The years 1962–63, which marked a high point after the crisis of 1957, were followed by a severe crisis in 1964–65, prompting cattle feed manufacturers to relaunch sales by encouraging farmers to develop pork. Other factors influenced the transition to greater industrialisation. As we have seen, farm-based production relied on potatoes and buttermilk. But it was at this point that conservation potato cultivation began to decline and the dairy industry changed over from collecting cream, leaving the farmer with the buttermilk, to collecting milk.

Table 4.2 French production and consumption of poultry (000 tonnes)

	1970	1975	1980[a]	1985	1988
French production	637	823	1,125	1,125	1,449
Domestic consumption	616	755	860	972	1,087

(a) Series break in 1980.

Source: Graph'agri 90, Service Central des Enquêtes et Etudes Statistiques (SCEES) Ministère de l'Agriculture, 1990, p.145.

Table 4.3 Deficit in French pork production, 1970–85

	1970	1975	1980	1985
Carcass equivalent (000 tonnes)	206·3	237·9	320·0	369·6
FFr million	1,310·9	1,722·9	3,156·6	5,557·9

Source: Direction Générale des Douanes; SCEES (revised series 2)

The poultry industry recovery was due essentially to diversification both in products, with the launching of turkeys and then of guinea fowl, and in export outlets. Table 4.2 illustrates the broad trends for the domestic market compared with the international market, showing that the turnaround occurred in the late 1970s. The export effort has not, however, eliminated overproduction crises, of which the most recent, in 1986, severely affected egg producers.

The broad context for the regional production of pork is altogether different since, as table 4.3 shows, France is in deficit. Despite the 1966 husbandry Act and the various special promotional campaigns, the deficit has increased over time. This suggests that if the right conditions for competitiveness were brought together, pig farming could be developed more extensively.

The lack of self-sufficiency in pig production is generally attributed to a number of objective factors, notably underemployment of agricultural manpower associated with particularly unfavourable agrarian structures in Brittany. Attention has also been drawn to the stimulus provided by the livestock feed companies engaging in production 'integration'. As Saunier and Schaller pointed out in 1978, although intensive rearing enterprises did originally develop on farms without land, they also developed on small, intermediate and even large farms, a trend more distinctive for pig farming than for poultry. Another point is that although farming structures are very similar in the four Brittany *départements*, the whole of the east, together with Ille-et-Vilaine, did not experience that evolution.

More subjective reasons have been advanced in the form of farmers' changing traditional attitudes towards schooling and the prolongation of studies, itself produced by the fortuitous encounter between deteriorating employment markets and a greater supply of education and training associated with transformations within the Church and with Catholic action movements (Grignon, 1981). The existence of a farming elite, especially in the lower Brittany area, seems to have been decisive in the start-up of intensive poultry breeding. Cultural forces certainly enabled the new techniques to spread more easily. Le Bihan (1961) notes the part played by migrants returning to their commune of origin, and the changing motivation of farmers held as prisoners-of-war on their return. Poultry breeders who were not farmers were also very active.

But educational factors are not enough to account for the spread of intensive rearing into low school attendance areas. In retrospect, we can see that poultry breeding was taken up in cantons affected by the dwindling or disappearance of some major activity which used to ensure that farms were in balance (during 1984). One example was Trégor, where serious problems arose when flax was discontinued in 1955; another was the Lamballe region for horses. But the potato plant provides the most telling example, where the decline in cultivation was almost simultaneously accompanied by the adoption of egg or table-chicken batteries. The technical and marketing discipline required to grow potatoes was an asset for the agro-food firms in implementing a strict contract policy. Conversely, the absence of potato culture in Ille-et-Vilaine accounts for the absence of intensive rearing in that *département*.

On pig production, Diry (1985) points out that its present distribution is associated with a previous background of traditional pig rearing. It was only natural for areas that already produced pigs to start doing so industrially in the mid-1960s. The livestock feed firms only exploited this trend in seeking to diversify their outlets after the 1964–65 poultry crisis.

The present situation
The distribution of intensive animal husbandry among *départements* can be seen from figs. 4.3 and 4.4, established by Diry in 1985 on the basis of the general agricultural census of 1979–1980. The fact that the data are not very recent is less of a disadvantage in that, as we have seen, a plateau was reached in the early 1980s. Trends between the 1970 and 1980 censuses show that there has been no real spatial redistribution at a sufficiently aggregated geographical area like a canton.

To simplify somewhat, a certain number of fairly typical production areas can be identified in the Brittany region. Finistère, lower Léon and the Châteaulin country, corresponding roughly to the Poher, make up a poultry zone for table chicken and egg farms. Another substantial poultry area is in Morbihan at Questembert and Haut-Vannetais, where table chickens and turkeys are bred. Côtes-du-Nord has three poultry/pig areas, in the north-east at Trégor-Goëlo, in the west at Saint-Brieuc, and south of Saint-Brieuc in the Loudéac country.

Environmental problems

The main emphasis on pollution problems arising from intensive breeding is on nitrate deterioration of drinking water. But other water and littoral eco-systems can be disrupted, and various other problems can arise.

Liquid manure pollution

No explanation is needed of the mechanisms for malodorous disamenity associated with the spreading of liquid manure. For other forms of pollution, the problems are associated with effluent composition, which in addition to organic matter will contain not only nitrogen, phosphorus and potassium but also germs and heavy metals. Spreading of liquid manures and leaks/over-flows from storm water tanks which may not be sufficiently watertight give rise to dispersion of pollution components into the environment through infiltration and run-off.

In contrast to phosphoric acid, whose losses through leaching are low, nitrogen is easily carried into infiltration waters, thereby contributing to higher nitrate and nitrite content, depending on the degree of oxidation, of ground water. A small proportion of the potassium is also leached. Soil contamination by trace elements, essentially copper and zinc, also arises because these are present in pig feed and occur in animal wastes. Spreading

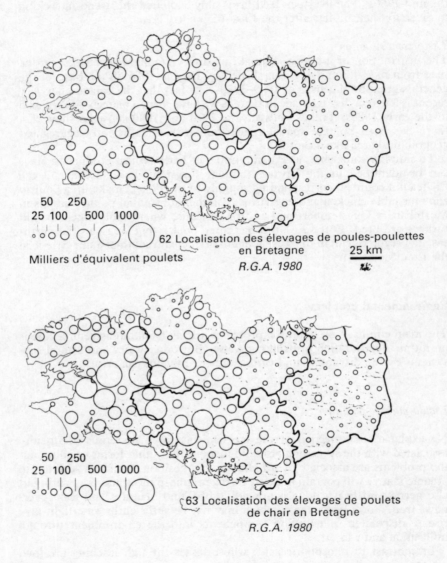

Figure 4.3 The distribution of table chicken and egg production, 1980.
Source: Diry (1985) p. 485.

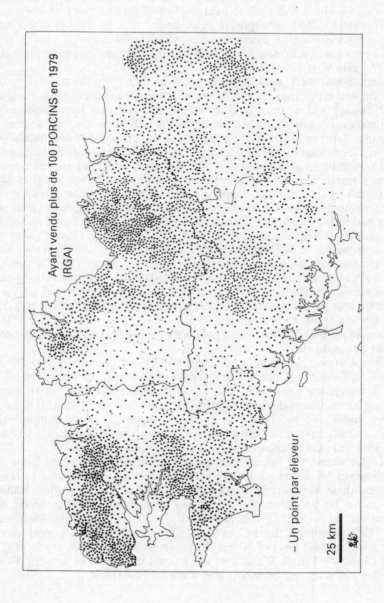

Ayant vendu plus de 100 PORCINS en 1979
(RGA)

– Un point par éleveur

25 km

Figure 4.4 The distribution of pig farms with sales of over 100 pigs, 1979. *Source:* Diry (1985), p. 483

on light, sandy soils increases the risk of direct percolation of all these constituents, and especially of nitrogen, into water.

Run-off water pollution will depend on climatic conditions (rain, frost), the extent to which the land is saturated, and on the slope of the land. Run-off waters include phosphoric acid, giving rise, as soluble orthophosphates in association with nitrogen, to eutrophication. Organic matter, producing ammonia as it breaks down, as do nitrite and nitrate reduction, is also carried by rain water.

Beyond certain thresholds, depending on the toxicity of the trace elements and on the self-cleansing capacity of the soils and waters concerned, environmental damage will occur. Slurry spreading in itself may have undesirable consequences, because the very heavy matter may degrade the soils and result in panning. Saturation which asphyxiates the soil encourages denitrification, and, when it dries again, nitrates are liable to find their way into lower layers.

Before looking in greater detail at the pollution effects of animal wastes, the heavy metals need to be considered. Scientific literature suggests that heavy metals (zinc, cadmium, copper, lead) accumulate and can make the soil irreversibly toxic. Some writers consider that large-scale spreading of pig slurries may bring this about because of the use of saline additives in animal feeds, leading to appreciable copper and zinc content in effluent. But research conducted in Belgium (Meeus-Verdinne et al., 1986), shows that plants grown under such conditions do not differ in copper and zinc content from plants grown with ordinary mineral fertilisation. Another point is that percolating waters take down only a tiny fraction of such metals. A problem could arise only in the event of lagooning. Slurry spreading limited to 2,500 kg per hectare seems to ward off any risk.

Potassium, of which a only small proportion is leached, at present presents no water pollution risk.

Table 4.4 summarises pollutants, their effects and their consequences for the medium concerned. The whole question of building and plant location is clearly a matter for regulation. The same applies to some extent to slurry spreading for such straightforward issues as disamenity in the neighbourhood and any impact on tourism. But as slurry spreading plays another role through infiltration and run-off water, it becomes a much larger issue. Other constraints and other mechanisms have to be considered.

Table 4.4 shows the special part played by nitrogen, whether from the decomposition of wastes or from the production of nitrites and nitrates, which find their way into drinking water. They can seriously affect human health in so far as there is a possibility that nitrites in the blood are transformed into carcinogenic nitrosamines (Fritch and Saint-Blanquat, 1985). This prompted the EC to set a guideline level of 25 mg/l and a maximum permitted level of 50 mg. Equivalents for ammonia are 0·5 mg/l, and for nitrates 0·1 mg/l. Water also acts as a saturation factor for such foods as radishes, lettuces and beetroot, which are 'nitrate traps'.

Table 4.4 Pollution associated with intensive rearing: sources, effects and consequences

Source	Effect	Consequence
Location of building, plants	Noise, appearance	Aesthetic, neighbourhood disamenity
Slurry spreading	Odour	Neighbourhood disamenity and tourism
Nitrogen infiltration	Production of nitrites and nitrates	Drinking water quality
Organic matter, nitrogen	Oxygen absorption in water courses Amonia production Nitrite, nitrate production	Drinking water quality
Nitrogen, Phosphorus	Eutrophication of standing waters and, possibly, estuary and coastal waters	Disruption of eco-systems Contribution to 'green tides'
Pathogens	Bacterial contamination of marine shellfish	Human health hazard

Table 4.5 Nitrogen transfers to the environment from animal wastes, diffusion through crops and contribution through fertiliser (per hectare usable agricultural area)

	Côtes-du-Nord	Finistère	Ille-et-Vilaine	Morbihan	Brittany
Diffusion through crops	150	144	146	146	146
Availability (animal waste)	149	148	114	118	134
Nitrogen from commercial fertilisers	81	95	129	62	93

The particular situation of Brittany

Amounts of animal waste
Applying conversion factors to livestock numbers available, the Centre d'Etude du Machinisme Agricole, du Génie Rural et des Eaux et Fôrets (CEMAGREF) has estimated the amount of nitrogen by source which flows through to the environment (table 4.5 and fig. 4.5).

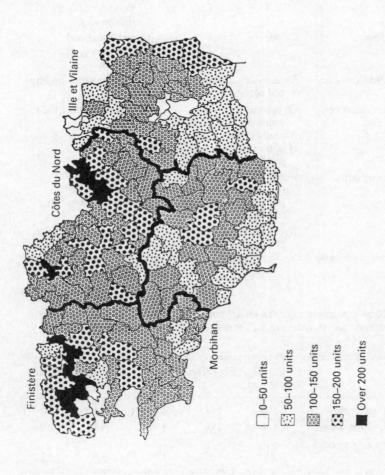

Figure 4.5 The average nitrogen load from animal waste in Brittany, by canton: nitrogen per hectare of usable agricultural area.
Source: CEMAGREF 1982)

Within the total theoretical nitrogen availability, cattle account for a noteworthy proportion (69 per cent regionally, varying from 59 per cent in Côtes-du-Nord to 84 per cent in Ille-et-Vilaine) and therefore represent considerable 'background noise'. This means that slurry pollution problems cannot be considered independently of the presence of cattle, and it is indeed the combination of the two types of production which accounts for the magnitude of the nitrogen load in Nord-Finistère, whereas in the east of the Côtes-du-Nord *département* pig farming is the main source.

Table 4.5 suggests at first glance that it would not take much to balance nitrogen units 'exported' through crops and those available on site through animal wastes. But nitrogen losses must be allowed for in storage, transport and effluent spreading. For the latter, some 6 per cent of nitrogen is thought to be volatilised in spring and autumn, but 25 per cent in summer.

There are also agronomic reasons for substituting animal waste nitrogen for only about one-third of mineral nitrogen. For health reasons no more than one-third of mineral nitrogen fertiliser can be replaced by manure in vegetable growing. Taking all these considerations into account, available nitrogen still exceeds exports for one canton in ten.

Pollution in Brittany

Because of its geological features, Brittany has few underground water deposits, tertiary basins being the only exception. Ground water represents some 20 per cent of total disposable drinking water, or 164,000 m^3 per day from 450 draw-off points. The 668,000 m^3 of surface water per day, from 108 draw-off points, are drawn direct from watercourses or from larger or smaller reservoirs (SRAE, 1982). In rural areas underground water accounts for a higher proportion (one-third in Côtes-du-Nord and Morbihan, three-fifths in Finistère and Ille-et-Vilaine).

The preponderance of surface water means that special importance attaches to run-off, especially since it gives rise to the autrophication of certain reservoirs such as Arguenon in Côtes-du-Nord. In this context the proliferation of green algae in Saint-Brieuc Bay has been mentioned. Run-off also brings pathogens (faecal streptococci) which have played some part in the bacterial contamination of mussels in Saint-Brieuc Bay, whose consumption is still banned.

Run-off is important but the effect of infiltration cannot be ignored. Many farmers take drinking water from wells and springs, often of less than ideal quality, whereas water obtained by drilling deeper is generally of better quality.

Looking only at nitrates in watercourses, the early 1986 water monitoring survey (SRAE, 1986) reveals a disturbing situation. Whereas in the early 1980s only Nord-Finistère showed nitrate concentration in excess of 50 mg/l, five other areas are now over the standard and twenty-one others are over 40 mg (fig. 4.6). The maximum, 70 mg, was noted in the river Frémur, in the eastern Côtes-du-Nord.

However, the samples are subject to seasonal variation, with a minimum during lower water periods and a maximum in January and February, when waters are higher and there is a greater degree of run-off, especially in the

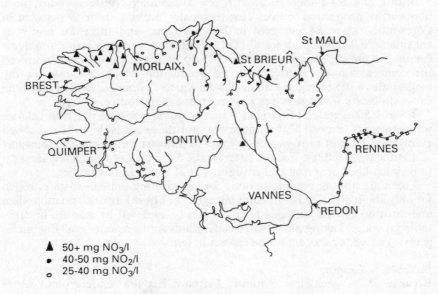

Figure 4.6 Nitrate concentration in watercourses, early 1986: Finistère, 14 April (CDAF and CDE); Côtes-du-Nord, February (CDAF); Ille-et-Vilaine, 2 and 4 March (SRAE). *Source:* Service Régional d'Aménagement des Eaux de Bretagne

absence of vegetable cover. The minimum/maximum discrepancy is about 50 per cent but depends on the geological background, schist basins being more irregular than granite basins, where the flow is regulated by underground water reserves. It also depends on the amount of precipitation.

However, the simultaneous presence of large-scale animal wastes and of pollution does not in itself demonstrate a link of strict causality. For nitrates in the market gardening areas of Nord-Finistère, the relationship has been clearly established between average nitrate concentrations in Léon watercourses and the share of market gardening and potato cultivation within the total area. The effect of poultry and pig breeding on water quality in that area, however, is not altogether clear (SRAE, 1986b).

Conversely, a study at commune level in Finistère has found that the quality of underground water in rural, non-market-gardening communes is positively associated with density of pigs and poultry, and negatively with the share of farmed area in use for forage crops. The latter make good use of animal wastes, limiting nitrate infiltration (André et Dubois de la Sablonière, 1983).

Technological solutions

There are four possible solutions which are not necessarily exclusive: slurry spreading on the spot, slurry spreading by neighbouring farmers (slurry bank), effluent treatment and the denitrification of drinking water.

ON-SITE SPREADING

In view of the fertiliser value of animal wastes, spreading at the point of production is by far the best solution, provided that each farm has sufficient land on which to spread the slurry produced. However, several factors limit the capacity of soils to accept sludge. These factors include the nature and condition of the soil, the type of crop and the need to protect water, etc. Soils are therefore distinguished for slurry suitability as being nil, poor, average or good. The classification has to be matched with weather conditions, which ultimately leads to three categories—those on which slurry spreading is unsuitable at any time of year, those on which it is unsuitable only in winter, and those on which slurry can be spread at any time of year.

Table 4.6 Surface available for slurry spreading in three communes

Availability	Pipriac (I. et V.)		Henanbihen (C. du N.)		Henansal (C. du N.)	
	Ha	% total surface	Ha	% total surface	Ha	% total surface
Totally unsuitable	2,239	46	1,208	39	1,525	52
Unsuitable in winter	1,434	29	1,299	41	1,029	35
Available throughout year	1,191	25	637	20	366	13

Source: CEMAGREF (1983), table 4.

Assessments at commune level in Côtes-du-Nord, where pig density is high, and in Ille-et-Vilaine, where pig density is 'average', show that between 40 and 50 per cent of the total surface is entirely unsuitable for sludge spreading (CEMAGREF, 1983). Moreover, only a fraction of the land can be used for manure spreading throughout the year (table 4.6).

In terms of total usable agricultural area the proportion of entirely unsuitable land naturally declines, to 38 per cent for Pipriac, 23 per cent for Hénanbihen and 39 per cent for Hénansal. Moreover, in winter, much less land is available for spreading and hence sludge storage facilities must be built.

Although regulations require storage capacity equivalent to forty-five days, it seems necessary to have capacity for at least six months, thereby also warding off risks of overflowing in high rainfall periods. The quality of the storage facility is crucial, especially as regards preventing infiltration, handling possibilities and the risk of accident.

Slurry spreading also depends on the suitability of the crop system in terms of time of year, soil condition and fertility. Consequently soil analysis can play an important role in improving soil fertility.

SLURRY BANKS

When intensive husbandry is on a scale which is disproportionate to the available spreading area in the immediate proximity, the question arises of making the fertiliser ingredients of the wastes available to nearby farmers. This is the slurry bank principle which presupposes a minimum of organisation within a fixed legal structure.

The slurry bank does not eliminate the storage problems, but in view of the cost of transport it requires greater attention to slurry quality and degree of dilution, which should not exceed 95 per cent for pig slurry and 90 per cent for egg-laying poultry. Hence the need for easy ways of ascertaining the main parameters of an effluent. Simple methods are available for pig slurry (Bertrand and Arroyo, 1984).

A pilot project has been undertaken in the Lamballe Hénanbihen region of Côtes-du-Nord. At the same time, cost studies based on reference models have been conducted (SOGREAH, 1985). An examination of thirty-two exporting communes and twenty-seven importing communes has found that slurry was being carried over distances averaging 61 km each way for liquid wastes and 75 km for solid wastes. This is a large-scale operation, since the surplus for carriage corresponds to 217,000 m³, requiring ten operators for loading and transport. The storage bank that was set up handled 3,500 m³ delivered over a distance of 30 km, yielding a cost per cubic metre of FFr 40. Since the slurry was purchased at a price varying between FFr 10 and FFr 12 per cubic metre, subsidies were necessary. This accounts for minimum costs of the order of FFr 2 million.

This is equivalent to FFr 9.4 per m³, while the theoretical fertiliser value of pig waste is estimated at FFr 30.1 per m³ when only 30 per cent of the total nitrogen is assumed to be effective (Deschamps *et al.*, 1984). Theoretically, then, the operation makes sense.

EFFLUENT PROCESSING

When slurries cannot be spread *in situ* or near by, processing units may be considered. Processing may be designed solely to deodorise the product by adding enzymes, blocking fermentation or adding other products. Aeration is also possible for slurries that are less than three days old. Deodorisation may also be achieved through methanisation, which also provides bio-gas that can be used for energy purposes.

Anaerobic fermentation or methanisation has been subject to an intensive feasibility study (Coillard, 1986) which found that when pig wastes were excessively diluted (97 per cent) a high proportion of the gas was consumed in the process (70 per cent), producing a rather negative return in view of the initial cost of FFr 900,000 (in 1982) for a rearing capacity of 2,400 pigs.

Accordingly, conventional processing techniques would appear to be more attractive. But, in either case, by-products still remain to be disposed of.

DENITRIFICATION OF DRINKING WATER

Here it is not the original pollution that is treated; rather, an attempt is made to remove the excess nitrate in drinking water. The elimination processes are physical/chemical or biological.

The biological processes, autotrophic when the carbon used for the reduction is of mineral origin, heterotrophic when it comes from an organic substrate, deploy bacteria which convert nitrates into nitrites and then into nitrogen. Although these processes provide water with under 10 mg of nitrates per litre, there seems to be a disinfection problem, especially for surface water, which adds to the processing cost. This is why up to now there have been only pilot units.

Physical/chemical processes depend on membrane technology (inverse osmosis, electro-dialysis) and ion exchange methods. The latter are especially used in full-scale applications, as in the commune of Plounevez-Lockhrist in Nord-Finistère which processes 150,000 m³ per year. The water passes over synthetic resins (an aromatic polymer) with a negative exchangeable ion which exchanges with the nitrate. But the ion nitrate requires resins with strong cation sites to be held back, with the result that other ions (chloride, sulphate) are also exchanged. This diminishes the theoretical capacity of the resin. When the medium has been exhausted it is regenerated with a concentrated solution.

The problem is how to eliminate the nitrate solution obtained after ion exchange. The solution or eluate contains, as we have seen, chlorides and sulphates and represents an appreciable volume (0·5 per cent of the volume produced for surface water and 1·5 per cent of the volume for underground water in optimum conditions). This makes it desirable to provide for eluates to be processed at a treatment station so as not to discharge them downstream of the water plant. As a result the real cost per m³ of the drinking water thus produced can amount to at least FFr 3.

Aspects of the regulations

As Helin (1983) points out, the legal background to pollution and disamenity problems associated with intensive rearing contains both general and more specific regulations. Those aspects are considered and we then discuss the farmer's liability.

General regulations

In principle the legislation concerned does not apply to agriculture as such but affects the ways in which any form of pollution or disamenity associated with intensive rearing has to be dealt with. The two broad categories involve town planning and water protection.

Intensive rearing, environmental protection and town planning
The installation of intensive rearing enterprises results, as we have seen, in environmental disruption and in neighbourhood conflicts, essentially with non-rural residents. Environmental protection in the broad sense, including landscape, and the incompatibility between production activities and residential areas are a matter for land use and thus come under town planning law, as Jegouzo (1983) points out. But, as he also points out, the problem is not simple in view of the principle of independence for town planning and environmental legislation. Very broadly, areas with no land use plans (POS) should be distinguished from those which have such plans.

Where there is no land use plan, problems associated with large intensive rearing installations are essentially governed by the national town planning regulations both as to relationships between farmers and private individuals, and in connection with permission to locate such installations in sensitive areas. Other more specific provisions apply, particularly the *département*'s health regulations. In each *département* the Commissaire de la République is empowered to set protection perimeters around classified installations, although he cannot entirely forbid the construction of dwellings around them. Lastly, since the 1983 Act giving communes new powers relative to other territorial collectivities, non-farmers cannot erect dwellings in a rural area. This provision has prompted communes to develop land use plans.

The land use plan sets out general rules for land use within a commune, allowing for specialisation of activities within the local territory. This it does on the basis of planned present and future use for land, the main distinction being between urban areas, designated by the letter U, and natural areas, designated by the letter N. Urban areas usually contain sufficient public infrastructure to warrant further building.

Apart from the two broad categories, there are also future urban areas, designated NA, and NB areas, which are already built-up but on which the possibility of extended infrastructure is not ruled out, and NC areas, where the quality of the land or the productivity of the soil or subsoil is such as to reserve it for agriculture. Lastly, ND areas are somewhat fragile, or contain nature areas or landscapes of aesthetic or ecological value. Intensive rearing installations are preferably located in NC areas, where agriculture is protected, or in ND protection areas when there is no special environmental sensitivity problem. When they are so located there is no possibility of habitat conflict.

In spite of some imperfections in the zoning system afforded by the land use plan it does make it possible from the town-planning standpoint to control the creation and extension of intensive husbandry facilities, especially if farmers continue to be associated with the formulation of land use plans. We merely draw attention to the lack of rigour apparently prevailing in the granting of permission for installations in protected areas, owing to insufficiently stringent ecological analysis when these areas are defined.

The 'Water Act' of 16 December 1964
The law of 16 December 1964 relates to the regime and distribution of water and to the control of water pollution. It was this Act which created the

present qualitative and quantitative management system for the six major hydrographic networks through advisory basin committees and financially independent executive basin agencies. Under the Act the agencies determine and define quality targets for watercourses. Quality and flow problems are viewed economically through a system of charges, but pollution discharges are also subject to a set of standards.

The Water Act provides the basis for regulations applying to all types of discharge, direct and indirect, of any kind of material and, more generally, of a kind liable to cause or increase water deterioration by modifying the physical, chemical, biological or bacteriological characteristics of any water, including surface water, underground water and sea water within the limits of territorial waters (Bady, 1981). Discharges are subject to quite strict regulations, and the government has certain powers in cases where safety or public health might be at risk. It can also regulate or prohibit certain types of discharge.

These provisions, designed to unify water policy in respect of discharges regardless of the legal status of the waters, naturally apply to intensive breeding facilities. Broadly speaking, the rule is that permission must be obtained for any kind of discharge and permission will not be granted unless the environment will not be adversely affected.

Specific regulations

These specific regulations, supplementing the general ones, concern classified installations and *département* health regulations.

Classified installations and intensive rearing

The Act of 19 July 1976 on installations classified for environmental protection purposes, and its implementation orders, supersede the 1917 Act on hazardous and other establishments, which did not cover intensive breeding activities. The 1976 Act sets out to protect the interests of neighbourhood amenity, health, security, agriculture, natural and environmental protection, and the conservation of sites and monuments (Lorvellec, 1981). An activity coming under this Act also has to be classified, i.e. listed within a given nomenclature. This applies to livestock husbandry installations of a certain size for some precisely enumerated species. The size depends both on the size of the flock or herd and on the weight of the animals (for pigs, only animals weighing more than 30 kg are covered). The species are those regarded as causing the most pollution: veal and beef fatstock, pigs and poultry. Dairy cows are not covered. The two criteria help to define the legal regime for any installation (table 4.7).

A reporting establishment must adhere to the rules laid down by the Commissaire de la République upon the advice of the *département* health board, which usually follows model decrees formulated by the Ministry of the Environment. They include a rule that the installation must be located at least 100 m away from third-party dwellings, camping and sports facilities and premises for professional use (in the case of pig installations), 35 m away from watercourses, 200 m away from bathing resorts and beaches, and 500 m

Table 4.7 Legal regime for classified establishments, according to species and size of herd or flock

Regime	Veal	Pigs	Poultry
Reporting	50	50	5,000–20,000
Permission	250	450	20,000+

away from fish farms. There are also rules for slurry spreading. For pig slurry a distance of 200 m is required from dwellings and business premises unless the slurry has been deodorised, in which case the distance is reduced to 50 m. Poultry slurry must be stored at least 500 m away from any dwelling. A reporting establishment must also adhere to water policy provisions.

The permission regime is more demanding, because the prefect's permission order is granted only after a public inquiry into the possible impact of the project upon the interests referred to in Section 1 and after consultation with the local municipal councils concerned and upon the advice of the *département* health board (Section 5 of the Act of 19 July 1976). This procedure does not prevent the Minister responsible for classified installations from imposing, by order, 'technical rules govering certain categories of installation subject to the provisions of the present Act. Such orders apply automatically to new installations' (Section 7).

The keystone of the permission regime is the impact study which the applicant has to provide. Section 3 of the decree of 21 September 1977 provides that 'the study must give details of the source, nature and magnitude of any disamenities liable to result from the installation concerned'. Disamenities include noise, the use and discharge of water, the protection of underground waters, and waste disposal. The applicant has to state how he proposes to remedy such disadvantages. To help him, especially in the case of pig farming, the profession has produced a document (Deschamps *et al.*, 1984).

Health regulation
Helin (1983, p. 447) notes that the health regulations are unique in that they are based on a local (*département*) standard, established on the basis of a model document formulated by the Ministry of Health, whose recommendations are minimum recommendations which the local authority may adapt by making them more rigorous. Various model documents have been formulated over time, the most recent to the author's knowledge dating from 1983, with the result that local regulations are somewhat disparate and variable.

Civil liability

In the last resort the farmer's civil and even penal liability may be invoked in the event of pollution.

On the basis of Article 1382 and those following in the Civil Code (anything a person does prejudicing another obliges the person whose fault gave rise to the prejudice to compensate him for it) the courts may require ecological damage to be made good. Traditionally 'making good' may take the form of equivalence, by restoring the situation to its original condition, or may take the form of compensation. For damage to watercourses, restoration to the original condition, which very often depends on the self-purifying capacity of the medium, presupposes delays, even if the first requirement for making the damage good involves eliminating the cause. The result in law is to 'look distinctly at the pursuit of the restoration of a "normal" situation [making good the damage] and add compensation for prejudice arising between the occurrence of the damage and its cessation, which will most often be compensation for disturbed possession' (Bady, 1981).

In these circumstances the court can theoretically rule that the polluting establishment be purely and simply closed down, since the courts have powers both of injunction and of compulsion. In fact they never go so far, because the polluting activity does have administrative permission and in most cases there has been no genuine fault. More basically, the courts hesitate to sacrifice economic interests to nature protection.

Lastly, it should be noted that civil proceedings cannot be taken by just anyone. Only a direct victim or an approved association acting on behalf of the general interest may do so. Of the 20,000 nature protection associations only 900, including fishery and fish farming associations, are approved.

Criminal responsibility

This is based largely on Article 434–1 of the Rural Code, which provides penalties of fines and imprisonment for anyone throwing, discharging or allowing substances potentially hazardous to fish to flow into watercourses. Fish protection was at one time and end in itself, but nowadays the provision serves to protect water, fish survival being regarded as one criterion of water quality. The kinds of discharge referred to apply to agriculture on the same footing as any other activity.

Discussing this article of the Rural Code, Bady (1981) notes that although it occupies a special position in the array of preventive measures available, there are several other regulations which establish criminal penalties for breaches of discharge regulations, in particular certain sections of the Water Act and of its implementing orders. Lastly, various instruments are designed to curb infringement of public health protection rules, while others deal with infringements of the classified installation laws. In contrast with Article 434–1 of the Rural Code, any penalties under the above-mentioned regulations cannot be subject of any compromise settlement.

An initiative to integrate agriculture and environment: the Côtes-du-Nord Agricultural Charter

In addition to regulatory measures, it should be mentioned that an effort was made in the Côtes-du-Nord area to encourage farmers to take the environment into account, on one hand, and local officials to take farming needs into consideration, on the other. This has resulted in the development in 1980, by the umbrella organisation for agriculture, land use planning and urbanism, environmental protection and public health, of a commonly agreed document called the 'Côtes-du-Nord Agricultural Charter'.

The charter, designed to safeguard agricultural activity, sought at the same time to resolve problems arising between farmers and non-farmers in the rural environment. It acknowledges that agriculture has historic precedence in terms of occupying the land. Accordingly, others residing in agricultural zones must accept any nuisances inherent to agriculture. On the other hand, the charter committed farmers to honouring existing environmental regulations, notably as regards slurry spreading and the construction of buildings in aesthetic harmony with the countryside.

Economic analysis

The economic analysis of pollution problems starts with the concept that nature can be exploited 'free of charge' and distorts resource allocation and the use of the environment. But, in this respect, economic mechanisms can be improved, especially by reliance on the 'polluter pays' principle. However, its application to any given case of livestock effluent pollution is not simple. Accordingly, theoretical outlines can provide no more than a guide for concrete proposals to improve the situation.

Optimum allocation of resources

Externalities

The availability of environmental goods free of charge prompts economic agents to shift (or transfer) the cost of the service obtained from the environment on to the community or to other users. This 'externality' process, (a negative one here) can be corrected by internalisation.

In an optimum situation each economic agent takes the best decisions, not only as regards himself but also, indirectly, as regards others. This presupposes certain conditions for producer technology, consumer preferences and the forms of interaction among the various agents. The assumption that economic agents interact via markets, and that a producer or consumer bears the costs resulting from the action he takes, is particularly questionable in the environmental context. Negative externality processes emerge in the sense that certain costs which ought to fall upon the individual agent are pushed back on to others.

When there are negative externalities—for example, when pig production adversely affects the quality of water which must then be denitrified to make it drinkable—the private costs of pig production are transferred to others.

The problem arises from the fact that the possibility of polluting, which is the equivalent of using the environment as a production factor, does not involve any cost. Accordingly there is no incentive, except through regulations, to 'economise' with nature. Nature is used as a production factor up to the point where further pollution would not permit any increase in output, i.e. when the marginal product from further pollution is nil. For an enterprise to reduce its use of nature as a production factor, either it must use more labour, more capital, more recycling, or it must reduce its own output, none of these possibilities being mutually exclusive. Resourse to other production factors means, for any given technology, an increase in cost, some part of which may be borne by the community (subsidy).

The internalisation of external effects

The effects of pollution can of course be limited through regulations, such as those examined in the preceding section, but the legal solutions have three major defects (Kneese and Lévy-Lambert, 1967). The first is that legal delays are such that even in the most favourable cases the plaintiff is liable to incur substantial losses while awaiting compensation. Secondly, there are so many kinds of damage, and so many individuals can be affected, that it is not realistic to imagine a proliferation of court proceedings in respect of all the existing forms of pollution and disamenity. Lastly, measures to be implemented only make sense at certain times, which are not fixed, whence the difficulty in enacting appropriate legislation. As there are no (sufficient) specific property rights for environmental goods from a standpoint of economic efficiency, one has to try to bring the external costs to bear upon the decisions each farmer makes.

In principle the aim should be that the producer should use the environment as a production factor up to the limit at which his marginal production cost equals the cost to society of the marginal unit of pollution. The firm will then set its production so that its contribution to collective well-being (its marginal product) is equal to what it costs society (marginal cost of pollution). This can be shown graphically by illustrating a situation when only the private cost is taken into consideration, and when the cost of pollution is included (fig. 4.6).

Figure 4.7a shows that at price $\bar{p}$ the entrepreneur, maximising his personal profit, produces volume $\bar{q}$, for which his private marginal cost equals the market price. But the real cost also includes the environmental deterioration that production imposes on other people. This brings the optimum volume of production for society back to $q^{\star}$. The same emerges from fig. 4.7b, which expresses the marginal income or profitability of the enterprise according to production, and the marginal cost of environmental deterioration according to the activity of the enterprise (curve C_{me}). The intersection between curve C_{me} and the cost to society of controlling the pollution (fig. 4.7c) gives the optimum social volume of pollution corresponding to level $q^{\star}$ for the enterprise.

Briefly, the internalisation of negative externalities can be defined as a combination of means which brings about, under the constraint of optimum factor allocation, a shift from quantity q to quantity $q^{\star}$. In theory there are three equivalent solutions:

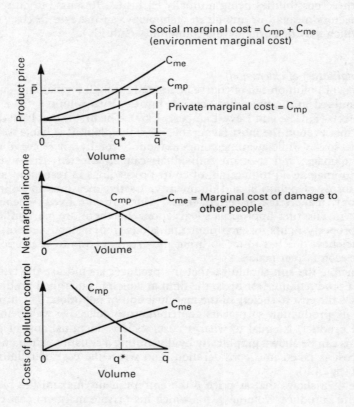

Figure 4.7 Private optimum and social optimum

1 A price per unit produced equivalent to social marginal damage as shown by curve C_{me}, bearing in mind that there is a direct link between the volume produced and the environmental deterioration.
2 Subsidizing the producer so that he reduces pollution to the point where the cost of reducing the pollution any further exceeds the subsidy he received, i.e. $q^\star$.
3 Creating a market in pollution rights corresponding to the optimum amount of environmental deterioration so that the total demand for pollution rights corresponds to curve C_{me} in fig. 4.7c.

The 'polluter pays' principle and its application (standards/charges)

Although in theory the three internalisation methods briefly mentioned above are equivalent, the pure lump sum subsidy system is ruled out owing to the poor allocation of resources it might bring about in the medium term. New firms might be tempted to enter the industry in order to benefit from the subsidies. As regards the market in pollution rights, there are both practical and psychological difficulties (Beckerman, 1975).

Internalisation is thus regarded essentially in terms of charging the economic agent responsible for the environmental deterioration (the 'polluter pays' principle). Equalising the marginal cost of damage (environmnental deterioration caused by the productive activity) and the marginal cost of controlling pollution allows for an optimum distribution of resources and an optimal amount of pollution which can be achieved by either setting a standard or levying a charge.

A charge allows the cost of pollution-associated damage to be fully internalised (Barde, 1980). This is because the polluter bears the charge which is equivalent to the cost of pollution control.

When one cannot ascertain the damage function representing environmental deterioration, no equivalence is available between the pollution standard and the charge, and a charge, if properly set, is more effective than the standard. This emerges especially when technological progress shifts the marginal cost of pollution control.

With a charge, the polluter is prompted to reduce pollution whenever pollution control becomes cheaper. If there is no charging system and only an emission standard system, the polluter has no incentive to reduce pollution.

As Barde points out (1980), the charge also constitutes an economic system when an environmental quality target is being set and when an overall reduction in the amount of emission is sought. Rather than seek a uniform abatement on the part of each polluter, it is better in this case to allow each agent to equalise his marginal treatment cost with the charge. This costs society less.

In fig. 4.8 polluter P_1, whose marginal pollution control cost is lower than the average marginal cost, will adjust his degree of treatment to the charge and thus considerably reduce his pollution, even if it was already below standard ON. Conversely polluter P_3, whose cost is high, will reduce his pollution less. But in total the expenditure of the three polluters will be lower than if everybody had been subjected to emission standard ON. The gain is equal to the difference between the hatched areas of P_3 and P_1.

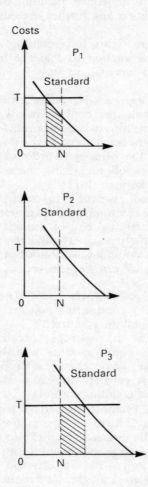

Figure 4.8 The usefulness of the charge in pursuing an environmental quality target. P_1 polluter with low treatment cost, P_2 polluter whose cost corresponds to the standard, P_3 polluter with high treatment cost

Apart from its optimisation and efficiency effects the charge has a redistributive role, with which either collective pollution control plants can be financed or polluters can be assisted to install more efficient equipment. This can be done by using the fraction of the charge corresponding to the tax on the environmental resource or by using the yield of the charges from those polluters with the highest costs. These economic agents pay more because, since they do less treatment than average, they pollute more and therefore pay the charge over a larger number of pollution units.

Application to slurry pollution

The above theoretical outline presents difficulties in application. The diffuse character of pollution in agriculture complicates the position. Having reviewed some of the obstacles we will see how financial agencies are tackling the question.

The difficulty of evaluating the damage function

Intensive husbandry produces manure which, when spread, often causes pollution associated with slurry production. Pig wastes are estimated to account for between 8 and 10 per cent of the increase in pollution growth. But when the processing capacity of the soil is used, by means of slurry spreading, which also provides fertilisation, the damage other than odour disamenity is significant only when a certain threshold is exceeded. As we have seen, the threshold varies according to the type of soil and weather conditions.

Slurry spreading and infiltration due to installations subject to heavy rain or which are insufficiently watertight give rise to pollution by diffuse effluents. Pollution flows are difficult to monitor consistently and reliably, and at a reasonable cost. Their stochastic nature may be partly taken into account by means of hydrological models based on basin soil characteristics, type of crop and climate. In fact such models barely reduce uncertainty as to actual pollutant flows. Damage evaluation is therefore intrinsically uncertain.

The very nature of the impact of animal wastes on the environment complicates the evaluation. The aesthetic disamenity due to buildings and plant, and the odour disamenity, are not perceived identically by country and city dwellers, which inevitably poses a problem of aggregation. Moreover, according to the type of damage, the marginal cost of environmental deterioration varies for different levels of production. For odours, disamenity appears as soon as slurry is spread, whereas the eutrophication of eco-systems requires a certain phosphate and nitrogen threshold to emerge.

Basin agency charges

The basin agencies are empowered to levy charges on pig farms which run to more than fifty-seven pigs. In fact the amount of the charge may be reduced if the farmer takes measures to limit wastes. The threshold of fifty-seven pigs is related to total effluent after treatment. The treatment is therefore more important since, if the quality of the slurry spreading is higher, there is less likelihood of pollution occurring. Generally the pollution is also less when there is sufficient storage volume and a satisfactory ratio of land to heavy livestock (taking all livestock present into consideration).

Table 4.8 indicates the proposed charges per pig produced in France in 1982 for pig farms with more than 208 places (Agence de Bassin Loire-Bretagne). This system, which has not yet been implemented, requires certain proofs as to the quality of slurry spreading. A simplified procedure is laid down for installations of under 1,372 places (reference details to be provided every five years) whereas for larger installations proof of good spreading practices will be required yearly and in greater detail.

The levying of charges, which can be extended to other types of intensive husbandry, makes it possible to contemplate subsidising and providing grants for animal waste pollution control operations individually or collectively, including the establishment of slurry banks. These grants are provided in advance, free of interest, covering up to 70 per cent of the cost excluding tax and repayable over ten years.

Conclusion

Conclusions can be drawn at two levels: Firstly, our understanding of the parameters for good, economic waste management; and secondly, the implementation of policy measures which limit the effect of discharges upon the environment.

Understanding the economic parameters

As a first step, an accurate estimate of the damage function for the environment based on the cost of denitrifying water is necessary. This is relatively easy to acquire and would be essential in setting charges.

At the same time, we need to know more about the cost of processing slurries on the farm and in collective installations in order to see how marginal cost curves can be worked out. This also presupposes an assessment of the fertiliser value of slurries in economic terms.

Lastly, a study of farmers' reactions to a levy or to the payment of anti-pollution premiums presupposes an estimate of the production function. This requires a study of the separability of production factors, of production elasticities and of elasticity in respect of input prices.

Policy measures

SRAE surveys in Côtes-du-Nord (1986b) in the area of very intensive pig farming show that hardly more than a third (37 per cent) of the farmers are aware of the fertiliser value of slurries, which results in further inputs of mineral fertiliser! Meanwhile storage containers are often too small, and receive rain water and run-off waters. At the same time animal wastes are badly used in terms of both space and time.

Table 4.8 Charge per pig produced, according to size of firm (number of places) and slurry spreading quality (FFr)

Quality of slurry	No. of pigs			
	208–686	687–1,372	1,373–2,744	2,745+
Poor	6·16	6·16	6·16	6·16
Good	0	1·86	1·86	1·86
Very good	0	0	0·93	0·93
Excellent	0	0	0	0–0·93

Accordingly, advisory activities and assistance towards a better understanding of slurry spreading capacities, soil qualities and crop requirements appear to be essential. Systems to improve the storage of slurry would also appear to be justified.

These are the only measures feasible in the short term, since the application of the 'polluter pays' principle would oblige farmers initially to absorb all the costs of protecting the environment. Under present economic circumstances and given falling commodity prices, this could mean the elimination of many farmers. Only in the long run would one see the impact of the rise in production costs on prices, and hence a transfer of the cost burden to consumers. Such a scheme does not seem realistic at this time, hence the preference for the above-mentioned measures.

References

Agence de Bassin Loire-Bretagne (n.d.), note relative à l'intervention de l'Agence du Bassin dans le cadre de l'amélioration de l'utilisation et la gestion des effluents d'élevage.

André, P, and Dubois de la Sabloniere, F. (1986), 'Elevage intensif et qualité des eaux souterraines dans un département breton. Réflexions pour une politique d'aménagement, Techniques et Sciences Municipales', L'eau, 5, 251–8.

Bady, J.P. (1981), 'La protection des eaux contre les pollutions d'origine agricole', *Agriculture et environnement*, 5ème Colloque de la Société Française pour le Droit de l'Environnement à Pau, Publications périodiques spécialisées, 105–44.

Barde, J.P. (1980), *La pratique des redevances de pollution*, OECD.

Beckerman, W. (1975), *Le principe pollueur-payeur: interprétation et principes d'application*, OECD, 35–68.

Bertrand, M., and Arroyo, G. (1984), *Méthode rapide d'appréciation de la valeur fertilisante des lisiers de porcs*, CEMAGREF, BI 321, 21–34.

CEMAGREF (1982), *Disponibilités en éléments fertilisants d'origine animale et évaluation des exportations des cultures en Bretagne*, 45 (Informations Techniques), 3.

CEMAGREF (1983), *Evaluation des surfaces agricoles disponibles pour les épandages de déjections animales*, 51 (Informations Techniques), 1.

Coillard, J. (1986), 'Contribution à la conception, à la réalisation et au suivi d'une installation de méthanisation de lisier de porcs dans un élevage industriel', *Bulletin*

Technique du Machinisme et de l'Equipement Agricoles (CEMAGREF), Hors série 54.

Coppenet, M., and Golven, J. (1984), 'Etude lisier-sol-plante. Bilan de suivi dans une soixantaine d'exploitations intensives du Finistère', INRA Quimper–SUAD Quimper, mimeo.

Deschamps, V., Heduit, M., Griperay, G., and Bessemoulin, J. (1984), *L'élevage porcin et l'environnement—Recommendations techniques*, ITP–GIDA.

Diry, J.P. (1985), *L'industrialisation de l'élevage en France—Economie et geographie des filières avicoles et porcines*, Ophrys.

Fritsch, P., and Saint-Blanquat (de), G. (1985), 'La pollution par les nitrates', *La Recherche*, 169, 1106–15.

Grignon, C. (1981), 'Les conditions sociales de l'intensification', *Economie Rurale*, 146, 3–13.

Helin, J.C. (1983), 'Elevage industriel et protection du milieu', *Revue de Droit rural*, 118, 445–53.

Jegouzo, Y. (1983), 'Protection du milieu élevage industriel et droit de l'urbanisme', *Revue de Droit Rural*, 118, 438–44.

Klatzmann, J. (1955), *La localisation des cultures et des productions animales en France*, *Paris*, Imprimerie Nationale.

Kneese, A.V., and Lévy-Lambert, H. (1967), *Economie et gestion de la qualité des eaux*, Dunod.

Le Bihan, J. (1961), 'Analyse économique du développement de l'aviculture moderne en Bretagne', Thèse Economie, Rennes.

Lorvellec, L. (1981), 'Elevage hors-sol et installations classées', *Agriculture et environnement*, 5ème Colloque de la Société Française pour le Droit de l'Environnement à Pau, publications périodiques spécialisées, 56–82.

Meeus-Verdinne, K., Scokart, P.O., and De Borger, R. (1986), 'Evaluation des risques pour l'environnement provenant des métaux lourds contenus dans les déchets animaux', *Revue de l'Agriculture*, 39(4), 801–16.

Saunier, P. and Schaller, B. (1978), *L'aviculture française vingt ans après, vol. 1, Quatre études sur la portée et sur les limites de lar 'révolution avicole': le cas des volailles de chair*, INRA–Série Econonmie et Sociologie Rurales.

Sogreah (1985), 'Etude d'un modèle de référence applicable à la création de banques de lisiers', guide méthodologique.

SRAE (Service Régional de l'Aménagement des Eaux de Bretagne) (1982), 'Réduction des nitrates dans les eaux potables distribuées en Bretagne', mimeo.

SRAE (1986a), 'Note sur les concentrations en nitrates relevées début 1986 dans les rivières bretonnes', mimeo.

SRAE (1986b), *Analyse descriptive visant à déterminer les mesures pour restaurer la qualité d'eaux conchylicoles (Bal de Saint-Brieuc—Bassin du Gouessant)*, rapport intermédiaire, April, 3 vols.

5. Dutch approaches to the management of pollution from intensive livestock production

Grontmij NV

Social developments after 1954 have led to an increasing demand for agricultural products, especially meat. The causes of this lie in the growth of the population, increases in employment opportunities in the industrial and service sectors and increases in purchasing power which in turn have resulted in increasing overall demand, especially in urban areas. Moreover, a shift in consumer behaviour has taken place, resulting in the substitution of highly processed animal products for vegetable products.

The agricultural sector has responded to these new demands and has taken full advantage of the market potential. Dutch agricultural policy and the European Community's Common Agricultural Policy have further stimulated developments in this respect. Both marketing and pricing policies and factors such as research, education and information linked to technological developments have also contributed to increased agricultural production. In some areas of the Netherlands all the above factors together have led to the considerable intensification of animal husbandry. Animal population per farm has increased to such an extent that it has become necessary to bring feed in from other farms and other countries.

The central location of the Netherlands in Europe, especially its position on the sea and its good infrastructure, has contributed to increased Dutch agricultural production. Animal husbandry especially profits from the relatively low costs at which animal feedstuffs can be imported. The good geographical position of the Netherlands is also of importance for the export of agricultural products. Europe's important centres of population in England, Germany and France are all within easy reach.

Similarly, relative to the land area available for cultivation, the size of domestic demand is large. The dense network of waterways, railways and roads makes all centres of population in the Netherlands easily accessible. As a result about 60 per cent of agricultural income is earned from exports. The value of agricultural exports amounted in 1985 to approximately Gld 51·5 billion, while agricultural imports were valued at approximately Gld 35·8 billion. The agricultural balance of trade was therefore a surplus of approximately Gld 15·7 billion.

Overall, these conditions, together with favourable climatic conditions, have led to a well developed agricultural sector, which in turn has helped to influence the development of technologies and methods relating to animal husbandry. There is a high level of expertise in the agricultural sector. The government provides sophisticated research, education and information services. These elements are closely linked organisationally and so provide the possibility of innovative activities and the rapid introduction of new techniques. Moreover the agricultural sector is highly organised and reflects strong relationships among agricultural contractors, suppliers and processing industries.

The possibility of importing low-priced feedstuffs and the use of by-products from the foodstuffs industry have contributed to the development of animal husbandry. High wage levels have also worked in favour of less costly methods of production.

Since the early 1950s the prices of agricultural products have declined by an average of a few per cent per year. For example, in relative terms the price of cereal grains, potatoes and milk has halved since the 1950s, with the result that farmers have had to produce more to offset lower sales prices. Productivity in agriculture has also been enhanced by technical improvements; a substantial increase in the use of agricultural inputs; and energy. Besides genetic manipulation, automation and improvements in the feed conversion rate have resulted in further intensification, mechanisation and increases in the scale of operation on many farms.

Strongly vertical integration in the agricultural sector is common in the Netherlands. Many processing companies work in close collaboration with intensive animal production units, and supply is often on a contract basis. In many cases these companies act as financial institutions. The financial dependence of the agrarian contractor on these institutions and the banks is very high in animal husbandry.

Dutch and European Community agricultural policies share the basic goal of stimulating self-sufficiency and ensuring a cheap supply of foodstuffs via marketing and price guarantees, market protection and supportive measures. Intervention prices exist for grains, sugar, beef and milk. But pork and poultry producers do not receive comparable financial support from the European Community. Nevertheless, market intervention, export refunds, price guarantees and supportive measures included in the framework of structural improvement have stimulated the increase in production and the continuing intensification associated with it.

The current situation of intensive animal husbandry

Historical developments in agriculture

During the period 1900 to 1985, particularly the period 1970–85, there has been:

1 An increase in the total surface area of the Netherlands from 3·3 million ha in 1900 to approximately 3·7 million ha in 1985.
2 An increase in the area under cultivation from approximately 2·1 million ha in 1900 (65 per cent of the total surface area) to 2·3 million ha in 1960 (70 per cent of the total surface area), and then a subsequent decline to approximately 2·0 million ha (54 per cent of the total surface area) in 1985.
3 A population increase from 5·2 million in 1900 to 14·5 million in 1985.
4 A decline in the number of farms from approximately 300,000 in 1960 to approximately 137,000 in 1985.

Table 5.1 Development of agriculture in the Netherlands

Input	1900	1950	1960	1980	1985
Total land area (000 ha)	3,255	3,483	3,616	3,731	3,730
Total land cultivated (000 ha)	2,116	2,336	2,310	2,020[a]	2,019[a]
Population (million)	5·18	10·20	11·56	14·21	14·40
Cattle (000)	1,656	2,726	3,509	5,266	5,248
Pigs (000)	747	1,864	2,955	10,138	12,383
Poultry (000)	4,300	23,500	42,400	81,200	89,900
Draught working horses (000)	295	256	187	47	46
Sheep (000)	771	390	456	858	814
Use of artificial fertiliser:					
P_2O_5 (kg million)	–	120	112	84	87
N (kg million)	–	156	224	486	478

(a) New method of compilation.
Source: CBS (Central Bureau of Statistics), The Hague, 1986.

Table 5.2 Distribution of farm size: cultivated area per farm

Area cultivated (ha)	1970		1985	
	No. of farms	%	No. of farms	%
0	3,494	2	3,650	3
0·01–2	31,483	17	64,563	48
2–10	67,169	37	–	–
10–50	79,960	43	64,082	47
50–100	2,283	1	4,139	3
100+	224	1	465	1
Total	184,613	100	135,899	100

Source: CBS (Central Bureau of Statistics), The Hague, 1986.

Table 5.3　Distribution of farm size by activity, 1985

Area cultivated (ha)	Farms with:					
	Fattened calves		Pigs		Poultry	
	No.	%	No.	%	No.	%
0	374	25·2	1,896	16·1	662	25·2
0·01–1	313	21·1	1,881	16·0	451	17·1
1–5	574	38·7	4,205	35·8	981	37·3
5–10	167	12·3	2,259	19·2	333	12·7
10–15	43	2·9	1,024	8·7	117	4·5
15–20	8	0·5	319	2·7	43	1·6
20–30	3	0·2	135	1·1	34	1·3
30–50	–	–	23	0·2	6	0·2
50–100	–	–	1	0·1	3	0·1
more than 100	–	–	–	–	–	–
Total	1,482	100	11,743	100	2,630	100
% with under 5 ha		85		68		80
% with under 10 ha		97		87		92

Source: CBS (Central Bureau of Statistics), The Hague, 1986.

Table 5.1 shows the total surface area of the Netherlands, the area under cultivation and the increase in the total animal population from the year 1900. The livestock population increased considerably, especially after 1950, while the amount of cultivated land has decreased somewhat. The number of cattle in 1985 is nearly twice that in 1950. The number of pigs has increased sixfold in this same period. The number of poultry has increased nearly fourfold.

Farm size

Table 5.2 shows the distribution of the total number of agricultural enterprises in relation to the amount of cultivated land. Farm size has increased in the period 1970–85. The average amount of cultivated land per farm was approximately 11·6 ha in 1970 and approximately 14·9 ha in 1985, an increase of more than 3 ha per farm.

Table 5.3 indicates the size distribution of farms used for intensive agricultural production. Nearly all the concerns in this category have an area smaller than 10 ha and the greater part have an amount of cultivated land smaller than 5 ha.

Regional distribution

Figure 5.1 indicates the distribution of intensive animal production in the Netherlands. Approximately 90 per cent of pigs and 75 per cent of piglets are concentrated in the provinces of Gerderland, Overijssel, Noord-Brabant and

Limburg. All these provinces contain many sandy areas, and intensive animal production tends to be concentrated on these areas.

Manure production

Amount of manure production

As a result of the increase in the quantity of farm livestock in the Netherlands, there has also been a big increase in the amount of manure produced. Manure produced by cattle, pigs and poultry increased from 76 tonnes in 1970 to 94 tonnes in 1985, corresponding to approximately 235,000 tonnes of phosphate. Owing to the spreading of the increased amounts of manure on the fields and also the use of artificial fertiliser, silt farm sewerage plants and compost, the concentration of minerals in agricultural land has greatly increased. In 1985 there was evidence of a doubling of mineral content (especially of nitrogen, phosphate and potassium) compared to 1970 (table 5.4).

Table 5.5 indicates the nature of animal manure production and, in particular, the significant quantities of nitrates, phosphates, potassium, copper and cadmium produced. Between 1970 and 1985 the total quantity of minerals, especially nitrogen, phosphate and potassium, applied from all sources to agricultural land doubled.

Manure spreading

The manure which is produced is spread primarily on agricultural land in the producing region. With the introduction of spreading restrictions, part of the manure production must now either be spread on agricultural land outside the region of production or treated. Some manure is now being exported outside the Netherlands.

In the future, the disposal of manure outside the concentration areas can be expected to become more frequent, notably in agricultural areas in the provinces of Zeeland, Noord-Holland, Groningen and the Veenkoloniën. Figure 5.2 gives a global indication of the areas between which manure must be transferred.

Exports of animal manure and vegetable compost in 1985 amounted to approximately 130,000 tonnes. Approximately 98 per cent went to Belgium and West Germany. The imports amounted to 150,000 tonnes, of which approximately 50 per cent consisted of horse manure for the preparation of black earth for mushroom production. Further export possibilities of animal manure to West Germany are very limited. In northern France there is further potential capacity for the export of manure.

Since 1972, in the provinces of Gelderland, Noord-Brabant and Limburg, three manure banks have been active.

Slurry can be disposed of at greater distances, aided by government subsidy (total turnover: approximately 440,000 tonnes).

There are also two installations in use for the pre-purification of fattening calf manure, with a capacity of 175,000 tonnes of fattening calf manure and 22,000 tonnes of pig slurry manure a year, which represents about 20 per cent

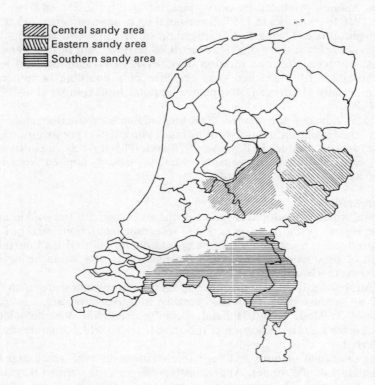

Figure 5.1 Areas where intensive animal husbandry is concentrated

Table 5.4 Total manure production in the Netherlands

Source	Tonnes	%
(Horned) cattle	72·0	77
Fattened calves	1·8	2
Fattened pigs	10·8	12
Breeding pigs	6·9	7
Laying hens (liquid manure)	1·8	2
Laying hens (dry manure)	0·2	0
Slaughter chickens	0·3	0
Turkeys (brooding/slaughter)	0·0	0
Total	93·8	100

Source: LEI (Agricultural Ecoonomics Research Institute), The Hague, 1986.

Table 5.5 Composition of the manure by type of animal, 1985 (kg/tonne, except Cu and Cd g/tonne)

Type of manure	Dry matter	Organic matter	N	P_2O_5	K_2O	Cu	Cd
Thin cattle manure[a]	95	60	4·4	1·8	5·5	3	0·03
Liquid manure (cattle)	26	10	4·0	0·2	8·0	–	–
Fattened pigs	80	50	6·8	4·4	6·5	48[b]	0·07
Chickens	150	90	9·6	8·8	6·9	18	0·11
Sows	60	40	4·0	4·0	4·0	25	0·05
Solid cattle manure	215	140	5·5	3·8	3·5	–	–
Pigs	230	160	7·5	9·0	3·5	–	–
Chickens (wet)	320	230	15·2	16·0	13·0	32	0·19
Chickens (dry)	600	370	24·3	28·3	22·2	57	0·35
Slaughter chickens	580	430	26·0	24·0	21·5	68	0·41

(a) Per stable period of 180 days.
(b) On the introduction of a new EC directive this will be 22 g per tonne of manure.
Source: Mesactieprogramma (Manure Campaign Programme), 1987.

of the total 13·8 million tonnes of manure to be disposed of. Current manure storage capacity in central installations is approximately 90,000 m³.

Productivity and national income

Free exchange of animal husbandry products has been possible within the European Community since 1967 and is now of considerable value. The value

Manure surplus area
Manure disposal area

Figure 5.2   Global indication of the main areas of manure production and disposal

of production in 1985 was almost twice that of 1970. Pork production has shown the largest growth. In 1982 60 to 65 per cent of the market was taken up by pork, poultry meat and eggs. The export share of total veal and fattening calf production amounted to between 80 per cent and 90 per cent. Of total exports of products from agriculture (approximately Gld 32,000 million in 1982), exports of animal husbandry products were the largest, followed by the dairy sector.

It has been estimated that one entrepreneur is fully employed at a production level of 400 fattening calves (approximately 800 animals supplied per year), 120 breeding sows, 1,200 fatteners, 50,000 table poultry (477,000 kg weight supplied per year) or 15,000 layers.

Animal husbandry and the environment

Animal husbandry has its consequences for the quality of the environment in the Netherlands, particularly where intensive animal husbandry is concentrated. The most important impacts on the environment are caused by the:

1 Leaching of nitrogen and phosphate in the soil, followed by diffusion into ground water and surface water.
2 Contribution to acid deposition by ammonia (NH_3).
3 Spreading of heavy metals and pesticides by way of manure.
4 Contamination of the soil and the water by potash (K_2O).

Apart from these primary effects, there are also important secondary effects, including:

1 Adverse effects on the visual quality of the landscape.
2 Adverse effects on flora, fauna and recreation.
3 Adverse effects on natural surroundings.
4 Adverse effects on drinking water supply.

The effects of manuring

Animal manure from animal husbandry can be described in terms of mineral content, and even of overdose. The quantity of nutrients far exceeds the requirements of the crops and is locally so high that damage to agricultural crops often follows.

The minerals P_2O_5, N and K_2O which are present in animal manure provide the environment with a richness of nutrients, which is then averaged out by natural variation.

Phosphate (P_2O_5) is fixed in the soil in the first instance. The buffer capacity is certainly lower in the sandy soils of areas of concentration than in areas of clay and fen land. In the Netherlands, continuing high levels of harmful manuring imply the threat that phosphate will pollute the ground water and surface water, and, ultimately drinking water.

Figure 5.3 gives an indication of the likely occurrences of phosphate pollution with no policy change in the next twenty-five years. In the first ten years polluted areas will amount to about 2,000 ha. In twenty-five years they could expand to approximately 20,000 ha.

Potash also contributes to the pollution of soil and water, leading to a decrease in soil fertility. The main source of potash is (horned) cattle. High quantities of K_2O on grassland can cause head illness in the case of (horned) cattle and through this decreased milk production.

Some of the gaseous nitrogen production from animal husbandry diffuses as NH_3 into the air, while most of the rest turns into NO_2 and after that into NO_3 (nitrate). Nitrogen in the form of nitrate is much more mobile than phosphate and drains into ground and surface waters much more rapidly. Figure 5.4 illustrates nitrate concentration in the ground water at 10 m below ground level. In the Netherlands two-thirds of the drinking water comes from ground water. In various wells the European Community guidelines of 50 mg of NO_3 per litre are exceeded. In the province of Brabant values of several hundred milligrams of NO_3 per litre are observed. At present the Dutch target value for NO_3 loading is 25 mg per litre; about 25 per cent of the population will, in the future, have to use ground water that approaches or exceeds this target value (KIWA, The Netherlands Waterworks Testing and Research Institute, Voorburg, 1984).

In parts of the water collection areas in Montferland, in the province of Gelderland, and Reuver, in the province of Limburg, the nitrate concentration exceeds 50 mg of NO_3 per litre. Measures are being taken for the purification of the drinking water and also to reduce the problem by diluting it with water from several other wells.

Acidification

Ammonia is released from manure produced by animals and amounts to 30 per cent of airborne acid deposition in the Netherlands. Acid deposition from the air accelerates leaching processes in forest soils, with a consequential impoverishment of elements (Cu, K and Mg) and absorption of heavy metals (Al, Mu, Cd and An) by plants. In the Netherlands approximately 130,000 tonnes of ammonia are excreted annually, and about 95 per cent of this comes from agriculture. Intensive animal husbandry (farms with fattened calves, pigs and poultry) produces 22,000 tonnes of ammonia from the production and storage of manure and 57,000 tonnes as a result of manuring.

Averaged out over the whole of the Netherlands, nitrogenous deposition from the air in the forms of NO_3 and NH_3 amounted to approximately 46 kg of nitrogen per hectare in 1984. In areas of concentrated intensive animal husbandry this load is much higher, occasionally exceeding 1,000 kg of nitrogen per hectare per year (Publicatiereeks Milieubeheer 23, Ministerie VROM—Publication series Environment, Ministry of Housing, Physical Planning and the Environment, The Hague).

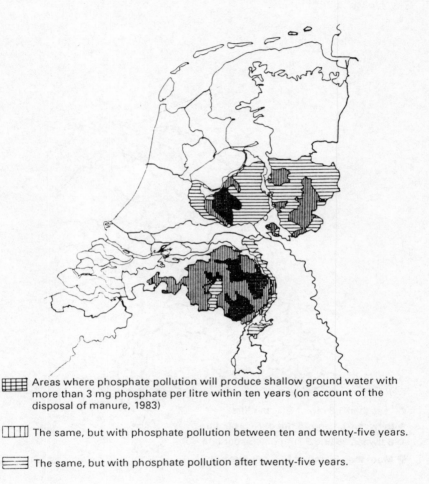

Areas where phosphate pollution will produce shallow ground water with more than 3 mg phosphate per litre within ten years (on account of the disposal of manure, 1983)

The same, but with phosphate pollution between ten and twenty-five years.

The same, but with phosphate pollution after twenty-five years.

Figure 5.3 Areas of likely phosphate pollution if there is no change in policy in the next twenty-five years. The phosphate content of the ground water can be expected to increase above 3 mg of PO_3 per litre within three time spans. *Source:* Rijks Planologische Dienst (Physical Planning), The Hague, 1985

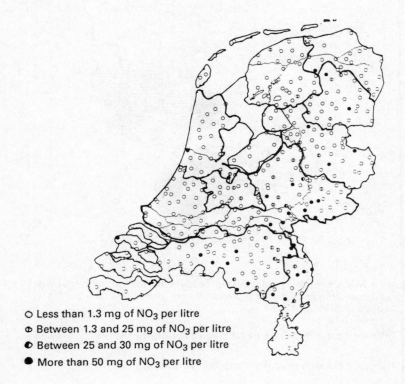

○ Less than 1.3 mg of NO_3 per litre
Φ Between 1.3 and 25 mg of NO_3 per litre
◐ Between 25 and 30 mg of NO_3 per litre
● More than 50 mg of NO_3 per litre

Figure 5.4 Nitrate concentration in ground water at 10 m below ground level, 1985. Twenty-five milligrams of NO_3 per litre is the Dutch target value for drinking water, 50 mg/l the limit. *Source:* RIVM (National Institute of Public Health and Environment Protection, Bilthoven) meetnet grondwaterkwaliteit

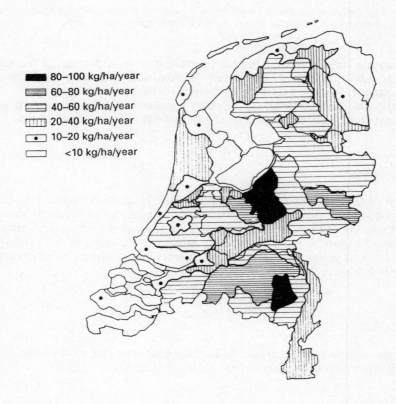

Figure 5.5 The emission of ammonia (NH_3) from animal manure in the Netherlands (kg/ha/year). *Source:* Rijks Universiteit (State University), Leiden, 1982

Table 5.6   Condition of trees in the Netherlands, 1985 (%)

Condition	Netherlands		De Peel area
	1984	1985	1985
Healthy	50·8	49·9	3
Affected	39·9	35·0	65
Sick	9·5	15·1	32

Sources: Staatsbosbeheer (Holland National Forest Service), Driebergen, 1985; Stichting Natuur en Milieu (Netherlands Society for Nature and the Environment), Utrecht, 1986.

Ammonia emissions are toxic for plants and trees, especially in the immediate vicinity of animal husbandries. Wooded and scenic areas near intensive animal husbandries have been severely affected (table 5.6).

Emissions of ammonia also cause smell nuisances, with adverse effects on flora and fauna and recreational amenities in areas of high concentration.

Heavy metals

Heavy metals such as copper, cadmium and zinc are introduced to the environment by way of feed. Considerable copper and cadmium is introduced into the soil by manuring, particularly in areas of high concentration. Continued inputs of these metals may affect certain crops in the medium term. Current policy aims to reduce the copper and cadmium concentration in cattle fodder.

Other effects

Intensive animal husbandry and the harmful emissions and wastes associated with it also exercise a number of secondary environmental effects.

Scenery
Owing to the increasing size of enclosed structures on farms, building density in certain areas has increased to the extent that more than 1 per cent of the usable area is covered by buildings, with the result that it is often difficult to discern the original landscape. The cultivation of maize, linked with (intensive) animal husbandry, also transforms the landscape from time to time.

Quality of life, recreation
The quality of life, work and recreation is negatively influenced in many places by unpleasant odours and the visual disamenity of the animal husbandry.

Natural beauty

Acidification and manuring can have deleterious effects on landscape amenity values through a decline in plant diversity, excess algae in surface waters and the grassification of moorlands.

As well as damage to woods, in the long run the fertility of the soil and the quality of the crops can be affected by manuring. The accumulation of minerals and heavy metals also poses health risks for cattle.

Policy aspects

Agricultural and environmental policy in the Netherlands is prepared and administered respectively by the Ministry of Agriculture and Fisheries and the Ministry of Housing, Town and Country Planning and the Environment. Policy measures related to the problem of manuring have been prepared and administered by these Ministries in collaboration. Further proposals have been submitted by the agricultural industry and the environmental movement to alleviate the problem of manure, and from a political standpoint. The measures to be taken have a broad basis of support.

The policy basis

Often more manure is produced at pig and poultry farms, and on cattle farms with high livestock density, than can be spread on the arable land belonging to these farms, and as a result of the regional concentration of livestock husbandry concern about the excessive manuring of agricultural land, legislation to control manure spreading was introduced. The government has opted for the following four measures to reduce the problem of manure:

1   Increasing disposal possibilities for manure by quality improvement.
2   Stimulating the treatment and processing of manure so that, for example, it may be exported.
3   Reducing excess minerals in manure by reducing the mineral content of feedstuffs.
4   Taking legal measures which will affect the production, use and disposal of manure.

These measures are being implemented through a Manure Campaign Programme and a series of legal controls.

The Manure Campaign Programme includes:

1   Apportioning the finances approved (Gld 25 million per year during the next four years).
2   A review of the measures taken and to be taken.
3   Monitoring the consequences for the livestock industry and the environment.
4   A review of the legislation.

In addition to the above, research within the framework of the mestonder-zoek (manure research) outline plan, which will run for five years, costing approximately Gld 65 million, has been established. Research activities include work on:

1 The reduction of the mineral surplus by feeding method and feed.
2 The distribution and quality of manure.
3 The treatment and processing of manure.
4 The identification of soils saturated with phosphates.
5 The technical development of manure storage facilities.
6 The injection of manure on grassland.

Publicity is organised with regard to the regulations developed, technical aspects of the manure problem and measures affecting the structure of the industry. These aspects are also covered by education programmes.

The legal measures planned are national, with no regional differentiation, and include the:

1 Wet bodembescherming (Soil Conservation Act), which contains a number of rules with regard to the use of animal manure.
2 Meststoffenwet (Fertilizer Act), which regulates the disposal of manure.

Table 5.7 gives details of the initial financial assistance to be provided by the government between 1987 and 1990 to assist with the implementation of the manure action programme. This will include assistance to establish a national manure bank, stimulate investment in manure storage facilities, to develop treatment and processing facilities and stimulate research.

Legal measures

Wet bodembescherming (Soil Conservation Act)
Amongst other things this Act contains a number of regulations which control the spreading of animal manure and came into force on 1 May 1987. Under the regulations it is forbidden to spread animal manure, measured in kilograms of phosphate (kg P_2O_5) per hectare, in greater amounts than those set out in table 5.8.

These new controls will be applied in three phases. The standards for 1987–91 and 1991–95 have already been laid down. In the first phase, although there will be not a surplus of manure nationally, there will be a surplus at farm level (13.8 million tonnes, base year 1985). In the second phase there will be an estimated 5 million tonnes of surplus manure at national level which may no longer be applied to the soil and so will have to be processed and/or exported.

The ultimate standard will have to be met by about the year 2000. For this third phase, however, only indicative standards have been indicated, but a final nationally uniform standard is envisaged. It is expected that the final

Table 5.7 Government expenditure on the control of problems associated with intensive animal husbandry, 1987–90 (Gld million)

Financial assistance	Year			
	1987	1988	1989	1990
Available from:				
Manure campaign programme	25	25	25	25
Pre-financing from previous years	–	–	5	6
MCB	15	–	–	–
Total	40	25	30	31
Destination:				
(Pre)financing national manure bank	7	a	a	a
(Pre)financing storage for national manure bank	4	a	a	a
Financing storage national manure bank	4	–	–	–
Stimulating manure storage	18	17	20	20
Treatment and processing	2	3	5	6
Research	5	5	5	5
Total	40	25	30	31

(a) Contribution from surplus levy.
Source: Mestactieprogramma (Manure Campaign Programme), 1987.

standards beyond the year 2000 will be equal in principle to that quantity which is extracted by plants on an annual basis.

In addition to the above limits on arable land, larger quantities of animal manure may be applied provided that no more than 250 kg P_2O_5 is applied per hectare in any two consecutive years.

The spreading of liquid fractions per hectare is:

1 Limited to 25 m^3 per year on building land and fodder maize land.
2 Limited to 50 m^3 per year on grassland.
3 Prohibited on nature reserves and other land.

It is forbidden to spread animal manure:

1 On sandy soils used for building land or growing fodder maize during the month of October unless the land has been cultivated at any time between the harvest and 31 October.
2 On arable land and fodder maize land unless the animal manure is worked into the soil within one day of application at the latest.

Table 5.8 Limits on the total quantity of phosphate in the form of animal manure which may be spread on land in the Netherlands (kg P_2O_5 per hectare)

Period	Grassland	Fodder maize land[a]	Building land	Nature reserves and other land
1 April 1987 – 1 January 1991	250	350	125	70
1 January 1991 – 1 January 1995	200	250	125	70
From 1 January 1995	175[b]	175[b]	125	70
From c. 2000	Final standard	Final standard	Final standard	Final standard
Crop extraction	110	75	70	–

(a) Average standards vary with soil type.
(b) Approximation.
Source: Mestactieprogramma (Manure Action Programme), 1987.

3　On grassland during October and November and on any soil covered with snow between 1 January and 15 February inclusive.

These restrictions, however, do not apply to land used for the cultivation of flower bulbs or land satisfying the conditions laid down in the regulations for combating potato root disease and the application of substances to combat potato eelworm.

Further efforts are to be made to bring about a general prohibition of application on snow-covered ground for the period 1991 to 1995.

Soils saturated with phosphates
In addition to the above, the provincial authorities may designate areas that are so saturated with phosphates that little capacity for fixing them remains. On these areas it is forbidden to spread animal fertilisers in greater annual quantities than: 70 kg P_2O_5 on arable land, 75 kg P_2O_5 on green maize land, and 110 kg P_2O_5 on grassland.

Soils deficient in phosphates
If it can be shown that agricultural land is deficient in phosphates, exemptions from the maximum permitted quantities on building land, green maize land and grassland may be granted by the Minister of Agriculture and Fisheries.

Fertiliser Act
On the basis of this Act, regulations have been drawn up relating to:

1   The need for farmers to keep manure production records and to establish a national manure bank.
2   The extending and establishment of new premises for livestock production.
3   Relocating manure production.
4   The implementation of a feed levy and a surplus levy.
5   The need for certain users, managers of storage sites and of processing plants to keep records of manure;
6   A simplified system of manure accountancy for certain categories of farmer, including:
    (a)   Stock keepers who produce less than 125 kg of phosphate per hectare per annum.
    (b)   Crop farmers and market gardeners (not being producers of animal manure) in what is termed the 'area of concentration' and arable farmers and market gardeners who have concluded a manure disposal agreement.

Prohibition of expansion

The Fertiliser Act contains a prohibition on the expansion of animal production which results in the production of more than 125 kg of phosphate per year per hectare on agricultural land belonging to the farm. However, it also:

1   Permits the relocation of intensive livestock production facilities to another site or to another farm, provided the total amount of manure produced is not greater than that produced on 27 October 1986.
2   Makes a number of exceptions to the rule that the production of manure must be reduced to 125 kg of phosphate per hectare when the surface area of the agricultural land belonging to the farm is reduced.

New financial measures

1   A general levy on stock feed, independent of demand or whether or not a manure surplus exists, to meet the costs of research related to solutions of the problem of manure surpluses; and measures to promote the quality and the determination of quality of animal manures. On the basis of 93,000 establishments producing manure, this levy amounts to Gld 65 per establishment and in total approximately Gld 6 million per year. The levy is raised on animal feed supplies and not on individual stock keepers.
2   A surplus levy imposed on farms which produce a specified amount of manure above a predetermined amount. The levy is imposed on producers of manure, according to the leviable amount of animal fertilisers, expressed in kilograms of phosphate per unit of time. The surplus levy should yield Gld 43·5 million per year. Approximately 48,000 manure-producing establishments are affected, at an average of Gld 910 per farm.

The levy is intended to meet the costs of:

1   Manure banks.

2 Infrastructural provision for the removal, treatment and processing of manure surpluses.
3 Surveillance and searches.

The surplus levy is implemented in proportion to the quantity of surplus.

1 No levy is due on an amount of 125 kg of P_2O_5 per hectare for the agricultural land belonging to the farm.
2 Gld 0·25 per kilogram of P_2O_5 is due on production in excess of 125 kg but no greater than 200 kg of P_2O_5 per hectare.
3 Gld 0·50 per kg of P_2O_5 for production in excess of 200 kg per hectare.

The levy-free amount of 125 kg of P_2O_5 is calculated on the basis of land availability. A company without arable land makes a direct payment of Gld 0·50 per kilogram. A reduction of Gld 0·25 per kilogram of P_2O_5 applies, however, for:

1 Dry chicken and turkey manure.
2 The disposal under a manure disposal agreement.
3 The export of manure to another country.

The total and average costs of the surplus levy by concern are shown in table 5.9.

Table 5.9 Costs of the surplus levy

	P_2O_5 produced (kg × 10^6)	No. of farms	Total costs (Gld 000)	Average costs per farm
Total production	235.0	135,899	–	–
Liable to levy	104.0	47,821	46,094	965
Liable to low levy	23.3	47,821	5,829	122
Liable to high levy	80.5	–	40,265	–
Estimated reduction	24.0	–	6,000	–
Total costs	–	47,821	40,000	836

Source: Mestactieprogramma (Manure Action Programme), 1987.

Manure storage provision
In order to facilitate the introduction of manure spreading regulations it has been necessary to introduce manure storage regulations. Manure storage facilities may be installed on individual farms or on a co-operative basis. Storage facilities are also necessary at national level.

During the first phase of the prohibition on application, storage capacity for at least two months is required. In later phases a storage capacity of five months' manure production will be required. The average cost per cubic metre of manure storage capacity is Gld 50 and at an industry level it is estimated that this will cost Gld 572·5 million.

Costs to the industry, i.e. annual costs and investment costs, are given in table 5.10 and vary widely. Disposal costs, for example, are highly dependent on the distance to the markets. Farms with little or no land of their own are faced with considerable outlay; poultry farms are especially hard hit, as they usually import most of their feed and have little land on which to spread the manure they produce. Table 5.10 indicates the nature of these costs.

Table 5.10   Investment costs of manure storage (Gld millions)

	Cattle	Fatteners	Total for layers	Pigs	Total capacity	Costs
Production per year (10^6m^3)	72·0	10·8	6·9	2·4	92·1	
Production per month (10^6m^3)	6·0	0·9	0·6	0·2	7·7	
Extension of storage capacity required						
Two months	3.0	0.04	0.04	0.01	3.1	155.0
Two to four months	7.4	0.4	0.4	0.06	8.25	412.5
Four to six months	10.1	1.0	0.8	0.2	12.10	605.0
Total six months	20.5	1.44	1.24	0.27	23.45	1,172.5
Total costs	102.5	72	62	13.5		1,172.5
Secondary costs						400.0
Total costs						1,572.5

Source: Mestactieprogramma (Manure Action Programme), 1987.

Accompanying measures

Other measures related to solving the manure problem consist, on the one hand, of regulations which influence stock-keeping concerns through other instruments of policy and, on the other, of support from the government in the form of subsidies and financing of research, etc. On the basis of another environmental Act, the Hinderwet, a draft decision has been taken to make possible the rapid construction of a large number of manure banks. This measure was taken to support the regulations concerning the spreading of manure. It is estimated that approximately 25,000 establishments will fall under the provisions of the order.

There are also guidelines for assessing the emission of ammonia from intensive livestock production. These guidelines apply to extensions and to new stock-keeping premises within 500 m of wooded areas or nature reserves on certain sensitive soils. They provide a method of assessment which indicates whether undesirable ammonia emissions occur in the immediate vicinity of woods and nature reserves. The method of assessment requires calculation of:

1 The amount of ammonia in the new situation.
2 The amount of ammonia already present, originating from premises within 300 m of the area.
3 The amount of ammonia already present within the area of a district council.

The resultant assessment is then reviewed and in certain cases additional restrictions can be implemented. These restrictions are more stringent for new premises than for the extension of established ones. Compulsory closure of existing premises will not take place. It is estimated that between 3 and 5 per cent of stock-keeping concerns will undergo some extra restrictions as a result.

Effectiveness of the new policy

The pig population increased by 9 per cent in both 1985 and 1986, and at 1 December 1986 there were over 14 million pigs in the Netherlands. This increase has occurred despite the temporary Act restricting pig and poultry-keeping concerns which came into force on 3 November 1984. Officially this Act prohibited the expansion of production facilities in all areas of surplus production, yet, in spite of this law, 1·1 million more pigs were added.

The Act prohibited extension of the stock population in the areas of high livestock density in North Brabant, Limburg, Overijssel and Gelderland, unless the pig and poultry keeper is able to show that he has undertaken investment commitments before 3 November 1984 or that he had applied for building or nuisance Act licences. Pig and poultry keepers outside the areas of high livestock density were permitted a 75 per cent extension in the number of livestock they carried, up to a ceiling. Inside existing buildings a maximum increase of 10 per cent in livestock number was permitted. There was also the possibility of exchanging one fattener for two breeding pigs or a breeding pig by two weaners. From 1 January 1987 the interim Act has been replaced by the Fertiliser Act, which has rendered further expansion almost impossible.

In connection with the registration of manure production on 31 December 1986 (the starting point for the determination of expansion) the number of pigs and cattle was extended just before the new year 1987 to obtain the largest possible manure quota.

The amount of copper that may be added to pig feed has been reduced from 1 September 1984 throughout the European Community and as a result the amount of copper in manure fell by 40 per cent in the period 1977–84. Moreover, because of the restrictions on the content of minerals and heavy metals in animal feed, the amount of phosphorus in various fodder sources was reduced by 30 to 50 per cent in the period 1973–85. The amount of cadmium was reduced by 70 to 90 per cent in the period 1976–85. A reduction of approximately 75 per cent in the emission of copper from the forage to the environment per delivered pig has been also achieved.

Possible developments for the future

The European Community threshold value of 50 mg of nitrate per litre in water for human consumption applies to ground and surface water, and a target of 25 mg of nitrate per litre has been set (EEC 75/400). These values, however, are exceeded on higher-lying soils in sandy areas and, as a consequence, the Provinces have to draw up a programme of intent before 1990.

Policy consequences

Consequences for the environment

Policy instruments that have been developed so far constitute a first phase; the aims for the final phase (around the year 2000) envisage setting standards for the application of animal manure so that the supply of phosphate does not exceed crop requirements. At the same time attempts are being made to reduce the emission of ammonia by a factor of two.

In the short term the first phase of the policy will lead to the:

1 Easing of environmental pressure from manuring in the areas of high livestock density.
2 Distribution of manure at national level so that present use of animal manure outside the areas of high livestock density.

Again, the leaching and drainage of phosphate and nitrogen will be limited by prohibiting spreading in winter, and the emission of ammonia will be reduced by requiring any manure which is spread to be ploughed into the ground. The prohibition on spreading in the winter, however, may lead to greater application of manure in warm weather, which could result in larger emissions of ammonia. Requiring injection of manure will limit the emission of ammonia, but the nitrate loading of the soil may increase.

In the short term the measures will bring about only a limited improvement in environmental conditions in poorly nourished woods and nature reserves. In the first phase, because of lag effects, an increase in the load on ground water is anticipated until the year 2000 as the nitrate now already in the soil leaches through to the ground water. As a result water treatment to remove nitrate will be required in some locations. In areas where the potential for the harmful accumulation of phosphate is considerable, additional stronger standards will be necessary. Until the year 2000 the accumulation of phosphate in the soil will continue. A problem arises if the soil does not have the capacity to hold this phosphate. The final standards will be applied first in areas which are saturated with phosphate. In the longer term, after the phased introduction of these measures has been achieved, the load on the environment from the emission of phosphate and nitrogen will begin to decline, with the result that eventually such measures as the extraction of nitrate from drinking water supplies may no longer be necessary.

Dutch legislation does not yet contain heavy metal standards for animal feed; however, legislation in this area is planned. Within the context of normal European Community negotiations, however, it has been proposed that Community-level animal feed standards should be set to reduce heavy metal pollution from animal feed.

Consequences for the industry

The approach to the manure problem described above has consequences for intensive animal husbandry at the industry level. The prohibition on expansion will seriously restrict the farms' capacity for development. There will also be financial consequences for the industry, the most important of which are:

Table 5.11   Farms with a surplus of manure, 1985

Type of enterprise	Total no. of farms	Farms with a surplus of manure		Average surplus of manure per farm
		No.	%	
Dairy cattle farms	57,710	3,572	7	250
Farms with fattened calves	1,482	1,252	84	764
Farms with fattened pigs	4,474	4,221	94	785
Farms with breeding pigs	5,755	4,911	85	687
Farms with chickens for slaughter	699	695	99	317
Farms with laying hens	1,839	1,797	98	858
Remaining intensive farms	1,606	1,487	93	1,030
Remaining animal husbandries	16,523	3,633	22	417
Remaining agricultural farms	51,811	1,065	2	273
Total	135,899	22,633	17	603

Source: LEI (Agriculture Economics Research Institute), The Hague, 1986.

Table 5.12 Costs of manure surplus disposal

Amount of surplus at industry level in tonnes	Number of farms	Total amount of manure to be disposed of (000 tonnes)	Total costs (Gld. 000)	Average costs per farm
0–100	4,253	209	1,881	442·25
100–250	4,900	829	7,461	1,522·65
250–500	4,810	1,751	15,759	3,276·30
500–750	2,772	1,714	15,426	5,564·95
750–1,000	1,877	1,624	14,616	7,786·90
1,000–2,500	3,422	5,065	45,585	13,321·15
2,500+	599	2,451	22,059	36,826·37
Total	22,633	13,644	122,787	5,425·15
Average per farm		0.60	5.425	–

Source: LEI (1986) Agricultural Economics Research Institute, The Hague, 1986

Table 5.13 Annual costs for the industry (excludes government costs)

	Total costs (Gld. 000,000)	Number of farms	Costs per farm (Gld.)
Annual costs			
Manure disposal	122·8	22,633	5,425·00
Surplus levy	40·0	47,821	836·00
Total	–	135,899	–
Upgrading facilities			
Manure storage	1,572·5	80,000	

1   Less work for builders of sheds, but more work for builders of elevators and for transport.

2   The replacement of fattened pigs by breeding pigs, as profits will be greater from breeding pigs.

3   Possible switches to other forms of production such as goats, fattened bulls, furred animals, fish farming and horticulture.

4   Higher land costs, since there can be no further expansion on a fixed area, and because land is now a factor in the cost of the disposal of animal manure.

5   Higher costs for the use and sale of manure: the 'polluter pays' principle
    is reflected in a trio of costing factors: the disposal of the surplus manure,
    the levy on surpluses, and the cost of storing manure.

Manure production in 1985 was 93·8 million tonnes (cattle, pigs, laying
hens, slaughter chickens and turkeys). In accordance with the norms applied
in the first phase, approximately 13·8 million tonnes of manure cannot be
absorbed by the manure-producing concerns themselves on their own land on
the basis of the situation in 1985. Within manure-producing regions 10·6
million tonnes can be sold to other concerns with a manure shortfall; 3·2
million tonnes will have to be transported over longer distances.

Disposal of surplus
Animal husbandry enterprises producing more manure than can be disposed
of on their own premises will have to accept responsibility for the disposal of
the surplus. Possibilities include disposal within one's own region, whether
or not via the national manure bank.
   Table 5.11 shows the number of farms with a manure surplus (22,633
concerns) by type of enterprise and table 5.12 the cost of disposing of this
surplus in a manner which is consistent with the new regulations.   The
Landbouw Economisch Instituut (Agricultural Economic Institute, or LEI)
calculated in 1986 that the average cost of disposal amounted to more than
Gld 9 per tonne of manure or an average of Gld 5,600 per farm. Transport
costs are particularly high, and when distances greater than 150 km are
involved disposal costs amount to Gld 30 per tonne. Table 5.13 presents an
overview of the costs of all the above regulations and levies for intensive
animal production in the Netherlands.

Part III
Dry-land farming, soil conservation and soil erosion

6. Impact of agricultural policies on soil erosion in two regions of Portugal

Alfredo Gonçalves Ferreira

Introduction

Erosion has always been a factor in the earth's evolution. The morphological alteration of the earth's surface is a result of the physical and chemical agents of the climate which modify the landscape as it evolves from youth to maturity and finally old age.

Widespread agricultural activity in Portugal began about 2,000 years ago with the Roman occupation, and until the beginning of this century agriculture and the environment were in equilibrium. The introduction of the mullboard plough that replaced the traditional Roman chisel plough, however, changed this equilibrium. Also, with the beginning of mechanisation and the use of fertilisers, more land has been brought under cultivation, triggering, as a result, more intense erosive processes, mainly in areas where the soil has no winter and spring cover.

Environmental impact of agricultural policies

Dryland farming of cereals

In Portugal a significant proportion of the country's soil erosion is associated with cereal farming. This is not only because of the large areas involved — around 44 per cent of the country's area — but also because erosion control practices are not commonly used.

Most of the precipitation occurs during autumn and winter and is concentrated in generally high-intensity rainfall events, coinciding with the period of lowest soil cover and hence the greatest risk of erosion.

Usually cereal grains are planted either in autumn or in early winter but it is only in later winter or early spring, with increased temperature and daylight, that the vegetation covers the ground to an extent that finally affords adequate protection. By this time, however, almost 80 per cent of the total year's precipitation has already occurred.

The most common annual cycle on a typical Portuguese cereal farm begins with the use of the mullboard plough during winter to turn the soil, and in the spring a row crop, such as sunflower, is planted. The following autumn the seedbed is then prepared for wheat, with a disc or a chisel plough, and seeding takes place in late autumn or early winter. The cereal crop is harvested in early summer. The goal of the winter ploughing and the spring row crop is to prevent natural grass development during spring, so as to avoid the need to use herbicides to reduce competition between natural grasses and the next crop.

Winter cereals, such as wheat, barley or oats, are seeded two or three years in a row and then the field is returned to pasture for two or three years. Each year the straw is baled because of its value (a bale of about 25 kg is worth about 120 escudos) and the remaining stubble is then grazed by livestock. A mullboard plough is usually used only once during the rotation period, and the number of years between each cropping phase depends on the soil characteristics and the amount of livestock on the farm.

Pricing policy

The price of cereals has long been subsidised by the government. The marketing system is also controlled, and this, together with easy access to short-term loans, encourages cereal farming in many areas where soil erosion and soil drainage are a problem.

This policy started during the 1930s and involved the provision of interest-free credit to help farmers purchase seed, price guarantees, and the establishment of a government monopoly over the wheat market. The aim of this policy was to decrease the country's dependence on imported cereals, mainly wheat. As a result there was a big increase in the area farmed, and dependence on cereal imports decreased drastically. With hindsight, this proved to be an important strategy, as during the Second World War cereal imports were difficult to obtain.

An advantage of this policy was that large brush areas were developed. As a side effect the area under cork oak increased, as cork tends to invade the cleared land. The main drawback was that many of these newly farmed areas were not suited to cereal farming, and soil conservation problems resulted.

Since the 1930s price support has remained central to agricultural policy. The minimum yield of wheat the farmer needs to cover his costs has remained around 1,200 kg/ha, and price support has remained around 42 per cent of the total price paid to the farmer. The variation of the area farmed under winter cereals has also been closely related to a succession of incentives for other crops or farming systems, mainly subsidies for reforestation programmes and livestock prices.

Soil conservation action

Owing to the climatic characteristics of the country, with wet winters and dry summers, the adoption of soil conservation practices is generally associated with reforestation and the establishment of permanent pasture on open land or under hardwood forest.

According to the Soils Service in Portugal, only 28 per cent of the country's total area is suitable for agricultural use and 57 per cent should be primarily under forest or permanent pasture. The actual distribution of land use is almost the opposite: 51 per cent of the area is devoted to agricultural use and 34 per cent is under forest or permanent pasture.

The implication is that 23 per cent of Portuguese agriculture is probably conducted on soils that are not suited to it, with the corresponding soil conservation problems. Conservation practices should thus ideally be targeted on these theoretically non-agricultural lands, to encourage reforestation and the development of permanent pastures.

Perennial vegetation, especially trees which have more developed and permanent root systems, is more able to increase the soil's water-holding capacity. Excess autumn, winter and early spring rainfall is stored for later use, so preventing surface run-off and erosion.

The correct balance between forest and grazing land is critical for a farm economy. With the intensification of livestock operations through pasture improvement and controlled grazing, as opposed to extensive herding, it is possible to prevent overgrazing, expand forest areas and prevent fires.

Reforestation programmes

Different types of programme have been implemented in order to encourage the reforestation of land cleared over the centuries for timber, firewood, grazing and cereal cropping.

Reforestation of communal lands

The first type of programme has been directed at communal land. About 500,000 ha of these communal lands, called baldios, are owned by parishes or municipalities. Of these lands 98 per cent are mainly located in the mountainous region north of the river Tejo and used in common by the community that owns the baldio. Most of these lands are upland and the lower land and the valleys are privately owned.

The baldio is used for gathering firewood, brush to make barn manure, for livestock grazing, and occasionally for a rye crop. Most, if not all, of these uses have tended to encourage the continuous degradation of the vegetative cover and consequently of the soil.

The first attempts to break this cycle were made at the turn of the century, but it was only in 1938 that a reforestation programme (Law 1971, 15/6/1938) was implemented by the National Forest Service. Under this programme the planting and the management of the forest were undertaken by the Forest Service and the cost was recovered at harvest time. The remaining money was returned to the communities. Along with the reforestation programme, pastures were established in order to maintain the livestock of these communities. Today, as a result of these programmes, the National Forest Service administers 600,000 ha of communal and State-owned land.

Reforestation of private land

The second type of programme has been directed at privately owned forest lands. On these lands the National Forest Service provides technical assistance and applies regulations to control wood and game harvesting.

From 1963 to 1980 reforestation and the establishment of improved permanent pasture on private lands (Decree-Law 45443, 16/12/1963) have been encouraged by the provision of low-interest long-term loans. Usually the farmer contracted the planning and execution of this with the Forest Development and Credit Agency that provided the loan. Repayment of the loan was required when the forest harvested, or five years after the new pasture was established. The repayments, however, never exceed 50 per cent of the total revenue collected from timber sales.

New trees were planted along the contour line in infiltration channel terraces or in bench terraces, which not only prevented erosion but also promoted the infiltration of running water. This enhanced the rate of weathering of the parent material and consequently the soil formation rate. When this programme ended in 1980, around 170,000 ha had been reforested.

In 1981 the encouragement of reforestation was given a new impetus by a joint programme of the Forest Development and Credit Agency with the World Bank. This programme was a continuation of the reforestation of private land programme and used the same type of incentives and techniques. When it ended in 1986, 54,000 ha of forest had been planted.

A reforestation programme with European Community funding commenced in 1987. This programme highly subsidises forest planting and the creation of pastures on both private and communal land, under the joint responsibility of the Forest Service and the Forest Development and Credit Agency.

Gully control programme

The third type of programme was directed at controlling gully erosion. The watershed of the river Lis and the southern part of the watershed of the river Mondego in the Coimbra region are on a Mesozoic formation of fairly deep sandstone. However, although most of the region was under pine forest, gully erosion was a very serious as well as spectacular problem.

In 1938 a gully control programme started as part of the Communal Land Reforestation Programme. It entailed the control and the reforestation of gullies. Usually the work was done by the Forest Service with the permission of the landowners, on the condition that the latter would harvest the forest only with the consent of the Forest Service and using techniques that ensured that the problem did not emerge again.

At present, all the gullies are under control and only maintenance and/or preventive work continues.

Soil loss quantification programme

An important parallel approach has been the establishment of experimental erosion stations. This programme started in 1960 with the objective of quantifying soil loss under different versions of a basic crop sequence. The first experimental erosion station was installed in Vale Formoso, south-east from Beja, near Mertola. In this region, although the slopes are steep and soils are shallow and derived from schist (layered rock with some degree of metamorphism), the natural vegetation of brush and green oak was cleared during the 1930s to farm cereals, causing erosion of the topsoil.

A second experimental station was established in the northern part of Portugal on the slopes of the river Douro, which have been under vineyards for centuries. The soils in this region are also derived from schist but they are so shallow and stony that hardly any erosion can take place.

Land management programme

The recognition of the importance of appropriate soil use in accordance with its natural capacity led to the implementation in 1965 of a land management programme (Decree-law 46595, 15/10/1965) with the objective of increasing agricultural production through appropriate soil use. It aimed to intensify the cereal yield on the soils most suitable for this crop and to adapt the cropping system to the natural soil characteristics in order to enhance the profit of the farmers and improve soil conservation. Long-term low-interest-rate loans with up to a 20 per cent grant were offered to participating farmers, and in order to qualify the farmer had only to agree to use the land according to its natural capacity.

These incentives were offered for almost all cropping systems, but some crops like rice and vineyards were excluded. An important feature was that both landowners and tenant farmers could qualify for loans under the programme. In 1974, owing to political changes, this programme stopped.

Drainage and soil conservation project in Alentejo

The last major approach to soil conservation in Portugal is being implemented in the southern part of the country known as Alentejo. The main problems in this region are poor drainage conditions and sheet erosion associated with winter rainfall.

In 1979 a joint programme between Portugal and the United Nations Food and Agricultural Organisation was launched. Its goal was to identify, characterise and establish guidelines for solving drainage and soil conservation problems in Alentejo. A series of experiments was set up to test the efficiency of different soil conservation and drainage systems. In a second phase of the project it was envisaged that the knowledge gained would be adapted to other regions of the country.

In 1981, following the identification of successful strategies, an experimental extension phase began with a view to the construction of drainage systems and to introducing successful soil conservation measures. In 1986 the joint programme with FAO ended, but the Portuguese team has continued with the partial support of European Community funds (Decree-Law 172-G/86, 30/6/86).

Other programmes of agricultural institutions related to the environmental impact of agricultural policies

In 1980 a technical assistance programme was developed involving collaboration between the Agriculture Research Service, the regional agricultural services and the US Agency for International Development with the objective of correcting soil acidity and promoting the rational use of fertilisers. As a side effect the correction of soil acidity will enhance aggregate stability, soil erodibility will decrease, soil degradation will be retarded and internal drainage will be enhanced.

In 1982 a law was passed in order to prevent land being taken out of agriculture owing to the spreading of urban and industrial areas. Under this law, soil classified by the National Soil Classification Service as agricultural soil cannot be taken out of agricultural use unless a special permit is issued allowing another form of utilisation.

Case studies

In order to convey a better understanding of the dryland agriculture/soil conservation and erosion interface in Portugal, two areas were chosen for more detailed description and analysis, according to the following criteria:

1 The use of the soil should be dryland farming.
2 Soil loss data should be available.
3 The problems of each region should be distinct and representative.
4 The agricultural activity in the region should be profitable enough to be maintained.

The two regions chosen were Beja or, more precisely, the region south of Beja, in Baixo Alentejo province, where the soils are derived from schist, and the region of Evora, in Alto Alentejo province, where the soils are derived from volcanic rocks. The most common soil of these regions is Mediterranean soil, which covers 40 per cent of the area. This soil presents an argilic horizon with a base saturation over 35 per cent, which increases or remains constant with depth from B to C or D horizons; no carbonates are found in the A or B horizons even when these soils are derived from calcareous materials.

The main activity in both regions on Mediterranean soils is dryland farming of winter cereals and cattle and sheep herding. In the region of Beja sheet erosion is more evident and the drainage problems are localised. In the region of Evora the soils have poor drainage characteristics coupled with erosion, which is, however, less evident.

The main agricultural policy still followed by the Regional and Central Agricultural Service is one established during the Dryland Farming of Cereals Programme. Credit for cereal farming is easy to obtain. The price of wheat during the 1986 crop year was around 46 esc./kg, so that a yield of 1,500 kg/ha was enough to cover farmers' production costs. This price is significantly higher than the European Economic Community wheat price, which was around 25 esc./kg during the 1986 cropping year.

Thus the cropping of winter cereals will likely continue, and we can estimate that each year around 20 per cent of the area in both regions will be under crop, while 8 per cent of the area will be fallowed under during late winter or early spring each year.

It is important to point out that, from the farmer's point of view, this type of land management is interesting because the production of 1,500 kg/ha of wheat is virtually certain and credit is easily accessible. In addition, the straw is available for feeding the cattle during winter, and the residues, left in the fields, constitute good grazing feed during the summer. In this way the farmer reaches an equilibrium between the cattle or the sheep and the small grain growing.

In prime soils for small grain, where the average yield of wheat is around 4,000 kg/ha, the farmers are essentially small grain producers and some of them have no livestock. In soils where the average yield of wheat is less than 2,000 kg/ha the farmers run livestock as a complement to cereal farming.

Beja region

The Beja region is part of the old massif of the Meseta formed during the Palaeozoic, and the main parent material is schist rock. The fundamental element of the geomorphology of this region is a large peneplain with an average altitude of 200 m rising gradually to the south owing to the tectonic emergence of the Caldeirao mountains that separate the Alentejo from the adjacent province, the Algarve.

This surface is, in some areas, covered by deposits called ranas but it is strongly desiccated, mainly in areas covered by shales and graywacks in the zone of influence of the Guadiana river. Hard rocks like quartzites, marbles and schists over the peneplain remain resistant to geological erosion, like the ridges of Ficalho for marbles, and Alcaria Ruiva for quartzites, both along the eastern border.

Along the western border the old massif-like Grandola, Cercal and Vigia mountain ranges separate the peneplain from the sea. Northwards of Beja, in the region of Portel, the peneplain is sharply broken by a fault scarp which separates the Baixo from the Alto Alentejo.

The potential evapotranspiration is maximum east of the Guadiana river, where it reaches about 950 mm per year. The average precipitation is 561 mm. November is the wettest month, with 80 mm, and January the coldest, with a mean average of 8°C. The hottest month is August, with a mean average of 24°C and almost no rain.

The most common Mediterranean soil in the Beja region is the red or yellow Mediterranean soil. It has a reddish or yellowish A and/or B horizon, with a loamy texture for the A horizon and clay loamy texture for the B horizon. Generally the parent material is schist rock, and in some families of red Mediterranean soils the presence of plinthite is common. The slopes generally associated with these soils are around 6 to 12 per cent, which favours sheet erosion and subsequent sedimentation in the bottom lands.

Soil management
Agriculture on Beja's red Mediterranean soils is usually based on a four-year rotation, starting wth bare fallow or a row crop, then one or two years of

small grain cropping, followed by one or two years of natural pasture or forage. The rotation usually begins with ploughing during late winter or early spring. The direction of this tillage operation, like all the others, is related not to the field slope but generally to the field shape. After seeding, the seed is buried with a chisel plough and the final topography of the field is a succession of small ridges on the top of which the cereal will stand. This practice is meant to protect against waterlogging during winter and spring.

As already discussed, this type of land use has been encouraged by cereal price policy and the control of its commercialisation. This means that, with a yield of 1,500 kg/ha of wheat, the farmer is able to cover his costs, so that there is little incentive to change.

Cattle and sheep production is important and well adapted to this type of farm management. During winter and spring, sheep and cattle graze the natural pastures on the fields which are under fallow. During summer and autumn, they have the cereal crop stubble and, if necessary, a supplement of straw or hay. In the Mediterranean climate the yield of hay in dry farming conditions is low so that the cereal straw is an important source of food for the livestock during the worst periods of the year.

In areas with shallow soils the effects of erosion are more evident. With soil loss the capacity of the soil to support the crop, mainly due to the lack of moisture during spring, decreases rapidly and cereal yields fall below the subsistence level. Once this occurs the cereal farming area is reduced and the number of years that a field will remain in pasture increases. In these cases, sheep herding becomes the main activity, supplemented by a cereal crop once in a while. Consequently the use of herbicides to maintain the natural grasses is not common and weed control of the cereals is done throughout the bare fallow.

On deeper soils and for soils associated with them such as the vertissols, waterlogging during autumn and winter is generally the limiting factor for cereal yields. Erosion decreases the soil's water-holding capacity and fertility, and is a problem that the farmer cannot perceive. When this occurs the usual response is to increase the proportion of the farm which is cropped and shift the crop sequence to a spring row crop, winter cereal pattern. In these conditions it is common to use heavy farm equipment and herbicides, and sometimes there is no livestock on the farm.

Estimation of soil erosion and its impact on soil productivity and off-site damage

The soil loss data for this region are available from the soil erosion Vale Formoso experimental station south-east of Beja. The soil in the experimental station is a red Mediterranean soil derived from schist. This experimental erosion station was set up in 1960 with different versions of the basic rotation, fallow and winter cereal cropping patterns in the region.

The orientation of the fields has proved to be a major factor in soil loss. The fields facing south or west exhibit the highest erosion rates, in the order of 2,800 kg/ha/year and those oriented north and east 1,200 kg/ha/year for a bare fallow winter cereal rotation. This is understandable because rain is mainly produced by frontal systems coming from a south-west direction.

Thus the slopes which face the storm are more exposed to the impact of the raindrops and hence exhibit the highest erosion rates.

It is interesting to point out that at this experimental station the measured amount of sediment trapped in a twenty-year-old dam with a catchment area of 192 ha gave a specific erosion rate of 3,700 kg/ha/year. The agricultural management and the slope of the catchment were similar to those of a plot with bare fallow winter cereal.

The measured soil loss rates are moderate to low when compared with worldwide standards. What is important is not the total soil loss but the quality of the sediment that has been produced, the actual soil depth and fertility and the deposition of sediment on bottom lands.

The quality of sediment is reflected in the sediment particle size. Sand is the first to settle out, the silt settles only in still waters, while the clay fraction goes to the estuaries, where if flocculates because of the increased sodium concentration in the water. Thus the exchange complex removed from the soil is lost for crop production.

The off-site damage due to erosion is a consequence of sediment deposition in the surface drainage system, waterways and dams.

The soil lost from the hills is deposited on the bottom lands, forming a fine-texture, poorly structured layer owing to the low organic matter content, common in the Mediterranean climate, and the presence of 1 : 1 clays, which are the most frequent in the red and yellow Mediterranean soils. This results in poor drainage conditions in the depressions during autumn and winter.

In soils like the red Mediterranean it is important to start with soil loss control measures by changing from cereals to permanent pasture and by building drainage systems mainly at the farm level.

Evora region

North from the fault scarp that breaks the Alentejo peneplain in the region of Portel the peneplain continues but with an average altitude of 300 m and the region of Evora commences. Here the old massif of the Meseta has many intrusions of volcanic rocks pushed up during the Palaeozoic period. This parent material was the one that influenced soil formation in the region.

The potential evapotranspiration is maximum near Evora, where it reaches about 800 mm per year and the average annual precipitation is 633 mm. November is the wettest month, with a mean of 100 mm, and January is the coldest month, with a mean of 8°C. The hottest month is August, with a mean temperature of 22°C and with almost no rain.

The grey Mediterranean soil which is derived mainly from granodiorites or diorites is common in the Evora region. Its A horizon is generally sandy and the B horizon clay loam to clay. The slopes associated with these soils are around 2 to 6 per cent, so poor internal and surface drainage is the main problem. During the normal rain season, perched water tables are formed and sometimes reach the surface. In these conditions, soil will be covered by a thin layer of water which enhances soil loss by transporting soil disaggregated and lifted by the impact of the rain drops on the surface.

Soil management

The most common rotation in the Evora region is an initial period of bare fallow or row crop, followed by two or three years of small grain and then one or two years of natural pasture. In the grey Mediterranean soils, winter drainage problems are more evident, owing to the heavier texture of the argilic horizon. The crop sequence adopted in the region is closely dependent on the soil drainage conditions, so that as the yield of small grain decreases this crop is replaced by pasture land.

The rotation starts with the mullboard ploughing of the soil during late winter or early spring. All the tillage operations are parallel to the slope in order to encourage surface drainage. After seeding, the soil is passed with a chisel plough, and the final shape of the field is a succession of small ridges where the plants swill stand. The main difference between this region and Beja is the incidence of waterlogging problem areas, which are the main feature to correct.

Owing to the deficient surface and internal drainage of this soil, a shallow layer of water is formed over the surface, enhancing the soil loss by creating an environment where the soil particles can be disaggregated, dispersed, maintained in suspension and transported.

As the excess of water during winter is the limiting factor for the crop yield, the problems caused by erosion are not generally recognised by most of the farmers, although they are responsible for the sedimentation of the surface drainage system and consequently exacerbating waterlogging problems. One way to solve these problems is to start with the implementation of surface drainage. When it is evident that it has become necessary to clean the drainage ways, owing to the deposition of erosion sediment, then the opportunity to implement the erosion control measures increases substantially.

Estimation of soil erosion and its impacts on soil productivity and off-site damage

The rates of soil loss for grey Mediterranean soils in the region of Evora are of a similar order of magnitude to those measured in the region of Beja. As an example, the soil loss measured on standard plots with 9 per cent slope, facing west, is 1,600 kg/ha/year for a rotation of winter wheat and forage. The soil loss in the grey Mediterranean soil results in decreased soil fertility and water-holding capacity of the first soil horizon and consequently aggravates the poor drainage conditions. Furthermore, the off-site damage due to deposition on bottom lands and water ways adds to the existing drainage problems.

The wheat yields from these soils can vary from about 1,000 kg/ha in a wet year with about 700 mm annual rainfall to 3,000 kg/ha in a dry year with about 450 mm annual rainfall. This means that, in order to maximise profits, it is necessary to turn every year into a dry one by draining from the soil about 250 mm of rainfall between November and March to avoid waterlogging.

In grey Mediterranean soils the main need is to build surface drainage systems to control the perched water tables. This is generally assisted through low-interest loans coupled with the provision of technical assistance.

Work begun under the Drainage and Soil Conservation Project in Alentejo is now starting to build the primary surface drainage system through a programme subsidised by the European Community. The next step is to establish a credit mechanism and a grid of technical assistance to permit the planning and building of the secondary drainage system needed at the farm level.

It is interesting to point out that the portion of the area in both regions that at the moment exceeds 3,000 kg/ha of wheat is around 10 per cent. With the implementation of drainage, mainly in the grey Mediterranean soils, this figure could be raised to about 30 per cent.

Possible agricultural policy changes in Portugal due to integration in the European Community and its relation to environmental changes

Integration in the European Community implies a change from the existing national policy to the Common Agricultural Policy. One of the big steps concerns the cereal price, which at the end of the integration period in 1996 will be gradually lowered from approximately 46 to 25 esc./kg at 1987 prices.

Recently farmers have been seeding as much as they can in order to take advantage of the high prices being offered, an option which is not advisable from a soil conservation point of view. The marketing of wheat, which has been a government monopoly since 1930s, was opened in 1990 to free enterprise, and a change is expected in the quality of the grain owing to free-market competition. Some modifications are also expected in small grain credit policy.

Until 1996 the main change in the agricultural system is expected to be the gradual reduction of the area farmed with winter cereal. If the present form of agricultural management continues, cereal production will be restricted to the areas where it would be possible to obtain a minimum wheat yield average of 3,000 kg/ha of wheat, which corresponds to 12 per cent of the country's total area. In the areas where the wheat yield is above 3,000 kg/ha the small grain farmers will remain in business and are likely to follow the general trend to increasing use of fertilisers, herbicides, heavy machinery and drainage works. All this will result in increased erosion unless control measures are implemented.

Land use in areas which yield less than 3,000 kg/ha can be expected to change more drastically and to shift largely to livestock production on permanent pasture or forest.

In both cases, programmes partially supported by European Community funds (Decree-Law 172-G/86, 30/6/86) which subsidise the restructuring of agricultural enterprises offer other alternatives, such as assistance for the construction of irrigation and drainage schemes in order to allow more intensive cropping and livestock systems to be established. Another possibility will be so-called rain supplement irrigation, in order to intensify rainfed agriculture by supplementing natural rainfall and so increasing yields and stabilising income. This alternative will result in a concentration of conservation problems in smaller areas and so they will be more evident to the farmers and consequently easier to correct.

The implementation of drainage systems to control perched water tables will be a major conservation work because it will increase production and so the protection given to soil. The expansion of forestry is also foreseen, as the European Community has a shortage of almost all forest products. As much as 23 per cent of the land area which is still under agriculture should be converted to forest use in a manner which will result in beneficial environmental effects.

In a general way, the rational use of the land through reforestation, the implementation of permanent pasture on non-agricultural land, control of perched water tables and soil loss on agricultural lands, and good crop management and the implementation of irrigation could be the main tools for improved land use and the restoration of degraded soils.

Conclusions

Conclusions from the case studies and their applicability to other zones

In a Mediterranean climate with an annual average precipitation of about 600 mm, and with layered soils, the main problem is the formation of perched water tables during winter, which is the limiting factor in dryland farming production and enhances soil loss. High, subsidised cereal prices have kept large areas in crop production which would have otherwise been under pasture or forestry use. Policy changes which result in the reforestation and the establishment of permanent pasture on these non-agricultural lands are the main conservation measures. The control of grazing in the pastures, to prevent their degradation, is another major conservation technique.

Only with this symbiosis between forest and permanent pasture is it possible to make forests economically attractive to the farmer and also prevent forest fires.

The main conservation practices on agricultural lands relate to:

1 The prior control of excess water during winter.
2 The implementation of irrigation.
3 The correct management of soil fertility in order to improve the soil aggregation and the plant cover.
4 The implementation of erosion control practices.

Identification of present difficulties and new possibilities in implementing agricultural and environmental policies

Under Portuguese conditions, erosion rates are rather small when compared with the values published for other countries. The annual amount of rainfall, its distribution throughout the year and the soil characteristics are the limiting factors for cereal yields, and these combined factors hide the effect of erosion in decreasing soil productivity and crop yields.

In the grey Mediterranean soils, for example, one way to overcome these difficulties is to start controlling excess water in the soil during winter by installing drainage systems. When these systems start to fill up with sediment the opportunity to pursuade farmers to adopt soil erosion control measures is increased.

To implement soil conservation measures, it is necessary to prove to farmers that soil conservation is efficient and necessary to maintain crop yields. Thereafter, one way to implement agricultural and environmental policies will be to merge them with a pricing policy. Prices define areas which cannot be profitably cropped, which is partially related to the soil capacity. Access to special credits and/or technical assistance for farmers who comply with conservation measures also has merits which should not be disregarded. This is similar to the land management programme discussed above, where conservation measures and production incentives are packaged together in a cross-compliance package which makes access to concessional credit and technical assistance conditional upon the use of the soil according to its natural capacity.

References

Araujo, E.B. (1974), 'A sedimentaçao na albufeira de Vale Formoso', *Separata da Revista de Ciencias Agrarias*, I (II), 157–70.

Bessa, M.R.T. (1969), *Possibilidades e limitaçoes dos solos da Regiao-Plano Sul*, I, Encontro sobre Desenvolvimento Regional da Regiao-Plano Sul, Evora.

Cardoso, J.V.J.C. (1965), *Os solos de Portugal, sua classificaçao, caracterizaçao e genese*, I, *A sul do rio Tejo*, Direceçao Geral dos Serviços Agricolas, Lisbon.

Conference on Mediterranean Soils (1966), Study Tour through Southern Portugal, International Soil Science Society, Portuguese Soil Science Society, Lisbon.

Diario da Republica (1972), Decreto-lei No. 451/82, I Série, No. 265, 16 November 1982.

Diario da Republica (1986), Decreto-lei No. 172-G/86, I Série, No. 147, 30 June 1986.

Feio, M. (1949), 'Le Bas Alentejo et l'Algarve', *Congrès International de Géographie*, Lisbon.

Feio, M. (1963), *Situaçao economica e perspectivas da cultura do trigo*, Federaçao dos Gremios da Lavoura do Baixo Alentejo, Publicaçao 10, Beja.

Feio, M. (1965), 'Clima e ocupaçao agricola de Portugal', *Separata da Revista Geographica*.

Ferreira, A.G. and Singer, M.J., (1985), 'Energy dissipation for water drop impact into a shallow pool', *Soil Sci. Soc. Am. J.*, 49, 15371542.

Ferreira, I.B.M., Ferreira, A.N.J.R., and Sims, D.A. (1984), *Analise preliminar dos dados dos talhoes de erosao do Centro Experimental de Vale Formoso, Alentejo (Portugal), em termos da Equaçao Universal de Perda de Solo, para os anos de 1962/63–1979/80*, Direcçao Geral de Hidraulica e Engenharia Agricola, Ministerio da Agricultura e Pescas, Lisbon.

Graauw, P.J. (1986), *Country Information Paper from Portugal for the ad hoc group on Agriculture and Environment*, Servico Nacional de Parques, Reservas e Conservaçao da Natureza, Ministerio do Plano e da Administraçao do Territorio, Lisbon.

Gomes, A.M.A. (1967), *Fomento da arborizaçao nos terrenos particulares*, Fundaçao Calouste Gulbenkian, Lisbon.

Instituto Nacional de Estatistica (1983), *Estatisticas agricolas do Continente Açores e Madeira*, Papelaria Fernandes, Lisbon.

Mendonça, J.C. (1961), *75 anos de actividade na arborizaçao de serras*, Direçao Geral dos Serviços Florestais e Aquicolas, Ministério da Economia.

Ministério da Agricultura, Comercio e Pescas (1982), *Plano de mudança da agricultura*, Lisbon.

Ministério da Economia (1965), *Regime cerealifero 1966–1970*, Diario do Governo, Decreto-lei No. 46595 de 15 Outubro 1965.

Ministério da Economia (1966), *Bases do apoio tecnico e financeiro à lavoura e da concessao das dotaçoes à cultura cerealifera*, Diario do Governo No. 89, I série, de 15 Abril de 1966.

Organisation for Economic Co-operation and Development (1975a), *Agricultural Policy in Portugal*, OECD Agricultural Policy Reports, Paris.

Organisation for Economic Co-operation and Development (1975b), *Review of Agricultural Policies*, OECD Agricultural Policy Reports, Paris.

Organisation for Economic Co-operation and Development (1978), *Regional Problems and Policies in Portugal*.

Rego, R.F. (1963), *Federaçao Nacional dos Productores de Trigo, subsbdio para o seu historial*, Editorial Império, Lisbon.

Sampaio, J.A. (1965), *Problemas de materia organica na agricultura alentejana*, Federaçao dos Gremios da Lavoura do Baixo Alentejo, Publicaçao 15, Beja.

Teran, M. (1952), *Geografia de Espana y Portugal*, I, Montaner & Simon, Barcelona.

Tropa, J.A. (1960), 'A lei dos melhoramentos agricolas como factor de valorizaçao agraria', *Separata de A Agricultura e o II Plano de Fomento*.

7. Integration of agricultural and environmental policies in the United States: the case of soil erosion and soil conservation in dry-land farming

P. Crosson

The pre-1980s

Perceptions of the soil erosion problem

Federal government policies to deal with soil erosion and its consequences date from the 1930s. Owing mainly to the persuasiveness of Hugh Hammond Bennett, the father of soil conservation in the United States, and the evidence of massive wind erosion in the Great Plains (the Dustbowl), the Congress in 1935 established the Soil Conservation Service (SCS) as part of the United States Department of Agriculture (USDA). Bennett was the first chief of the SCS, a position he held for some fifteen years.

From the beginning, the main focus of soil conservation policy in the United States was the protection of soil productivity. *In Soil Erosion: a National Menace* (1928) Bennett vividly described the damage erosion had done to soil productivity across the wide stretches of the southern piedmont, the middle south, the Middle West and the Great Plains. In summing up his account Bennett wrote:

> To visualize the full enormity of land impairment and devastation brought about by this ruthless agent [erosion] is beyond the possibility of the mind. An era of land wreckage destined to weigh heavily upon the welfare of the next generation is at hand.

This perception of the nature of the erosion threat found expression in SCS statements of soil conservation policy from the 1930s (e.g. see *Soils and Men*, the USDA yearbook, 1938) to the 1980s (e.g. see USDA, 1985). The SCS, and conservationists generally, recognised that soil erosion contributes also to off-farm sediment damage, but primacy was always given to the threat to soil productivity.

Emergence of agriculture/environment concerns

It was not until the 1960s that much public attention was given to the relationship between agriculture and the environment—defined here as the soil, water, air and biological context within which agriculture operates. The main focus of that attention was the effect of pesticides on wildlife and human health. Rachel Carson's book *Silent Spring* (1963) eloquently described these effects, as she perceived them, and gave impetus to the emerging concern about the environmental consequences of agriculture. But Carson paid little heed to soil erosion, nor, initially, did the environmental movement which her book inspired.

This began to change with the passage of the Clean Water Act amendments of 1972. Section 208 of this legislation dealt with non-point pollution, and recognised that farmland erosion is a major source of such pollution. Under section 208 the Environmental Protection Agency (EPA) was charged with primary responsibility for non-point source pollution policy, and the individual states were required to come up with region-wide plans to deal with this kind of pollution. The plans had to meet the approval of the EPA Administrator, but EPA delegated responsibility for implementation of the plans to the states. Erosion control was prominent in many, if not most, of these plans.

The off-farm consequences of erosion thus came into focus as one of the principal elements in the agriculture–environment relationship. For the USDA, however, the main concern with erosion continued to be its impact on soil productivity, not its off-farm effects.

Conservation policy

Objectives and instruments
The principal objective of federal soil conservation policy is and always has been to reduce erosion to socially desirable levels. Until the 1985 Farm Bill the main instruments of policy were an offer to share with farmers the cost of erosion control practices, technical assistance in the design and implementation of these practices, education about the nature and consequences of erosion, and persuasion of farmers, emphasising the ethical obligation of each generation to protect the interests of future generations in the productivity of the soil. Most funding for cost-sharing practices was provided by the Agricultural Stabilization and Conservation Service (ASCS), and technical assistance (and minor funding) by the SCS. The Management of these programmes was through some 3,000 soil conservation districts (SCDs), roughly one for each county in the nation. The ASCS, SCS and local farmers were represented on the managing boards of the SCDs.

The pre-1985 Farm Bill instruments adopted for soil conservation policies reflected the sense of the Congress, and presumably of the American people, that soil conservation objectives must be achieved by inducing the voluntary compliance of farmers in the adoption of soil conservation practices.

Regulatory approaches, e.g. of the sort employed by the EPA in respect of pesticides, were never adopted.

Performance in reducing erosion
The socially desirable levels of erosion sought by soil conservation policy are defined by T values. T stands for 'tolerable' levels of erosion per acre per year, that is, the maximum amount of annual erosion an acre can sustain indefinitely without impairment of soil productivity. T is five tons on most soils, but on some shallow soils with unfavourable subsoils it is two tons.

Until 1977 no data existed by which to judge the performance of soil conservation policies in achieving T values on the nation's farms. For 1977, however, the National Resources Inventory (1977 NRI) showed that wind and water erosion exceeded T on 242 million acres of agricultural land, about 18 per cent of the total. T was exceeded on 32 per cent of the crop land. The 1982 NRI, which included more comprehensive estimates of wind erosion, indicated that wind and water erosion exceeded T on 21 per cent of total agricultural land, and on 44 per cent of crop land.

Because comparable erosion estimates do not exist for years before 1977, it is impossible to know for sure whether land eroding in excess of T was more or less in those years. Indirect evidence suggests, however, that total cropland erosion declined from the 1930s to the late 1970s. The evidence is presented in some detail in Crosson and Stout (1983, pp. 15–18). In summary it is as follows:

1   The total amount of land under crops was almost exactly the same in the late 1970s as in the late 1930s.
2   Land in soya beans, a highly erosive crop, increased sharply, but this was approximately offset by a decline in land under cotton, also a highly erosive crop. On balance, this shift in crop mix probably had no significant effect on per-acre erosion. Land under corn, a crop almost as erosive as soya beans, declined.
3   Yields of maize about tripled and soya bean yields increased about 75 per cent. These yields increases would tend to reduce per-acre erosion by increasing both crop canopy, protecting the soil from raindrop impact, and crop root mass, tending to bind the soil more tightly in place.
4   Conservation tillage, which on erosion land reduces erosion 50 to 90 per cent compared with conventional tillage, was used on 25–30 per cent of the nation's crop land in the late 1970s. In the 1930s conservation tillage was virtually unknown.

The combination of no change in the amount of crop land, no factors suggesting an increase in per-acre erosion and several factors suggesting a decrease points to the conclusion that cropland erosion, and probably total erosion from agricultural land, declined from the 1930s to the late 1970s. It does not necessarily follow that in this case the percentage of land eroding in excess of T would also decline, but this is a plausible inference.

It is also plausible that because of soil conservation policies both total erosion and erosion in excess of T were less in the late 1970s and 1980s than they would have been otherwise. Between 1936 and 1980 some \$43 billion (1977 prices) of private and public funds were spent on on-farm soil conservation programmes (Pavelis, 1985) and many tens of thousands of people in government, in private conservation organisations and on farms devoted themselves diligently to erosion control. While there is no quantitative measure of the pay-off to these efforts, it could hardly not have been positive.

Any judgement of the performance of soil conservation policy, however, must note other factors tending to reduce erosion over the last fifty years. One is the increase in crop yields. The increase reflected in good measure public investment in research on higher-yielding crop cultivars, such as hybrid maize, and on crop management practices, e.g. optimum per-acre seeding rates, as well as on extension activities, both public and private. Soil conservation policy played no role in this. Similarly, farmers' adoption of conservation tillage, at least until the early 1980s, was spurred primarily by the economic advantages of the technology, not by soil conservation policy.

It must also be noted that prior to the 1980s much of the effort expended on soil conservation programmes was focused on land with little if any erosion problem. This has been documented in studies both by the US General Accounting Office (1977) and by the ASCS (1980). The reasons for this scattering of soil conservation resources are complex. Pressures from the Congress to spread funds widely was surely part of it, as was the policy of the local ASCS and SCS offices of offering cost-sharing on a 'first come, first served' basis. The first farmers to come were not necessarily those with the most severe erosion problems.

In the early 1980s the USDA took steps to concentrate soil conservation resources more closely on the land with the greatest erosion hazard. This is dealt with more fully in the next section. In assessing policy performance prior to the 1980s, however, the scattering of resources cannot be ignored.

It might be argued that more reduction in erosion could have been achieved had policy relied more on regulatory approaches and less on voluntary inducements. Whatever the force of this argument in principle, it has little relevance to the situation of American agriculture in the period since soil conservation was placed on the nation's policy agenda. That situation included a deep commitment throughout the social and political structure of the country to the concept of fee-simple property rights in farmland. Given this, and the commonly held perception that the main threat of erosion was to soil *productivity*, a regulatory approach to soil conservation would have aroused violent opposition, not only from farmers, and virtually no support. This was so obvious that regulatory approaches were never seriously considered by either the USDA or the Congress.

Performance in reducing erosion costs

Although T values serve as the standard of performance in soil conservation policy, it is clear that T values in fact are a proxy for the correct standard: the economic costs of erosion effects on soil productivity. The rationale for T values demonstrates this. If T values are exceeded over a substantial period, soil productivity will decline and this will impose higher costs of food and fibre on future generations.

This view of T values raises a host of questions. If one of the objectives of agricultural policy is to avoid imposing higher costs on future generations, then *all* factors bearing on costs, not just soil productivity, must be in the policy arena. In this case optimal policy may dictate increasing the productivity of one or more of these factors, e.g. new technology, and accepting some loss of soil productivity. This is to say that the proper objective of soil conservation policy is to hold loss of soil productivity to that amount consistent with the objective of avoiding higher long-term production costs, taking into account all the options available for meeting the objective. This amount of loss may be, but is not necessarily, zero, as the T value criterion assumes. (This view of T values is discussed at length in American Agricultural Economics Association, 1985, and in Crosson, 1987).

A second question is how well T values serve to indicate limits of soil loss beyond which erosion reduces soil productivity. The history of T values is obscure. There is some evidence that the five-ton limit is owed to an observation by Bennett that topsoil is formed at that rate (McCormack *et al.*, 1982). Whatever the scientific validity of that proposition—not much—it is clear that T values have been set by a variety of criteria, the effect on soil productivity being only one. (For an account of the evolution of T values see Schertz, 1983.) The fact is that T values are not reliable indicators of erosion-productivity relationships.

It does not follow from this that soil conservation policy failed to contain losses of soil productivity. On the contrary, there is evidence that policy may have had considerable success in this. Studies of the effects of erosion on soil productivity in the late 1970s to early 1980s indicate that, if then prevailing rates of erosion continued for 100 years, crop yields at the end of the period would only be 5 to 10 per cent less than they would be otherwise (Pierce *et al.*, 1984; Crosson, forthcoming). Assuming crop prices of the early 1980s and a 4 per cent discount rate, the annualised present value of this loss was only 1–2 per cent of the total annual crop production costs (Crosson, 1987). These estimates indicate that the productivity effects of present rates of erosion do not pose a serious threat to the interests of future generations in maintaining low costs of food and fibre.

Of course, soil conservation policy cannot claim all the credit for containing the effects of erosion on production costs. As already indicated, advances in technology tended to reduce erosion, and this must have contributed to the relatively small impact of erosion on production costs. Nonetheless the contribution of policy likely was also significant.

A third question concerns T values as indicators of off-farm costs of sediment damage. It is fair to say that T values bear no relationship to these costs. This is not surprising, since T values were developed to measure erosion consistent with zero loss of soil productivity, not zero off-farm sediment damage. In the early 1980s the chief of the SCS appointed a high-level committee of SCS scientists to consider modification of T values to make them serve as a guide to off-farm sediment damage. This proved exceedingly difficult, not surprisingly, because the relationship between erosion on the land and downstream sediment damage is highly variable, both in space and in time. This is discussed further below.

Perhaps in part because soil conservation policy has been aimed primarily at protecting soil productivity, the off-farm costs of sediment damage are roughly half to one order of magnitude higher than the on-farm costs of soil productivity loss (Crosson, 1987). That off-farm costs should be higher than on-farm costs should not be surprising. Not only has policy been aimed primarily at the on-farm problem, but farmers have far more incentive to protect the productivity of their land from erosion damage than they do to reduce sediment damage off the farm.

The discrepancy between on-farm and off-farm erosion costs, combined with the difference in the way farmers view the two kinds of costs, suggests that soil conservation policy should be more concerned with reducing off-farm sediment damage than traditionally it has been.

Integration of agricultural and environmental policies

Policy integration as discussed here has two dimensions. In one, agricultural and environmental policies are said to be integrated when each set of policies is formulated with attention to its consequences in the policy domain of the other set. I call this substantive integration. The main agricultural policies of interest are those dealing with agricultural output, income and prices. For brevity these are called macro-agricultural policies. In the other dimension of integration, policies are said to be integrated when either formal or informal arrangements exist to assure that government agencies responsible for agricultural and environmental policies maintain reciprocal exchanges of information (and perhaps people) among themselves.

Since this study is concerned with soil erosion and soil conservation policy, the substantive integration issue can be stated as follows: to what extent do macro-agricultural policies take into account their effects on soil erosion, and to what extent do soil conservation policies take account of their effects on agricultural output, income and prices? The institutional issue is the extent to which the exchanges of information and people among agencies actually result in closer harmony among agricultural and environmental policies than would occur in the absence of the exchanges.

Substantive integration
There may be an asymmetry in the substantive relationship of agricultural and environmental policies. That is, each policy set will not necessarily give the same amount of attention to consequences in the policy domain of the other set. This seems to have happened with regard to macro-agricultural policies and soil conservation policies in the pre-1980 period. There is no evidence that in the formulation and implementation of policies to support farm prices and income any serious attention was paid to the consequences of the policies for soil erosion. Soil conservation policy, however, was always sensitive to the impact of soil conservation on farm income. It could not have been otherwise. Soil conservation policy relied on farmers voluntarily adopting erosion control measures. Clearly adoption on the desired scale would not occur if it entailed a loss of farm income.

The erosion consequences of the failure to integrate macro-agricultural policies with soil conservation policies are not obvious. One of the instruments of macro-policies was inducements to farmers to take land out of production as a way of reducing crop supply, making it easier to support commodity prices and farm income. This probably resulted in less erosion than if land retirement had not been part of the policy. Farmers were encouraged to put the retired land to environmentally beneficial uses but there were no sanctions if they did not. And the contracts for land retirement were only for one year, so whatever erosion reduction benefits were achieved in one year could be lost in the next. Moreover, since the scale of commodity programme benefits the farmers received was based in part on the number of their 'eligible' crop acres, they had an incentive to convert to crops some highly erodible land that otherwise would have been in grass or trees. In this respect, failure to integrate policies may have resulted in more erosion than would otherwise have occurred.

Macro-agricultural policies also were not integrated with EPA policies for pesticide regulation or non-point pollution. That is to say, policies to support commodity prices and farm income appear to have been developed and implemented with little if any attention to their consequences for farm use of pesticides and fertilisers. There are a couple of reasons for believing that these policies may have encouraged increased use of these materials. By encouraging farmers to withdraw land from crop production the policies probably induced more per-acre use of pesticides and fertilisers to extract more output from the land remaining in production. In addition, higher commodity prices increased the pay-off for pesticides and fertilisers, also encouraging greater use of them. To the extent that greater use of these materials meant more environmental damage, the failure to integrate macro-agricultural policies with pesticide and fertiliser policies imposed uncompensated environmental costs. This possibility is worth noting here even though the main focus of this study is on the integration of macro-agricultural policies with soil conservation policies.

Institutional integration

There are a number of agencies involved in macro-agricultural and environmental policies in the United States. At the national level the Congress is a principal actor, as is the executive branch, operating primarily through the Department of Agriculture and the Environmental Protection Agency. The Council on Environmental Quality (CEQ) has policy co-ordinating responsibilities but little authority, and other agencies, e.g. the Departments of the Interior, Defense and Energy play peripheral roles. The dominant federal agencies in the implementation of agricultural and environmental policies are the USDA and EPA. In the formulation of policies these agencies share responsibilities with the Congress.

State and local agencies are also active in the policy process, in many but not all instances operating within advisory or regulatory guidelines set by the USDA and the EPA. Some states have imposed more stringent environmental policies than EPA or USDA, e.g. California on the use of pesticides, and Iowa on permissible soil erosion where off-farm sediment damage results.

Because policy for the use of non-federal lands is a state responsibility in the United States, state and local agencies are particularly active in the implementation of soil conservation policies. Local soil conservation districts, established under state law, are the principal agencies for doing this. Most states have agencies devoted to the protection of water quality and other environmental values. To a greater or lesser extent these agencies are involved in environmental aspects of soil conservation policy.

In 1974 the Secretary of the USDA and the Administrator of the EPA signed an agreement providing for co-ordination of the activities of the two agencies. Under the agreement the agencies have undertaken numerous joint research, training and project implementation activities. In addition, the agreement provided a useful mechanism for frequent meetings between top-level officials of both agencies where issues of mutual concern could be discussed.

The legislation establishing the Rural Clean Water Programme (RCWP) states that at the national level the Secretary of Agriculture will administer the programme in consultation with the Administrator of the EPA. It is stipulated that the EPA must concur with USDA in the selection of the best management practices (BMPs) which under the programme farmers are encouraged to adopt, e.g. conservation tillage and the conversion of crop land to permanent vegetative cover to reduce erosion and integrated pest management to reduce environmental damage by pesticides. Day-to-day administration of the RCWP was delegated to the ASCS and co-ordination of technical assistance to the SCS.

Although the potential range of collaboration between the USDA and the EPA is wide, it is fair to say that most of the interaction between the two agencies has been in respect of pesticide policy. In matters touching on agriculture this was, and is, the EPA's principal concern. It perforce was, and is, a main concern of the USDA as well. In contrast, the two agencies have had much less to say to one another about soil conservation policy.

The main policy area in which the two agencies have common responsibilities is in connection with non-point pollution. As indicated above, section 108 of the Clean Water Act amendments of 1972 assigns the EPA principal policy responsibility for this kind of pollution. The EPA has chosen to delegate this responsibility to the states, which for the most part have seen the non-point problem as a problem in erosion control. Accordingly, they have relied heavily on the SCS, and state soil conservation agencies, to deal with the problem. The EPA has played a role in this, but a minor one. According to the *United States Country Information Paper*, EPA has accepted SCS personnel in regional EPA offices to help co-ordinate EPA and USDA work on non-point pollution control policy. In this connection EPA has given USDA people access to its data base on water quality. For the most part, however, interaction between the two agencies on soil conservation policy has been very limited.

The present situation

Erosion and its consequences

As noted above, recent studies indicate that the continuation of late 1970s and early 1980s rates of erosion would have small impact on crop yields and crop production costs. Should erosion increase by 50 to 100 per cent and be maintained at the higher level, this judgement would have to be reconsidered. Such an increase now seems unlikely. The amount of land under crops now is less than it was a few years ago, indicating that erosion is also probably less. Should foreign demand for US grain and soya beans recover and continue to grow, more land may be brought under crops, in which case erosion also would likely increase. But if current rates of increase in crop yields continue, the rise in exports would have to be rapid and sustained before enough additional land would be cropped to raise erosion 50–100 per cent above the levels of the late 1970s and early 1980s. Such an expansion in export demand now seems unlikely. Some future increase in erosion is possible, even probable, but that it should be on a scale posing a significant threat to soil productivity is not.

The situation for off-farm sediment damage is different. According to Clarke *et al.* (1985) this type of damage now costs between $4 billion and $16 billion annually, five to ten times the annualised cost of soil productivity loss (Crosson, 1987). It is likely that more resources should be devoted to reducing off-farm damage. There are two reasons. One is that, acting individually, farmers have little incentive to reduce the damage, since by definition the damage occurs off the farm where the sediment occurs. For any farmer to reduce his erosion on this account is a matter of all cost and no benefit. The other reason for doing more to reduce off-farm damage is the relative neglect of this damage in soil conservation policy, noted above.

Conservation policy: recent initiatives

In the last few years the USDA, acting on its own and in response to the Congress, has begun to take some initiatives in soil conservation policy which, if carried through, would constitute major shifts in emphasis, all of them, in the judgement of this writer, in the right direction. Three initiatives in particular stand out: targeting, more attention to off-farm sediment damage, and better integration of soil conservation policies with macro-agricultural policies.

Targeting
Partly as a result of the GAO and ASCS studies showing a scattering of effort and partly, perhaps, because of tight budgets, the USDA under-secretary decided to target a rising share of the soil conservation budget on those soils around the country where erosion is highest. Beginning at 5 per cent, the share was to rise to 25 per cent in annual 5 per cent increments.

Targeting makes so much sense that the USDA's failure to target from the beginning can be explained only by the political pressures, mentioned earlier, to spread soil conservation spending widely. That these pressures remain strong is evident from the fact that after a couple of years Congressional opposition succeeded in blocking, at least momentarily, the increase in targeted funds. Should the blockage prove permanent, an opportunity for a major improvement in soil conservation policy will be lost.

Should the move towards targeting resume, two ideas for setting the targets deserve more attention than they have so far received. One is to come up with a better guide than T values to the soils where erosion poses a significant threat to productivity. The limitations of T for this purpose have already been discussed. Models developed by soil scientists at the University of Minnesota (the PI model) and by USDA people at Temple, Texas (EPIC), promise to do a much better job than T in identifying soils under threat of serious productivity loss. The SOILEC model developed at the University of Illinois also has promise in this respect. These models need further refinement, but even now they are a marked improvement over T as targeting guides.

The other idea that should inform targeting policy is that off-farm sediment damage deserves more attention than it has so far received. This is relevant to targeting because the best targets for reducing off-farm sediment damage are not necessarily the best for reducing losses of soil productivity. In fact the targets may be different. The reasons for this become clear in the next section. The point here is that recognition of the importance of off-farm erosion damage requires rethinking the criteria for identifying targets. The threat to soil productivity will not suffice.

Off-farm sediment damage

There are signs that in formulating soil conservation policy the USDA is beginning to pay more attention to this type of damage relative to loss of soil productivity. The latter continues to get top priority, but it is less dominant than formerly. Or at least so it seems to this observer, based largely on informal discussion with SCS personnel rather than on formal statements of policy.

Any shift towards relatively more attention to off-farm sediment damage would clearly be in the right direction, since it would begin to reflect more closely the relative importance of the two kinds of erosion damage. However, reducing off-farm sediment damage is likely to pose more difficult policy problems than reducing damage to soil productivity. The site where erosion threatens productivity is the site where that erosion occurs. The policy imperative is obvious: to protect soil productivity at that site, reduce erosion at that site.

The situation with off-farm sediment damage is quite different. The site where the sediment originates may be tens, hundreds, perhaps even more miles from the site (a stretch of river, a lake, reservoir or habour) where the damage occurs. There is spatial discontinuity between the place where soil is detached and the place where, as sediment, it does harm.

There is also temporal discontinuity. The time between soil detachment and sediment damage can vary from a few hours, following a heavy rainstorm, to centuries as sediment slowly and haltingly makes it ways through a watershed. Trimble (1975) found that 90 per cent of the soil eroded since colonial times in twelve watersheds of the southern piedmont is still stored somewhere above the fall line of the watersheds. Other studies of other areas in the United States come to similar conclusions.

Because of the spatial and temporal discontinuities between site of soil detachment and site of sediment damage one cannot be sure that controlling erosion at the site where it is high will give proportional and timely reductions in sediment damage downstream. The emphasis on timeliness is important. Even if the spatial relationship is known, a long delay between initiation of erosion control and realisation of reduced sediment damage will make it difficult for the control project to pass a cost–benefit test. This is not to say that erosion control will always fail to yield satisfactory and timely reductions in sediment damage. Where the watershed is small and the damaging sediment readily traced to its source on the landscape, erosion control can be the optimal policy. These conditions seem to have been met, for example, in the Blue Creek basin of Illinois and the Prairie Rose Lake area of Iowa. The lakes in both areas were undergoing accelerated sedimentation, with loss of recreational value. Erosion control projects were undertaken in the watersheds upstream from the lakes. Studies of the projects (Davenport and Lowry, 1985, for Blue Creek and Agena *et al.*, 1985, for Prairie Rose Lake) indicate that targeted reductions in sedimentation were achieved within a few years.

But in large watersheds the spatial and temporal discontinuities between site of soil detachment and site of sediment damage are likely to be great enough to call into question whether erosion control will yield satisfactory reductions in sediment damage. This has important implications for targeting policy. More attention should be given to targeting *sediment management* at the point where damage occurs as an alternative to targeting *erosion control* at sites where erosion is high. The issue is an empirical one. There must be circumstances in which erosion control is the optimum policy for reducing sediment damage. But there likely are others, probably many others, where sediment management at the point of damage would be optimum. In many instances the optimal policy would no doubt be some combination of erosion control and sediment management. As the USDA moved towards more emphasis on control of off-farm sediment damage and targeting, the sediment management alternative may prove to have high pay-off.

Soil conservation policy: integration

SUBSTANTIVE INTEGRATION

The 1985 Farm Bill contains a number of features which have the effect of integrating policies to support commodity prices and farm income with policies to conserve soil and reduce off-farm damage from sediment and sediment-related pollutants. These features are the Conservation Reserve and making participation in price support and other programmes conditional upon adoption of soil conservation measures (cross-compliance). ('Swampbuster' features of the Farm Bill are designed to protect wetlands from conversion to other uses and are outside the scope of this report.)

The objectives of the Conservation Reserve (CR) are to reduce erosion; preserve land productivity; reduce sediment-related pollution; limit the supply of surplus commodities; protect wildlife habitat; and maintain farm income (Dicks and Reichelderfer, 1986). Under the programme farmers submit bids to the Secretary of Agriculture to take some, or all, of their highly erodible land out of crop production and put it in the CR for ten years. The bids include a plan for putting the land under some vegetative cover that would control erosion. In autumn of 1986 'highly erodible' land was defined as that eroding at three times T. The secretary establishes maximum per-acre amounts that the government will pay for land put in the CR. Bids which do not exceed that amount are accepted. The government also pays half the cost of the erosion control plan which accompanies the bid.

By the 3T criterion 69 million acres of the 421 million acres of US cropland are eligible for the CR. Texas, with 11·5 million eligible acres, leads the nation in this respect. Iowa is second, with 6·6 million acres, followed by Montana (5·0 million acres), Missouri (4·1 million acres) and Colorado (3·7 million acres). Between them these five states had 24 per cent of the nation's crop land and 45 per cent of the acres eligible for the CR. Texas, Colorado and Montana are among the top five states in eligible acres because they have exceptionally high wind erosion.

The USDA aims to have 40 million to 45 million acres in the CR by 1990, with 5 million in 1986 and 15 million in 1987. In the autumn of 1986 8·9 million acres had been accepted. Texas and Colorado each had 12·7 per cent of the accepted acres, followed by Minnesota (7·4 per cent), Kansas (6·9 per cent), Oklahoma (5·0 per cent) and Iowa, Montana and Missouri (4 per cent each). The prominence of Plains and Mountain states (Texas, Oklahoma, Kansas, Colorado and Montana) is owed to high wind erosion. On a per-acre basis erosion by water in these states is well below the national average.

Judged by number of acres in the CR, the programme performed well in the first year of its existence. Judged by attainment of the several goals of the programme, its performance is more problematical. Some 5·6 million acres of the 8·9 million in the reserve would have been eligible for deficiency payments under commodity programmes. Dicks and Reichelderfer (1986) assert that direct and indirect savings from not having to make these payments may be sufficient to offset the cost to the government of putting the 8·9 million acres in the CR. If this is correct, and if the CR also makes progress in achieving its various environmental goals, then the programme will be a distinct improvement over earlier ones which failed to integrate macro-agricultural policies with environmental policies.

Progress towards the goal of reducing erosion seems clearly to have been made. According to Dicks and Reichelderfer, the conversion of crop land to permanent cover in the reserve is estimated to reduce erosion on CR land by an average of 25 tons per acre per year (t.a.y.), with the greatest average reduction in Texas and Oklahoma (46 t.a.y.). New Mexico, the state with the highest percentage of eligible acres in the CR, is estimated to achieve a 56 t.a.y. erosion reduction, more than in any other state.

Dicks and Reichelderfer provide no information about progress towards other environmental objectives: protection of soil productivity, reduced sedimentation, improved water quality and improved animal habitat. Measurement of movement towards these objectives is inherently more difficult than measurement of erosion reduction. Indeed, since less erosion *per se* conveys no benefits, one suspects that erosion reduction was included among the CR objectives only because it is measurable and may be a proxy for the achievement of erosion-related objectives, especially protection of soil productivity, reduced sediment damage and improved water quality.

But there is a problem here. Erosion reduction may be a very poor proxy for these objectives. As noted, much of the land in the CR is there because it is subject to high wind erosion. This is troublesome on two counts. One is that there are serious technical questions about the accuracy of the wind erosion equation. As part of an analysis of erosion data from the 1982 NRI, the Board on Agriculture of the National Research Council commissioned a special study of the wind erosion equation. The author of the study concluded that, outside the area in Kansas where it was developed, the wind erosion equation tends systematically and substantially to overestimate wind erosion.

The prominence of high wind erosion land in the CR is troublesome also because very little is known about the effects of wind erosion on soil productivity. The PI, EPIC and SOILEC models were developed to measure the relationship between water erosion and soil productivity.) And the contribution of wind erosion to sedimentation and impaired water quality is surely less, probably substantially less, than the contribution of water erosion. One must question, therefore, whether the inclusion of so much wind-eroded land in the CR is optimal for achieving the objectives of reduced sedimentation and improved water quality.

But reduction of water erosion may also be a poor proxy for progress towards other environmental objectives. The PI and EPIC models show that on deep soils with favourable subsoils, e.g. the loess soils found widely throughout the Mid-west, erosion at 3T can continue for many years without significant impairment of soil productivity. And, for the reasons given above, control of water erosion may contribute little to the reduction of sediment damage in many watersheds around the country.

On balance, the CR clearly represents an important step towards better integration of macro-agricultural policies and environmental policies. However, exclusive reliance on erosion control to achieve a variety of environmental objectives is probably not wise policy. The highly uncertain relationship between erosion control and these other objectives suggest this. It is to be hoped that as the CR idea matures, and experience is gained with managing it, ways can be found to make the CR a sharper instrument for achieving the objectives of reduced sediment damage and improved water quality.

So long as farmers have an incentive to participate in price and income support and other programmes, cross-compliance has potential for bringing about greater integration of macro-agricultural policies and environmental policies. Under the cross-compliance provisions of the 1985 Farm Bill, farmers who crop high erodible land without an approved conservation plan would be ineligible for price support or disaster payments, farm storage loans from the Commodity Credit Corporation, federal crop insurance, or loans from the Farmers' Home Administration if the secretary determined that the loan would finance an activity resulting in excessive erosion on highly erodible land.

The 1985 Farm Bill stipulates that until 1990 the cross-compliance provisions will apply only to potentially erodible land converted to crops. In 1990 and thereafter, however, all highly erodible land will be covered. The result is that since the Farm Bill was passed the cross-compliance provisions have had little bite. The reason is that markets for farm commodities have been so weak that farmers have had little incentive to convert any non-crop land to crops. In 1990, however, cross-compliance should become an important element in soil conservation policy, assuming the relevant provisions of the Bill remain intact.

Like the Conservation Reserve, cross-compliance is an important step towards better integration of macro-agricultural policies and environmental policies. Unlike the CR, cross-compliance does not necessarily have to rely on land retirement to achieve environmental objectives. There is no reason in principal why cross-compliance could not be used to induce environmentally desirable changes in pesticide, feriliser and irrigation practices, as well as in land management. Cross-compliance thus seems a more flexible instrument than the CR for achieving policy integration.

INSTITUTIONAL INTEGRATION

The advance in substantive policy integration incorporated in the 1985 Farm Bill have not been accompanied by comparable movement towards stronger institutional integration. Pesticides continue to dominate the EPA's concern with agriculture. Other aspects of non-point pollution, and soil conservation in particular, remain well down on the agency's list of policy priorities. Consequently EPA–USDA interaction on soil conservation issues remains at a low level.

Should targeting re-emerge as a USDA objective, then some shifts might occur in USDA relations with other institutions, particularly the soil conservation districts. If targeting is taken seriously, and federal budgets remain tight, then some, probably many, soil conservation districts would find themselves with less and less to do. Others, on the other hand, would find their workload much heavier. Adapting to this changing situation while maintaining effective delivery of soil conservation measures to farmers would require adroit manoeuvering by the USDA. Its relationship with the conservation districts has evolved over a fifty-year period under the operating principle that federal soil conservation resources should be widely dispersed, with virtually every county in the country participating through its conservation district. Targeting runs squarely against that principle.

The movement towards greater emphasis on off-farm sediment damage also may introduce new elements in USDA's relations with other institutions. This is particularly likely if USDA decides on a strategy which increases the weight given to off-farm sediment management relative to on-farm erosion control. The conservation districts are adapted to do the latter. Sediment management, at least in large watersheds, is a regional problem requiring a regional approach, for which the soil conservation districts would be inappropriate. The main agency involved in sediment management on the scale of large regions is the army Corps of Engineers. SCS involvement in sediment management occurs under authority of PL 566, and management is confined to small watersheds. It seems likely that a major thrust by the USDA towards sediment management could raise difficult questions of how best to integrate its activities with those of the corps.

Soil conservation policy: untried alternatives

Cost-sharing, education, technical assistance, moral persuasion and more recently the Conservation Reserve and cross-compliance: these are the instruments of soil conservation policy in the United States. Underlying all these approaches is reliance on securing the voluntary participation of farmers in soil conservation programmes.

It might be argued that cross-compliance departs from the principle of voluntarism, and perhaps it does to some extent. Under cross-compliance no farmer is required to adopt conservation practices, but he will lose some important programme benefits if he does not. Is compliance in these circumstances voluntary?

Be that as it may, voluntarism is still the guiding principle of soil conservation policy. Should it continue to be? From time to time measures have been suggested that would require farmers to adopt conservation practices under sanction of severe fines or even legal proceedings. Particularly with respect to off-farm sediment damage, or any other kind of off-farm damage, it can be argued that the 'polluter pays' principle ought to apply to farmers just as it applies to industrial and municipal polluters. Farmers might be required, for example, to reduce erosion to some specified level under threat of legal action if they do not; or made subject to a tax, perhaps a progressively rising tax, for every ton of erosion above the specified level.

In principle it is difficult to see why farmers should not be subject to the same sanctions as other polluters. In practice, however, there are some strong reasons for doubting that the principle could be equitably applied with existing knowledge of erosion and its off-farm consequences. One reason is a question about the ability of the Universal Soil Loss Equation (USLE) to predict erosion with sufficient accuracy to justify its use in taking legal action or levying taxes on excess erosion. The USLE was developed from experimental data reflecting the soil and climate conditions in the Mid-west. Its accuracy elsewhere may be quite good enough for a general assessment of rates of erosion but this likely would not be sufficient to pass a strict legal test. Moreover the USLE is designed to measure average annual erosion over a long period, not that which might occur as a consequence of a single storm. Yet such events likely contribute more to off-farm sediment damage than the same amount of erosion over an extended period of time.

A stronger reason for doubting the feasibility of the 'polluter pays' principle in soil conservation policy is the highly tenuous relationship between erosion at a given place on the land and sediment damage downstream. Because of this it would frequently, if not usually, be difficult to identify the farms which contribute the damaging sediment. Even if a group of farms could be reliably identified, assigning responsibility among them in a way sufficient to pass a legal test would likely be difficult.

This is not an argument against the sort of law now in force in Iowa and some other states under which farmers can be held legally accountable if sediment from their land damages someone else's land. These laws apply in circumstances where sediment damage can be assessed and those responsible for it clearly identified. These are not the circumstances, however, under which most sediment damage occurs.

It is highly desirable that research be undertaken to improve understanding of the processes by which sediment moves through a watershed. Such understanding could contribute importantly to the improvement of policies for dealing with off-farm sediment damage. If the research is undertaken, then in time it may be possible to do a better job in linking erosion on the land with downstream sediment damage. This, in turn, would remove one of the main practical obstacles to applying the 'polluter pays' principle to farmers. Until then, however, there seems little alternative to voluntarism in soil conservation policy. But with a *caveat*: that cross-compliance should continue to be accepted as consistent with voluntarism.

Erosion consequences of reduced price support

At a meeting of OECD Ministers in the spring of 1987 it was agreed in principle that OECD member countries would move towards 'a progressive reduction of assistance to and protection of agriculture'. Should the United States act to reduce commodity price support and other forms of assistance to agriculture, the amount of land under various crops, and hence erosion, likely would be affected. However, the magnitude, perhaps even the direction, of the effects is not entirely clear. Wheat, maize, cotton and soya bean are the crops responsible for most cropland erosion in the United States. Of the four, per-acre erosion is highest on soya bean land. In 1986 the amount of harvested land under soya beans, wheat, maize and cotton was 24·0 million, 24·6 million, 28·0 million and 3·4 million hectares respectively. Support prices for soya beans are and always have been low relative to support for the other three crops. It is plausible to believe, therefore, that incentives to grow soya beans would be less affected by the elimination of all price support than incentives to grow wheat, maize and cotton. Indeed, the relative profitability of soya beans would rise, suggesting that some land now under these three cops might be shifted into soya beans. It is not certain, therefore, that the total amount of land in soya beans would decline with the elimination of price support, in which case the change in erosion from soya bean land would also be uncertain. The amount of land under the other three crops would likely decline, assuming that demand for them would not grow enough to offset the decline in prices following the elimination of support programmes. Elimination of the programme, therefore, likely would reduce erosion from maize, wheat and cotton land.

On balance, there would probably be some reduction in total cropland erosion if price support and other assistance programmes were eliminated, with most if not all the decrease occurring on land under wheat, corn and cotton. Nothing more conclusive can be said in the absence of detailed analysis of the relationship between the programmes and farmers' incentives to put land under these crops.

It should be noted that in the 1985 Resource Conservation Assessment the USDA projected that by the year 2000 land in crops would be several million hectares less than in 1982, reflecting faster growth in crop yields than in crop demand. Total erosion is projected to decline proportionately more than the decline in crop land, reflecting the adoption of erosion-reducing tillage practices on an increasing percentage of land. The projected reduction in erosion because of this combination of economic and technological factors would dwarf whatever decline might occur with the elimination of price and other assistance programmes.

The case studies

The discussion thus far has dealt with national-level issues of soil erosion, soil conservation policy and policy integration. The rest of the paper focuses on these issues as they are revealed in a couple of small regions. Erosion and its consequences vary widely around the country, depending upon climatic, soil and topographical conditions, the commodities produced, and the farm practices used to produce them. This heterogeneity of conditions suggests that focusing on quite specific circumstances in a small region will provide insights into problems of soil conservation policy not obtainable within a national or large regional perspective. This is the rationale for the case studies.

The criteria for selecting the two regions were that they should have significant problems with soil erosion and that there should be an existing body of data that would permit analysis of the problems and policy options for dealing with them, including policy integration options.

Reelfoot Lake, Tennessee

Reelfoot Lake is located in the north-west corner of Tennessee. It was formed by the New Madrid earthquake of 1811–12. A shallow lake, it now covers about 14,000 acres and drains some 154,000 acres, mostly in Tennessee, some in Kentucky to the north. The total land area, excluding that surrounding the lake where sediment deposition occurs is 117,000 acres. Of this, over 60,000 acres are in row crops, mostly soya beans. Some 15,000 acres are under a variety of non-row crops. Virtually all the rest of the land is under trees and grass, somewhat over half of it under trees. About 33,000 acres of crop land directly north of the lake are in the Mississippi river flood plain. This land is generally level, and much of it drains directly into the lake. Most of it is in continuous row crops (Park and Dyer, 1986).

Another 30,000 acres of cropland lie east and north-east of the lake. This land is hilly, with elevations rising up to 300–400 feet. The soils are thick, silty loess and generally are highly productive, although subject to high erosion when under crops, especially row crops. Most of this land in fact is continuously row-cropped.

The drainage area of the lake consists of nine watersheds. By far the largest (almost two-thirds of the drainage land area) is Reelfoot Creek, which enters the lake at its north-east corner. The next largest watershed, Big Sandy Creek, covers 9·5 per cent of the drainage area.

Apart from its value as a source of agricultural output, the Reelfoot Lake area is a recreational resource of regional significance. The watershed includes a state park, a national wildlife refuge and a state wildlife management area. These features and recreational uses of the lake attracted an average of over 500,000 visits annually to the area in 1979–84.

Problems
According to the Reelfoot Lake Rural Clean Water Programme annual progress report for 1985 (LCC, 1985, p. 3) the main problems in the Lake are sedimentation, eutrophication from nutrient run-off and contamination of aquatic life by pesticides. The progress report for 1986 (LCC, 1986) reaffirms this statement of problems. It goes beyond the 1985 report, however, by citing surveys showing that non-agricultural sources of pollutants delivered to the lake are minor relative to agricultural sources.

In support of this assessment of the nature of the problem affecting the lake (LCC, 1985, 1986) present data on erosion in the nine watersheds drained by the lake and the resulting delivery of sediment to the lake. Eighty-nine per cent of total erosion in the drainage occurs in Priority Area IE to the east and north-east of the lake. (This is sheet and rill erosion only. Wind erosion is not significant. Ephemeral gully erosion is believed to be high, perhaps as high as sheet and rill erosion, but there are no measurements.) Priority Area IE also accounts for 89 per cent of total cropland erosion, 78 per cent of gully erosion (not ephemeral gully erosion) and 98 per cent of all other erosion. Cropland erosion in the area is estimated at 60 tons per acre per year, or 12T. For all land in the area the per-acre erosion rate is 32 t.a.y.

Priority area IW (W means west of the lake) accounts for only 2·4 per cent of total drainage area erosion and 2·9 per cent of the erosion from crop land. Per-acre erosion on the 16,500 acres of crop land in this area is 3·5 t.a.y., 0·7T.

Priority Areas II and III (II is 13,500 rolling acres in the far north-east of the region; III is 18,000 acres of Mississippi flood-plain land lying west of and adjacent to Area IW) together account for 8·4 per cent of total drainage area erosion and 7·9 per cent of cropland erosion. Per-acre cropland erosion in Area III is 3·5 t.a.y. In Area II it is 8·5 t.a.y.

Prior to 1980 an estimated 851,000 tons of sediment was delivered to the lake each year, 85 per cent of it from Reelfoot Creek watershed. The implied sediment delivery ratio (SDR) for the whole drainage is 0·36, and for Reelfoot Creek watershed it is 0·35. SDRs for the other eight watersheds vary from 0·5 to 0·37.

LCC (1986) asserts that sedimentation is the largest single problem affecting the lake, attributing the sediment to high erosion of upland areas in the drainage. Sedimentation is especially notable in the east and north-east sides of the lake where it is entered by Reelfoot Creek and Indian Creek. An extensive delta area has been formed in this area which has practically cut off passage between the northern and southern portions of the lake. Sedimentation is apparent also in many other areas of the lake, in some cases impeding boat traffic.

In contrast to the relatively abundant data about erosion and sedimentation, little is known about feriliser and pesticides as sources of lake pollution. LCC (1986) provides estimates of nitrogen, potassium and phosphorus fertilisers applied to maize, wheat and soya bean land in the drainage, but no information about how much of these nutrients reaches the lake. No data on pesticide use are presented.

Assessment of problems

There are large unanswered questions about both the nature and the magnitude of the problems affecting the lake. There can be no doubt that sedimentation is a problem, but its significance and origins are unclear. The estimates of erosion in the drainage appear somewhat artificial. All the crop land in Priority Area IE is said to erode at exactly 60 tons per acre per year (t.a.y.), regardless of watershed. All the crop land in Priority Areas IW and III erodes at exactly 3·5 t.a.y. across all watersheds, while that in Priority Area II erodes at exactly 8·5 t.a.y. across all watersheds. Similarly, the t.a.y. erosion rates for woodland, grassland, gullies and other land are fixed for each category of land use. All woodland and all 'other' land erodes at exactly 1 or 0 t.a.y., depending on priority area, all grassland at either 2·4 or 9·2 t.a.y., and all gullies at either 170 or 929 t.a.y., again depending on priority area. SCS people in the Reelfoot area defend these estimates as reasonable, and accurate enough for the uses to which they are put.

The 60 t.a.y. estimate for crop land in Priority Area IE is especially important. At 60 t.a.y. the 29,943 acres of crop land in this area produce 1,796,580 tons of eroded soil annually, 75 per cent of the total from all sources in all nine watersheds, and 74 per cent of the estimated amount of sediment delivered to the lake.

LCC (1985, 1986) does not indicate how the per-acre erosion estimates were arrived at, nor does the Reelfoot Lake RCWP Plan of Work (revised October 1980). Other sources suggest that the 60 t.a.y. estimate may be high. The 1982 National Resources Inventory (NRI) showed that in Obion County, which is drained by Reelfoot Creek, per-acre erosion averaged 22·8 tons on 107,700 acres of crop land eroding in excess of 5 t.a.y. According to Park et al, (1986) the NRI also showed that 438,000 acres of west Tennessee crop land in Land Capability Classes IV, VI and VII (land ill suited to crop production because of high erosion potential and other limitations) were eroding at rates of 10 t.a.y. or higher.

Other county-level data also suggest that the 60 t.a.y. estimate is high. Lauderdale County, bordering the Mississippi river, in central west Tennessee, is much like the counties in the Reelfoot Lake drainage with regard to cropping patterns and soils. most crop land in Lauderdale County is under soya beans, wheat and cotton, and the soils are deep, silty, highly productive and highly erosive. Park *et al.* studied the adoption of conservation practices on 5,980 acres of crop land in Lauderdale County, making use of data collected by SCS technicians on rates of erosion on this land prior to adoption of the practices. The data showed that the average pre-practice erosion rate was 29 t.a.y. (personal communication with William Park of Park *et al.*, 1986.)

A question arises also concerning the sediment delivery ratios (SDRs) used to estimate the amount of soil eroded from various sources which enter the lake. The question concerns the appropriateness of SDRs for estimating sediment contributions to deterioration of water quality. At the core of the problem are the spatial and temporal discontinuities in the movement of sediment through a watershed, discussed in an earlier section. The SDRs used to estimate sediment delivery in the Reelfoot Lake drainage (and those in general use elsewhere) obscure these discontinuities by 'lumping' them in a single number. They implicitly assume sediment movement to be smooth and orderly across space and time when in fact this movement reflects a complex halting process combining stochastic weather events with highly area-specific watershed geomorphology, soil type and vegetative cover. After a review of the issues involved, Novotny and Chesters (1986, p. 28) conclude that:

> The spatially and temporally lumped estimates of DRs [Delivery Ratios] are not suitable for non-point pollution-water quality studies. The delivery process and the values of parameters to which the magnitude of DR is related represent a hydrological stochastic process that should be treated as such.

Yet another question arises about the LCC (1985, 1986) estimates of sediment delivered to Reelfoot Lake. In the late 1950s a study was undertaken of how to protect the flood plain land around Reelfoot Lake against flood damage and the lake itself against sediment damage (Reelfoot–Indian Creek Watershed District *et al.*, 1960). The study focused on the Indian Creek and Reelfoot Creek watersheds. It estimated that the two watersheds were delivering 120 acre ft of sediment per year to the lake. According to LCC (1985) they were delivering 568 acre ft of sediment to the lake in 1980. The two watersheds had 22 per cent more·land in crops in 1980 than in the late 1950s (43,300 acres, up from 35,495 acres), which would tend to produce more erosion and perhaps more delivery of sediment. However, the LCC (1985) estimate of sediment delivered is 4·7 times the earlier estimate. This difference is the more remarkable because between 1960 and 1980 six floodwater retention dams were built in Indian Creek–Reelfoot Creek watersheds, and these would have trapped much of the sediment previously delivered to the lake.

Of course, these comparisons do not necessarily mean that the sediment delivery estimates in LCC (1985, 1986) are too high. Those in the earlier study may have been too low. But the difference, taken together with the questions about the per-acre erosion rates and SDRs used in LCC (1985, 1986), suggests great uncertainty both about amounts of erosion in the Reelfoot Lake drainage and about amounts of sediment delivered to the lake.

As noted above, LCC (1986) gives estimates of the amounts of fertiliser used on soya bean, wheat and maize land in the drainage, but nothing about amounts of nutrients delivered to the lake. Hence evaluation of the threat of fertilisation practices to the quality of lake water is difficult. Eutrophication of the lake appears to be a threat to recreational values, but it is not clear that current fertilisation practices are the main cause. LCC (1985) states that historically nutrients carried to the lake by sediment and in run-off 'undoubtedly' were the main source of nutrients stimulating eutrophication. At present, however, nutrients recycled from lakebed sediments may be the main source. LCC (1986) adds yet another piece to the puzzle, stating that sedimentation may be promoting eutrophication by making the lake more shallow, thus permitting more sunlight to penetrate to the lake bottom, where it can stimulate plant growth.

Pesticide poisoning of fish and other aquatic life in the lake may be less of a problem than it was thought to be at the planning stage of the RCWP. In 1983–84 the United States Fish and Wildlife Survey did a study of contaminants in fish taken from the lake and in lake sediments. No contaminant levels of concern were found in the fish. Most metals were below detection levels or not high enough to be of concern. Three organochlorine pesticides were found in fish tissue (BHC, DDE and DDD), but at barely detectable levels. No PCBs were found.

In the sediment samples heavy metals were in the normal range found in many lakes in the United States. The only organochlorine pesticides detected in the sediment were DDD and DDE. The Fish and Wildlife Service summed up the results by saying that:

Although the Reelfoot Lake NWR [National Wildlife Refuge] is located in the Mississippi River Valley and surrounded by agriculture, there is little evidence that contaminant levels of concern to fish and wildlife resources are present in either the fish or sediments of the refuge. [LCC (1985), appendix C, p. 2.]

In summary, it appears that sedimentation and eutrophication of Reelfoot Lake are a genuine threat to recreational and ecological values provided by the lake, but the extent of the threat is unknown. The physical dimensions of the problem—amounts of sediment delivered annually to the lake and the rate of eutrophication—are quite uncertain; and evidently there are no dollar estimates of the recreational values, let alone the ecological values, under threat.

The absence of such estimates greatly complicates the problem of how to protect the quality of the lake's water. It is not just that policy-makers lack clear guidelines as to the extent of the threat to the lake, although that is serious enough. Perhaps more serious is that the lack of damage estimates makes it difficult to gain a consensus among the various users of the lake about how best to protect it. LCC (1985) indicates that action to this end has long been impeded by controversy among users about the nature and extent of the threat to the lake and about how best to deal with it. Quantitative, objective and widely understood estimates of damage would not necessarily eliminate such controversy, but they likely would do much to reduce it.

The Rural Clean Water Programme
Concern about the lake prompted the designation of the Reelfoot Lake drainage as an RCWP project in 1980. A first step was to identify 'critical' areas within the drainage. Two criteria of 'criticality' were applied.

1 Intensive use of fertiliser and pesticides where these would be easily transported to major drainage ways or direct to the lake,
2 Areas where erosion exceeded T and which contributed to sediment loads of major streams and directly to the lake. As of October 1986 the total critical area was a little over 52,000 acres (LCC, 1986).

Two goals were set for the RCWP project, both aimed at achieving a 'desirable level of water quality in Reelfoot Lake'. One goal was to 'adequately treat' 80 per cent of the critical acres in this drainage area. The other goal was to reduce the annual delivery of sediment to the lake by 75 per cent over the years 1981 to 1990.

The 'adequate treatment' of the land was to be achieved by encouraging farmers to adopt best management practices (BMPs) on their land. Sixteen BMPs were identified (LCC, 1986, table 17).

The LCC (1986) 'felt' that the sediment reduction goal could be achieved by reducing per-acre erosion rates to 5 t.a.y. on 80 per cent of the land in the drainage that exceeded this in 1980 (34,388 acres).

Performance indicators of the RCWP project are: number of RCWP contracts written in the area and number of critical acres they cover; number of critical acres 'inadequately treated'; number of acres previously eroding in excess of 5 t.a.y. which have been brought down to that level; reduction in erosion, in number of tons annually; and reduction in sediment delivered, assumed proportional to the reduction in erosion.

None of the performance indicators tells anything directly about improvement in the quality of the lake's water. This reflects the absence of pollutants. Evaluation of performance, accordingly, is difficult. This is especially true in respect of fertilisers and pesticides. BMP 15 has to do with fertiliser management and BMP 16 with pesticide management. For both BMPs the RCWP goals were set in terms of number of acres to be covered by RCWP contracts. Table 26 of LCC (1986) shows that through 1986 the goal for the fertiliser management BMP had been exceeded by 73 per cent and the pesticide management goal by 62 per cent. But there is no way of knowing in either case how much the quality of lake water was improved, if at all.

Table 26 also credits the RCWP with reducing erosion by 490,000 tons from 1981 to 1986. Eighty-two per cent of this was attributable to one BMP, conversion of land to permanent vegetation. Other erosion control programmes in the area brought the total reduction in erosion to 655,000 tons over the six years. Given the uncertainty about how much of this soil might reach the lake, and when, the contribution of the erosion reduction programmes to improve water quality is quite uncertain.

In 1986 farmers in the Reelfoot Lake drainage, as elsewhere in the country, were given the opportunity to put some or all of their most erodible land in the Conservation Reserve. As of mid-February 1987, somewhat more than 3,000 of these acres were in the reserve. What the effects might be on reducing erosion or damage to the lake were uncertain at this writing.

The Reelfoot Lake RCWP, like the Conservation Reserve, is based on the same fundamental principles as all USDA soil conservation programmes: offer farmers financial inducements to adopt conservation measures voluntarily (including taking land completely out of crops). None of the material reviewed for this paper, nor any of the discussions held with SCS people in the Reelfoot Lake area, gives any indication that a regulatory approach has ever been considered.

This is not surprising, since a regulatory approach would conflict with long established principle and practice. It is worth noting, however, that if a regulatory approach were considered it would face severe obstacles, quite apart from the defiance of tradition. The lack of even rough quantitative information about the nature and extent of damage to the lake, and the inability to assign responsibility for sediment and other pollutants, would make design, let alone enforcement, of a regulatory programme most difficult. Farmers and others on whom the burden of regulation would fall likely would find good political, if not legal, grounds on which to resist.

Policy integration

SUBSTANTIVE INTEGRATION
The Reelfoot Lake area provides examples of the consequences of failure to achieve substantive integration of macro-agricultural policies with environmental policies. In the mid-1970s crop prices were high, exports were booming, and farmers were being encouraged by the USDA to expand land under crops to take advantage of the favourable market conditions. In the Reelfoot Lake area farmers responded, like farmers elsewhere, by converting to crops, particularly soya beans, some land that had been under grass or trees. The material reviewed for this report does not indicate how much land in the Reelfoot area was converted, or how much erosion may have increased as a consequence. The point here is that little if any thought seems to have been given to such consequences at the time farmers were being encouraged to increase crop production.

Experience with the Conservation Reserve in the Reelfoot Lake area also provides evidence of a failure of substantive integration. In 1986/87 the price and income support programme for cotton, and to a lesser extent for maize, was so generous that farmers planting those crops had little incentive to put their land in the Conservation Reserve. Since cotton and maize are two of the three most erosive crops (soya beans being the third). this failure to achieve substantive integration likely resulted in higher rates of erosion and more environmental damage than would otherwise have been the case.

INSTITUTIONAL INTEGRATION
The low level of EPA–USDA contact on soil conservation–environmental issues, noted at the national level, appears characteristic of the Reelfoot Lake RCWP and related programmes as well. In a trip to the Reelfoot Lake area made as part of the preparation of this report, the author was accompanied by four USDA representatives but no one from the EPA. The trip was organised by USDA people, and our hosts in the area were SCS staff. No one from the EPA participated.

Although the RCWP PL 566 activities (small watershed flood control and sediment management projects) and the Conservation Reserve in the Reelfoot area are all run by the USDA, predominately the SCS, the agency works closely with other federal, state and local institutions. Under a memorandum of understanding between the USDA and the Soil Conservation Districts of Fulton County, Kentucky and Lake and Obion Counties, Tennessee, technical assistance provided under the RCWP was to be delivered to farmers (Reelfoot Lake RCWP Project Plan of Work, 1980, p. 28). The executive committee of the RCWP Local Co-ordinating Committee (LCC) is responsible for the day-to-day operations of the RCWP (Reelfoot Lake RCWP Project Plan of Work, 1980). The executive committee, as originally constituted, included representatives of the SCS, ASCS, county extension services, and the Tennessee State Department of Public Health, Division of Water Quality Control. In 1985 the ASCS entered into an agreement with the Tennessee Department of Health and Environment (evidently the earlier Department of Public Health) under which the latter was to provide the RCWP with data on the quality of lake water. Although the SCS was given overall responsibility for technical assistance in the RCWP, the Agricultural Extension Service, the United States Forest Service, and the Tennessee and Kentucky State Divisions of Forestry, and Wildlife Resources Agencies also were to contribute technical assistance in their respective fields (RCWP Project Plan of Work, 1980, p. 26).

This is not a complete listing of the inter-agency agreements which exist to carry out the RCWP and other soil and water conservation projects in the Reelfoot area, but it is enough to suggest a high level of inter-agency involvement. Whether this involvement results in close integration of conservation efforts could not be determined in the brief time available to do this report. That integration is less than optimum, however, is suggested by several passages in LCC (1985), p. 25.

> . . . 1985 saw a return of Reelfoot's old friend . . . controversy . . . And the issue of management of Reelfoot Lake once again ended up in court . . . The prolonged debate and courtroom battle over a proposal of the Tennessee Wildlife Resources Agency to drastically draw down the level of the Lake to combat aquatic plant growth once again exposed long-term problems that have yet to be completely resolved. Such as:
> To what extent will local residents have a hand in management of the lake? . . . Federal and State responsibilities remain unclear.

What is or should be the priority of lake management efforts? Sedimentation control? Water quality? Fisheries management? It is not now nor will it ever be sufficient to merely assume lake management priorities. They should be developed, prioritized, justified, then implemented in the form of a lake management plan.

Summary

The lack of substantive policy integration has pretty clearly hampered the achievement of soil and water conservation goals in the Reelfoot Lake area, as it has elsewhere in the country. This is an issue that must be addressed at the national level. There is little that the people or institutions in the Reelfoot area can do about it.

There is an extensive inter-agency (federal, state and local) involvement in conservation efforts in the Reelfoot area, but the extent to which this results in more integrated efforts is not clear. The quotations above from LCC (1985) suggests less than complete integration.

The last paragraph quoted appears to support an argument made earlier in this report. The lack of credible, quantitative estimates of the damage from agricultural pollutants to economic and ecological values provided by the lake is a major obstacle to getting consensus about the nature and extent of the problems and about how best to attack them. The absence of consensus, in turn, is an obstacle (not the only one) to better institutional integration.

If there is a lesson here it is that attempts to foster closer integration among conservation institutions in the Reelfoot area are not likely to improve conservation policy greatly so long as knowledge of the contribution of agriculture to the problems of the lake remains so limited.

Oakwood Lakes–Poinsett, South Dakota

An RCWP project was established in this area in 1981. The area consists of 106,193 acres in east central South Dakota, near the border with Minnesota. Some 65,000 acres are under crops, 52 per cent of this in maize, soya beans and sunflower, 24 per cent in small grain, and 24 per cent in hay and other crops. Three lakes in the area, the largest of which is Lake Poinsett, cover 10,532 acres. There are no urban centres in the project area, but its recreational advantages attracted 240,000–300,000 visits per year in 1981–85 from several South Dakota cities and several small communities.

Problems

People in the area are concerned about the presence of nitrates, and possibly pesticides, in ground water and about eutrophication of the lakes and the resulting loss of recreational value. Sedimentation of the lakes is not itself a threat but sediment carries phosphorous fertiliser, and phosphorous is the nutrient most limiting aquatic plant growth.

The Big Sioux aquifer underlies the area and is the principal source of water for the population. A study of nitrate levels in eighty private wells drawing on the aquifer, done in 1978 as part of a Section 208 non-point pollution study, showed that in 24 per cent of the wells concentrations of nitrate–nitrogen exceeded the 10 mg/l standard recommended by the United States EPA for potable water. The mean and median concentrations were 9·2 mg/l and 2·85 mg/l respectively. Subsequent studies showed varying results. Samples taken from forty-seven wells between July 1980 and March 1983 showed mean concentrations of nitrate–nitrogen varying from 3·1 mg/l to 1·1 mg/l, well below the 10 mg/l standard. As of July 1985, however, 26 per cent of 238 samples taken from sixty-six monitoring wells exceeded 10 mg/l of nitrate–nitrogen.

Two hundred samples of ground water were taken from forty-eight wells during 1983–85 and analysed for the presence of twenty-five pesticides commonly used in the area. Traces of one or more of these materials were found in 12 per cent of the samples. The authors reporting this result concluded that it showed that pesticide contamination of ground water is not a problem in the area.

A national eutrophication study done by the EPA in 1977 showed that among thirty-one lakes in South Dakota ranked for eutrophication (with 1 being least eutrophic), Lake Poinsett was nineteenth, East Oakwood was twentieth and West Oakwood was twenty-second. Other indicators of eutrophication of these lakes are not given in the material reviewed for this report. However, economists at the USDA's Economic Research Service did a study of the annual value of recreational uses of the lakes currently lost because of eutrophication. The study used so-called 'contingent valuation' techniques to analyse data collected in a sample survey of people using the lakes for recreation. One of the questions asked of these people was how many recreational visits they would make to the lakes if 'algae blooms would no longer be a hindrance to recreation' (Piper et al., 1986, p. 4). The results showed that the net value of the additional visits would be $3·5 million annually, i.e. this is the value of the visits after subtracting travel and all other costs incurred to make the visits. The $3·5 million is thus an estimate of the annual cost of lost recreational value because of eutrophication damage to the lakes.

Solutions

The RCWP aims to deal with the problems of nitrate and pesticide contamination and eutrophication by inducing farmers in the area to adopt one or some combination of three BMPs: conservation, tillage and pesticide management on 65,000 acres considered to be critical to ground and surface water quality; feriliser management on 70,000 critical areas (most, if not all, of the 65,000 acres are included in the 70,000). The project also aims to reduce the amount of animal waste entering waterways, lakes and ground water by applying waste management systems on ten livestock operations.

A first step towards achieving these objectives was to evaluate parcels of land within the area according to their contribution to water quality degradation. Two criteria were applied to each parcel: (1) amount of sediment delivered to the lakes: (2) impact on quality of ground water. The parcels of land then were assigned to one of the three priority groups. These priorities provided guidelines to RCWP personnel in deciding which lands and which farmers were most in need of BMPs.

As in all RCWPs, farmers in the Oakwood Lakes–Poinsett area were induced to adopt BMPs by an offer from USDA to share the cost of adoption. Farmers were made aware of the opportunity through their regular contacts with SCS, ASCS and Extension Service personnel, by meetings called to explain the programme, and by notice through USDA-provided newsletters, news columns and radio programmes. No doubt word of mouth also played a role. By the end of the 1985 fiscal year 136 farms with 41,253 acres were under RCWP contract. This was 52 per cent of the land in the total project area.

Monitoring of soil and water conditions is an important part of the Oakwood Lakes–Poinsett RCWP. Monitoring is carried out under a Comprehensive Monitoring and Evaluation Plan (CMEP). The SCS is formally responsible for the CMEP, but the agency contracted administration of it to the state Department of Water and Natural Resources.

The objective of the CMEP is to systematically collect data which will reliably track quantities of pesticides and nutrients in the soil and ground and surface waters of the area. The data are collected in ten 20–40 acre fields, nine of which are being, or will be, farmed and one of which is not. Two of the nine farmed fields will not employ BMPs, thus providing information for comparison with those fields where BMPs are used.

The data collected include the rate and volume of water moving through the soil profile to the ground water system, using tensiometers, neutron probes and infiltrometer testing; chemical analyses of the soil; volume of run-off and its chemical characteristics as well as sediment load; and climatic conditions such as precipitation, solar radiation, soil and air temperatures, and wind speed and direction.

This intensive data collection activity is designed to serve several purposes. One is to help evaluation of the effects of the RCWP in improving ground and surface water quality. Another is to provide inputs to the Agricultural Non-point Source Pollution Model (AGNEPS) developed by scientists at the Agricultural Research Service (ARS) station at Morris, Minnesota. AGNEPS can track movement of sediment and nutrients carried by sediment and run-off. One of its proposed uses is to simulate soil and water responses to storm events and to changes in land use and conservation treatment in the Oakwood Lakes–Poinsett area. The outputs of the exercise with AGNEPS should be useful in identifying places and farming practices which are main contributors to water quality problems and in assessing the usefulness of alternative BMPs in ameliorating the problems.

The data collection, monitoring, and modelling effort undertaken by the Oakwood Lakes–Poinsett RCWP is impressive in its scope and obvious attention to careful scientific procedure. The 1985 annual report of the project devotes some seventy pages to an account of this effort. The impact of the effort on the performance of the RCWP is not clear from the material reviewed in preparing this report. The USDA's Economic Research Service did an analysis of improvements in ground and surface water quality in the Oakwood Lakes–Poinsett area since inception of the RCWP. The data available indicated no trend in nitrate concentrations in ground water, nor in annual recreational visits to the lakes. The ERS analysis does not mention pesticides. Recall that monitoring results reported above suggested that pesticides in ground water are not currently a problem. All this suggests that the pay-off to the CMEP still is to be realised.

Implementation

As in all RCWPs, the SCS and the ASCS are the agencies responsible for administering the Oakwood Lakes–Poinsett RCWP. As in the Reelfoot Lake area, however, other agencies—federal, state and local—are also involved in selected aspects of the project. The earlier noted agreement between SCS and the South Dakota Department of Water and Natural Resources assigning monitoring responsibilities to the DWNR is an example of this. The Water Resources Institute at South Dakota State University along with the Extension Service is engaged in work to determine the fertilisation practices of farmers in the area and to provide farmers with information about nutrient management alternatives likely to reduce nutrient loadings in run-offs. As noted earlier, the USDA's Economic Research Service is responsible for economic analysis of the project.

Integration

The material reviewed for this report provides no information about changes in cropping patterns and management practices over the last decade or so, nor about the amounts of land in the project area eligible for and in the Conservation Reserve. When the information becomes available it will likely duplicate the situation noted in the Reelfoot Lake RCWP: lack of substantive integration of macro-agricultural policies and soil and conservation policies. This is likely because these policies are drawn up at the national level, and their incidence in the Oakwood Lakes–Poinsett area will be much as in the Reelfoot Lake area.

The two areas appear similar also in the degree of inter-agency involvement in the respective RCWPs. As in the Reelfoot Lake area, it is difficult to discern in the Oakwood Lakes–Poinsett RCWP the extent to which inter-agency involvement results in institutional policy integration.

Perhaps the most notable management difference between the two RCWPs is the vastly greater attention given in Oakwood Lakes–Poinsett to data collection and monitoring, hydrological modelling and economic analysis. As noted, the pay-off to this effort is not yet apparent. In time, however, it is likely that the information base for performance evaluation of the Oakwood Lakes–Poinsett RCWP will be much richer than that for Reelfoot Lake. The economic study of damage to recreational value at Oakwood Lakes–Poinsett in particular provides important evaluative information. Such a study is badly needed for Reelfoot Lake. Were it available it likely would contribute importantly to sorting out just what is at stake in the management of the lake and the agricultural practices in its hinterland. Such a sorting out would not likely eliminate the conflicts now hindering achievement of RCWP objectives, but it should contribute to that end.

References

Agena, U., Wnuk, M. and Lawyer, C. (1985). 'Prairie Rose Lake Rural Clean Water Programme Project', in *Perspectives on Rural Non-point Source Pollution*, EPA 440/5–85–001, EPA, Washington, D.C.

Agricultural Stabilization and Conservation Service (1980, *National Summary Evaluation of the Agricultural Conservation Programme—Phase I*, USDA–ASCS, Washington, D.C.

American Agricultural Economics Association (1985), *Soil Erosion and Soil Conservation Policy in the United States*, Report of the Soil Conservation Policy Task Force, Ames, Iowa.

Bennett, H. (1928), *Soil Erosion: a National Menace*, US Department of Agriculture, Washington, D.C.

Carson, R. (1963) *Silent Spring*, Hamish Hamilton, London.

Clark II, E., Haverkamp, J., and Chapman, W. (1985), *Eroding Soils*, Conservation Foundation, Washington, D.C.

Crosson, P., with Stout, A.T. (1983), *Productivity Effects of Cropland Erosion in the United States*, Resources for the Future, Washington, D.C.

Crosson, P. (1985), 'New perspectives on soil conservation', *Journal of Soil and Water Conservation*, July–August.

Crosson, P. (1987), 'Soil erosion and policy issues', in T. Phipps and P. Crosson, eds., *Agriculture and the Environment*, Resources for the Future, Washington, D.C.

Crosson, P. (forthcoming), *Land and Environment in American Agriculture*, Resources for the Future, Washington, D.C.

Davenport, T., and Lowrey, J. (1985), 'Watershed water quality programmes: lessons learned in Illinois', in *Perspectives on Rural Non-point Source Pollution*, EPA 440/5–85–001, EPA, Washington, D.C.

Dicks, M., and Reichelderfer, K. (1986), *Implementing the Conservation Reserve Programme: a draft report*, US Department of Agriculture, ERS, Washington, D.C.

General Accounting Office (1977), *To Protect Tomorrow's Food Supply Soil Conservation needs Priority Attention*, CED–77–30, Washington, D.C.

LCC (Local Coordinating Committee) (1985 and 1986) *Reelfoot Lake Clear Water Project Annual Progress Report*, Reelfoot Láke, RCWP.

McCormack, D., Young, K., and Kimberlin, L. (1982), 'Current criteria for determining soil loss tolerance', in Schmidt, B., Allamaras, R., Mannering, J., and Popendich, R., eds., *Determinants of Soil Loss Tolerance*, ASA Special Publication 45, Madison, Wisconsin, American Society of Agronomy and Soil Science Society of America.

Novotny, V., and Chesters, G. (1986), *Delivery of Sediments and Pollutants from Non-point Sources—a Water Quality Perspective*, Water Resources Center, University of Wisconsin, Madison.

Park, W., and Dyer, E. (1986), 'Off-site damages from soil erosion in west Tennessee', in T. Waddell, ed., *The Off-site Costs of Soil Erosion*, Conservation Foundation, Washington, D.C.

Park, W., Keller, L., Manard, R., and Monteith, S. (1986). *Strategies for Conversion of Highly Eroding Cropland in West Tennessee*, Tennessee Water Resources Research Center, University of Tennessee, Knoxville.

Pavelis, G. (1985), *Natural Resource Capital Formation in American Agriculture*, USDA–ERS, Washington, D.C.

Phipps, T. (1987), 'The Conservation Reserve: a first year progress report'. *Resources*, winter, Resources for the Future, Washington, D.C.

Pierce, F., Dowdy, R., Larsen, W., and Graham, W. (1984) 'Productivity of soils in the Cornbelt: an assessment of the long-term impacts of erosion', *Journal of Soil and Water Conservation*, 39 (2).

Piper, S., Richards, M. and Lundeen, A. (1986) The Recreation Benefits from Improvements in Water Quality at Oakwood lakes and Lake Poinsett, unpublished study, US Department of Agriculture, Washington, D.C.

Reelfoot–Indian Creek Watershed District *et al.* (1960), *Watershed Work Plan Reelfoot–Indian Creek Watershed*, Obion County, Tennessee, and Fulton County, Kentucky.

Schertz, D.L. (1983) 'The basis for soil loss tolerance', *Journal of Soil and Water Conservation*, 38 (1). 10–14.

Trimble, S. (1975), 'Denudation studies: can we assume stream steady state?', *Science*, 188 (4194) 1207–8.

US Department of Agriculture (1938). *Soils and Men: the Yearbook of Agriculture*, Government Printing Office, Washington, D.C.

US Department of Agriculture (1985). *Soil and Water Conservation Research and Education Progress and Needs*, Soil Conservation Service, Washington, D.C.

Part IV
Changing landscapes, land-use patterns and the character of rural landscapes

8. Maintaining alpine landscapes in the Austrian Tyrol: policies for the maintenance of the rural cultivated landscape in the Tyrol

W. Puwein

Definition of the problem

The Tyrol is a readily accessible alpine area with relatively densely populated valley floors. The rural cultivated landscape is mainly conditioned by mountain farming. This landscape, in addition to being used for productive purposes, fulfils many public objectives such as protecting settlements and roads from avalanches, floods, landslides, falling rock and other natural disasters. The landscape also serves recreational purposes for both the indigenous population and tourists. The place of the alpine area in agriculture has become less and less important in the last few decades, a development related to surpluses in the dairy and meat markets. Because of the difficult natural conditions, production in the alpine areas is not very amenable to further rationalisation; production costs can be covered only with the help of high government price subsidies. In the Tyrol the cultivated landscape is intimately linked with peasant farming. The recreational function is of great economic value because of intensive tourism.

The objectives of government policies for preserving and improving the agricultural landscape have so far focused on preserving the existence of alpine farms. It was assured that the 'external' effects of agricultural production would help to preserve the cultivated landscape. Some of the measures undertaken to improve the production base of the farms, however, such as levelling of the land and drainage, worked to the detriment of the natural scenery. In the last few years, policy emphasis has shifted to direct payments, independent of production, and premiums for certain kinds of maintenance work. In this chapter the efficiency of these measures is evaluated and the areas requiring further attention if the protective and recreational value of the rural cultivated landscape is to be enhanced are indicated.

Characteristics of the Tyrol

Geography, population and the economy

The Tyrol is a province of the Republic of Austria covering 12,647 km², about 15 per cent of the total federal territory. The climate of this mountainous area is continental with local Mediterranean and oceanic influences, and its ranges rise to 3,800 m above sea level. Yearly precipitation ranges from less than 600 mm in the dry south-western valleys to more than 1,500 mm in the north-eastern part around Kitzbühel. Average duration of the snow cover in settled areas is fifty-five days in the Innsbruck region, and 153 days per year in the Galtür region. Accordingly, the growing period in settled areas ranges from 266 days in Innsbruck to only 134 days in Obergurgl.

Permanent agricultural settlements from 500 m up to 2,000 m above sea level cover 36 per cent of the area of the Tyrol, 33 per cent is covered by forest and 31 per cent is unproductive, mostly alpine waste land (glaciers, rocks, etc.). With 586,633 inhabitants, the population density was forty-six persons per square kilometre in 1981. By comparison the population density for the republic as a whole was ninety persons per square kilometre. The density within the Tyrol reflects the following factors:

1 One third of the area is uninhabitable high alpine waste land.
2 Only 8 per cent of the area is under intensive agricultural use and suitable for settlement.
3 Only 3 to 4 per cent of the area is more or less flat valley bottom.

Table 8.1 Total population change in Austria and the Tyrol region

Year	No. (000s)		Index (1869=100)		Agricultural population as % of population	
	Austria	Tyrol	Austria	Tyrol	Austria	Tyrol
1869	4,498	236	100	100	–	–
1910	6,648	305	148	129	31·3	44·2
1934	6,760	349	150	148	27·3	35·1
1951	6,934	427	154	181	21·9	25·6
1961	7,074	463	157	196	16·3	18·6
1971	7,492	544	167	231	10·6	10·8
1981	7,555	587	168	249	6·7	6·1
1984	7,553	598	168	253	–	–

Source: Osterreichisches Statistisches Zentrlamt, *Statistische Nachrichten*, Wien, relevant years.

After World War I population growth in the Tyrol was substantially greater than in the other provinces with high birth rates and immigration the main factors (table 8.1). The Tyrolean economy also recorded a markedly higher growth rate than the rest of Austria. The tertiary sector, the tourist industry in particular, expanded very rapidly. Cultural attractions, favourable transport routes but above all the beautiful scenery and the excellent winter sports facilities make the Tyrol the No. 1 tourism region in Austria. The Tyrol accounts for more than 40 per cent of all overnight stays by foreign tourists, and two-thirds of foreign exchange earnings originate in this sector (Creditanstalt, 1986). Over the last few years summer tourism has stagnated, but the number of overnight stays in the winter season has kept growing at a healthy pace. While in 1970 the winter season accounted for less than 30 per cent of total overnight stays, by 1985 nearly half the total overnight stays occurred in winter. Earnings from tourism in the winter season are substantially higher than in the summer, however, because winter tourists tend to spend more than summer tourists.

The secondary sector, too, grew at a faster pace than in other provinces. The generation of electricity from hydro power, which is very abundant in the Tyrol, is also of great importance.

The share of agriculture and forestry in gross value added is below the Austrian average, while until 1971 the share of the population engaged in agriculture was above the average recorded for Austria, because farmers in the mountain areas tended to have large families.

Environmental problems

The observations made so far already imply certain possible environmental problems:

1   The increase in population suggests strong pressures for land development in the few valleys suitable for settlement.
2   The large tourist industry requires the construction of roads, hotels, and other infrastructure.
3   Winter sports facilities, including ski lifts and wide downhill runs through forests, can endanger changes in the natural landscape and the vegetation can be mechanically damaged by ski edges and rollers for ski trails and slopes.
4   The utilisation of water for power generation dramatically changes the scenery.
5   Industrial pollution, on the other hand, is limited; local environmental damage can be observed near the few mining sites, limestone quarries and cement factories.

The Tyrol's geographical position makes it one of Europe's important transit routes. The heavy transit traffic on Tyrolean roads, with its noise and air pollution, represents one of the severest environmental problems.

For the first time in 1985 the Tyrolean Assembly asked the provincial government to produce a report on the state of the environment in the Tyrol and on the activities of the government in this area. The first annual report, published in 1986, contains the following assessment:

> *Soil and vegetation.* As a result of the expansion of settled areas and the construction of roads more and more agricultural land is being lost to other uses. Several

legal measures (Real Estate Transaction Act, Land-use Planning Act, Building Code) attempt to control the loss of agricultural land.

As far as the flora is concerned, the rapid deterioration of the quality of the forest is cause for alarm. According to the 1985 forest inventory conducted by the Federal Ministry of Agriculture and Forestry about one third of the trees in the over-sixty-year age class are damaged. This damage cannot be attributed to any single factor: the main causes are air pollution from local sources (heating, manufacturing, transport), emission of pollutants from other countries and natural stress factors such as drought, insects, grazing in the forest, and game. Trees high in the alpine region as especially susceptible to disease. At present most of the measures taken concern monitoring and analysing the forest damage. The state of many of the slopes used as ski runs is also cause for concern. In the alpine pastures the sod has been damaged and erosion has occurred.

Water. The Tyrol as an alpine area is fortunate in having ample sources of fresh spring water: 82 per cent of all communes are supplied exclusively by spring water. The quality of ground and surface water is in general very good. At the end of 1985 55 per cent of waste water was cleaned biologically, and by 1990 it is expected that 82 per cent of communal sewage and industrial waste water will be treated by biological plants.

Air. The effects of air pollutants on the forest have been pointed out above. Several regulations at the federal and provincial level attempt to reduce air pollution. Sulphur dioxide pollution peaked at the end of the 1970s, and has been falling off since. Concentration of NOx has, however, increased in recent years, mainly because of the increase in road traffic.

Noise. Road traffic on the transit routes is the most important source of noise pollution. The situation is aggravated by the fact that in the relatively narrow valleys the whole inhabited area is affected by noise pollution. Attempts have been made to alleviate this problem by erecting protective walls.

Waste disposal. At present, 60 per cent of all waste is disposed of in central waste disposal facilities. There are also several recycling programmes in operation.

Nature and landscape conservation. At the end of 1985, 17 per cent of the total area of the Tyrol was designated as a protected area in the form of a nature reservation, recreational area, protected landscape or protected region. Valuable landscapes and biotopes are particularly endangered by the construction of dwellings and winter sports facilities. Generally, agriculture and forestry practices do not cause significant environmental damage, and in the Tyrolean Environmental Report no adverse effects of agriculture and forestry production on the environment were noted. The agricultural areas are maintained almost exclusively as permanent grassland, which tends to safeguard the environment. Mostly organic fertilisers are used. While the average rate of commercial fertiliser used on Austrian agricultural land was 150·7 kg per hectare, the quantity of commercial fertilisers applied to agricultural land (excluding alpine pastures, alpine hay meadows, derelict arable land and pastures) amounted to only 33·1 kg per hectare in the Tyrol in 1985.

Agriculture and forestry

For 1981 the population census recorded 16,117 persons active in agriculture and forestry on a full-time basis, 6·3 per cent of Tyrol's total active population (Austria 8·5 per cent). Only 35 per cent of the 20,912 owners of agricultural and forestry enterprises are active on a full-time basis in agriculture and forestry (Austria: 39 per cent). The average family size on an agricultural holding was 4·6 persons, somewhat higher than the Austrian average, but the number of full-time workers per family (1·2 persons) was slightly lower. The average cultivated area per farm in the Tyrol is 35·5 ha, about one third higher than Austria overall, but because of the high proportion of alpine pastures and extensive hay meadows, the actual agriculturally cultivated area per farm is lower by a quarter than the Austrian average.

Table 8.2 Structure of agriculture in Austria and the Tyrol, 1980

Indicator	Austria	Tyrol
Family size per holding	4·3	4·6
Permanent labour force per holding	1·3	1·2
Cultivated area (ha per holding)	21·6	35·5
Cultivated area per worker (ha)	17.0	28.5
Agricultural area in use as percentage of the cultivated area	53·6	55·0

Source: Amt der Tiroles Landesregierung, *Bericht über die Lage der Tiroler Land- und Forstwirtschaft, 1984/85*, Innsbruck 1986. Bundesministerium für Land- und Forstwirtschaft, Berichte über die Lage des österreichischen landwirtschaft 1985, Wien 1986.

Tyrolean agriculture is strongly dominated by cattle farming, and the number of head per hectare is high. This branch of agriculture accounts for more than half of total production in agriculture and forestry (Austria: 35 per cent). Within this branch, dairy farming dominates, but the production of veal has also gained in importance because of the good sales potential to the tourism sector (table 8.2). The raising of breeding cattle is more important than meat production. Other types of animal and crop production are insignificant for the market; they basically serve the personal consumption of the farmers. The strong predominance of cattle raising is a result of the natural production conditions as well as of the agricultural market system. Because of sales guarantees and more or less fixed prices dairy farming is the safest and most profitable source of earnings for the farmers in the alpine regions. There is a long tradition of raising cattle for breeding and meat in the Tyrol but, because of the European Community market regulations, the Tyrolean farmer is at a disadvantage in nearby markets in northern Italy and southern Germany. Recurring market problems have to be offset by public subsidies.

Agricultural income per family worker on farms with accounting records was 35 per cent below the average in Austria. This average, however, masks even greater income disparities. Agricultural income per family on farms in the extreme alpine zone is as much as 55 per cent less than that in the low valleys (table 8.3). There are attempts to offset these disparities, caused by adverse production conditions, through direct income payments.

Table 8.3 Income of Tyrolean agriculture and forestry, 1984 (Sch)

Results of holdings with accounting records (full-time farmers)	Austria	Tyrol
Agricultural income per family worker		
Alpine valley holding		121,813
Mountain farm zone I		97,259
Mountain farm zone II		90,558
Mountain farm zone III		54,518
Total	117,729	76,856
Working income per family worker		
Alpine valley holding		137,534
Mountain farm zone I		124,505
Mountain farm zone II		119,030
Mountain farm zone III		86,700
Total	131,441	105,152
Total income per family worker		
Alpine valley holding		154,740
Mountain farm zone I		145,054
Mountain farm zone II		140,171
Mountain farm zone III		105,176
Total	147,449	124,322
Average net income of industrial workers		121,752

Source: As for Table 8.2

Agricultural income, direct public payments and income from other non-agricultural sources together constitute total working income. In 1984 total income per household member in the Tyrol was 20 per cent below the average in Austria; in alpine zone 3 average total income per household member was 37 per cent lower than that in the low valleys even when grants are included (table 8.3). Farms in the valleys and alpine farms on favourable sites have greater opportunities for earning extra non-agricultural income, opportunities which are limited for the inhabitants of extreme alpine regions because of their remoteness.

As a result of the relatively high number of children, farmers in the Tyrol tend to receive more direct income support through family allowances than the average in Austria. Thus total incomes in alpine zone 3 are only one third lower than in the valleys in the final analysis.

The rate of abandonment of farms is strongly affected by the income differential *vis-à-vis* industrial workers. The average net income of a Tyrolean worker in manufacturing in 1984 was Sch 121,752, which is similar to the net working income per family farm workers. For all farms, working income was 14 per cent below manufacturing wages.

The Tyrol's landscape and the impact of agriculture and forestry

A history of the landscape

All of Tyrol is alpine country. The special attraction of this landscape is the variation between cultivated areas, meadows, pastures, forests and farm buildings that are architecturally appealing and in harmony with the surroundings and untouched wilderness with rugged rocks and glaciers. Changes in wilderness areas above the timber line are limited to tourist facilities, such as paths, refuges, ski lifts and cable cars and land levelling for ski slopes and development of dams, stream diversions, electric power lines for power generation.

The cultivated landscape below the timber line contains only a few original nature relics, mainly wetlands such as swamps, riparian forests, and moors; much of the vegetation pattern and the courses of waterways have been determined by human beings. In addition to the settlements, traffic, industry, energy production and tourism, agriculture and forestry also shaped the appearance of the cultivated landscape. The clearing of forests was the most important change in the alpine ecological equilibrium brought about by man.

Cultivation in the Tyrol by the indigenous Illyric population began in a warm period around 1500 B.C. At first the alpine meadows above the timber line were used for cattle farming. The area of the alpine pastures was then extended by clearing the forests in regions below the timber line. Owing to a deterioration of the climate, the permanent settlements in the mountains had to be abandoned: further clearing of the forest in lower sites was the next step in development. In most cases the bottoms of the valleys were not settled before the end of this stage; brooks and rivers had to be dammed first, and swamps drained (Bätzing, 1984).

The forest area was not cleared for agricultural purposes only; the enormous quantities of wood needed for mining, foundries and salt mines also led to extensive felling. Grazing on these areas prevented reforestation. This trend ended in the second half of the nineteenth century, when the forests were placed under rigorous protection. These measures were prompted to some extent by the dangers of avalanches and landslides, which forest clearance on steep slopes poses. In the course of agricultural and industrial development in the twentieth century much arable land in the mountainous areas, dominated by subsistence agriculture, was abandoned; the farming of permanent grassland came into being. As a result of rising labour costs after World War II the intensive utilisation of grassland by hand was progressively reduced. Land levelling and improvements helped to expand the areas suitable for mechanisation and more intensive utilisation. The least suitable areas were either reforested or allowed to return to natural woodland when grazing and haymaking was discontinued.

Besides these changes in agriculture, road traffic, industry, the construction of settlements, the production of energy, and tourism also had an impact on the landscape. The construction of roads, power lines, lift facilities and ski runs required in some cases the clearing of forests. This substantially changed landscape features; good arable land was converted to residential construction sites, industrial plants and restaurants and hotels were built; and the construction of dams flooded whole mountain valleys. The present pattern of the cultivated landscape comprises:

1   A chain or urban agglomerations (Innsbruck, 117,287 inhabitants; Hall, 12,614 inhabitants; Schwaz, 10,928 inhabitants; Kufstein, 13,118 inhabitants) with many expanding suburban settlements in between, linked by high-capacity transport routes (e.g. the Inn valley between Telfs and Kufstein).
2   Several smaller industrial centres, expanding in the flat valley bottoms (e.g. Landeck, Imst, Reutte, St Johann, Lienz.
3   Regions with intensive tourism; some of the most recent have been developed just below or above the timber line, while others are heavily built-up towns on the valley floors, with tourism facilities on the slopes (e.g. St Anton, Ischgl, Seefeld, Ötztal, Stubaital, Zillertal, Kitzbühel).
4   Side valleys with few tourism facilities, where the landscape continues to be shaped mainly by farming (e.g. Virgental, Defreggen).

Positive and negative effects of agriculture and forestry
on the functions of the rural cultivated landscape

The open spaces on the floors of the densely populated main valleys, the mountain slopes and the side valleys are shaped by agriculture and forestry. Basically, agricultural areas and forests fulfil two functions: production and the supply of public welfare goods.

The productive function
The productive function of alpine agriculture and forestry consists of the production of cattle, meat, dairy products and timber. The present surplus situation in agricultural markets and the high subsidies necessary for marketing the dairy surplus mean that the agricultural production of the mountain farms is not essential. Timber is, however, an important raw material for the domestic manufacturing sector, while some 60 per cent of wood products (sawn wood, paper) are exported.

The welfare function
While the owners of agricultural and forestry plots can always derive income from production, they are compensated for the other benefits they provide in only a few cases. These benefits, as public goods, are formally valued in the market only in rare cases. Now, however, they are more highly valued by society than agricultural production. The additional public value of alpine agriculture and forestry derives from the following factors:

1   Protection against floods, erosion, landslides and avalanches.
2   The positive influence on the micro-climate and water system.

3   The provision of a refuge area for native flora and fauna.
4   The recreational value.

The protective function. The protective function is an important physical welfare function in the high mountains, a function which on steep slopes is most effectively ensured by the forest, as agriculture aggravates the risk of avalanches and erosion through clearing of forests and grazing. A halt to the cultivation of steep slopes can pose further dangers: fires can spread quickly in the high and dry grass and then jump to adjoining forest areas.

In the literature it is often noted that the long grass stems, when flattened, create slides for snow slips and avalanches. But studies in the Auberfern (Greif and Schwackhöfer, 1979) have shown that there is only a tenuous link between the occurrence of avalanches and the abandonment of alpine hay meadows in areas where avalanches originate. According to the opinion of avalanche experts it is irrelevant whether a grassy area is mown or grazed, because almost all avalanches slide on a snowy chute. Only a few per cent of avalanches slide on a grassy underground. The long grass, shrub and tree seedlings can, however, also be frozen into the snow cover; ground avalanches tear off the sod, and erosion cracks form which in turn can lead to landslides. The tending of the grass cover by farmers protects the soil from erosion in as much as mowing or grazing keeps the grass short, barren spots are reseeded and new tree growth is cleared away (Schemel and Ruhl, 1980).

But perhaps the most important welfare function is fulfilled as the farmer continually strives to ward off dangers to cultivated areas, farm buildings and roads by continuously repairing damage and devastation.

Many extremely steep slopes were abandoned by agriculture because they could be mown only by hand and were not well suited to cattle grazing because of the danger of cattle falling and erosion from cattle trails. If these areas need to be kept open for aesthetic reasons, this can be accomplished by grazing sheep and goats. For optimal protection of the environment, light afforestation is certainly the best solution. Many people consider that the Tyrol has insufficient forest cover (Kreisl, 1982), as it has a large proportion of wilderness land and substantial unforested areas with slopes inclined at more than 35°.

The impact on the micro-climate, water table, flora and fauna. The effect of the forests on the micro-climate and the water system is much more favourable than those of agricultural use.

Heavy rainstorms pose a special danger of landslides and floods for low-lying valleys; while water masses run off quickly from grassland, wooded areas absorb part of the rainfall directly through litter and forest soil is capable of storing more water than the sod. Moors and wet meadows also play an important role in storing water. But most of these wetlands have been drained for agricultural use and their stabilising function in the water system has been lost.

The mountain areas are an important habitat and refuge for flora and fauna (Brugger *et al.*, 1984). Agriculture has strongly altered the original habitat, and its genetic diversity has been reduced. As already mentioned, the indigenous vegetation survives only on wetlands. Ending agriculture in some areas would allow valuable plants to grow again and provide a suitable habitat for many species of animals.

The recreational function. The recreational function of the alpine landscape derives from its provision of tranquillity, clean air, various sports facilities, and the aesthetic visual attractiveness of the rural cultivated landscape. While tranquillity, clean air and sports areas are also provided in the alpine waste land and in the forest, the aesthetic appeal of the rural landscape derives from agricultural use, which 'provides for a harmonic variation of forest and open areas, intensively and extensively used grassland, grazing cattle, farms, and huts on alpine meadows' (Schemel and Ruhl, 1980). Farmers tend and conserve the landscape by keeping certain areas open through mowing and grazing, by maintaining farm buildings, fences, dry stone walls, and similar facilities. Abandonment of agriculture would change the landscape into 'monotonously wooded slopes, on which ski runs constitute the single lively element' (Bätzing, 1985), and into alpine meadows covered by shrubs and damaged by landslides. Cessation of cultivation in such areas also implies the abandonment of farm buildings and other structures. Yet it is exactly these buildings with their characteristic style that form an important part of the aesthetics of the landscape.

Rationalisation in agricultural management has to some extent adversely affected the landscape. The abandonment of crop farming has certainly reduced its diversity. A special problem is posed by modification to the landscape for agricultural purposes, notably the elimination of trees and bushes at the edges of fields and land levelling to facilitate the use of farm machinery. This 'adjusted' tractor country (Krysmanski, 1971) can certainly be farmed more easily and safely, but its aesthetic appeal is small.

The drainage of wetland has also destroyed valuable smaller features in the landscape. In many cases farm buildings have been renovated or razed and rebuilt in a style inappropriate to the landscape. Silos, and the noise of hay drying and tractors, diminish the recreational value of the landscape. The construction of forest and service roads is essential for modern management of forests and alpine pastures, but some of the roads are built in a way destructive of the landscape.

Leisure activities for relaxation and recreation are very diverse. They range from contemplative walking, the viewing of nature and the intensive experience of nature to sports activities where nature is merely a backdrop. The question arises as to how much people seeking recreation value the 'well-tended cultivated landscape'. Leisure activities vary from season to season, implying changes in the shape of the cultivated landscape desired by visitors. A profile of the vacation motives of German tourists, the largest group visiting the Tyrol, reveals their preferences in the summer season. Above all they value environmental factors such as a beautiful scenery, clean air, forests, quietness, cleanliness and unspoiled nature (table 8.4). They want to gather new strength and to experience nature. The main activity of summer vacationers is walking and hiking. This survey clearly indicates the importance of the cultivated landscape for tourism in the Tyrol, with agriculture and forestry assuming a special role. In addition to hiking and mountain climbing, other summer sports which do not impair the landscape include

Table 8.4 A profile of vacation motives of German tourists to Austria, expressed as a percentage of possible responses

Motive	All visitors	German tourists
Reasons for selecting vacation destination		
Beautiful scenery	68	51
Clear air	50	37
Many forests	45	19
Quiet	44	27
Cleanliness	43	21
Unspoiled nature	36	21
Reasons for travelling		
To gather new strength	65	51
To experience nature	53	40
To escape a polluted environment	36	30
Vacation activities		
Walking	75	71
Hiking	77	47
Photography	58	50

Source: Studienkreis für Tourismus, *Reiseanalyse '85*, Starnberg, 1986.

horse riding, white water canoeing and rafting. Even the establishment of golf courses can be evaluated positively if the natural features of the land-scape are preserved.

The winter vacationer visits Austria mainly for downhill skiing; but increasingly cross-country skiing is also practised. While the cross-country skier can experience intensively the interchange between forests, open areas, and farm buildings, nature is more or less only a backdrop for the alpine skier for whom lift facilities without long queues, and wide, well manicured *pistes* with good snow conditions are important. The construction of lift facilities, mountain restaurants, and downhill runs often engenders serious changes in the landscape, notably the broad trails in the forest for ski runs and the corridors for ski lifts. Even in open areas, single trees or groups of trees, protruding rocks, gullies and ditches, which contribute so much to the diversity of the cultivated landscape, are removed or levelled because they pose a danger to the skier.

Planning the future development of recreational areas will have to take into account that summer tourism, with a less predictable impact on the environ-ment, will at best stagnate in the Tyrol, while there is still growth potential for the winter season.

In our society the recreational function is gaining in importance: even in economic terms, it has by far surpassed the productive function in the Tyrol. In 1985 agriculture and forestry contributed 2·3 per cent to the total added value of the Tyrolean economy; the share of hotels and restaurants was 11·2 per cent. Farmers gain substantial income from the recreational function by letting guest rooms and apartments, or by running snack bars. They also

receive payments from the communes, ski lift operators and tourist associations for the use of ski slopes and for tending the slopes. In many cases
farmers are owners or co-owners of inns, hotels and ski lifts.

It is a remarkable characteristic of the tourist sector in the Tyrol that it has
been essentially built up by the local population. The farmers seized their
chance as landowners and as the dominant group on the local councils, and
have not left the development of tourism to outside investors. Even in the
highly developed tourist centres such as Obergurgl the tourist facilities are
almost without exception owned by local farmers or their children.
According to a recent survey (Meleghy *et al.*, 1982) in 1970 practically all
farmers received income from tourism, as owners of hotels and inns, owners
in ski lift companies, or as ski instructors, mountain guides, etc. This source
of income has helped to secure the existence of the mountain farmers, which
in Obergurgl since 1900 have fallen in number from twenty to seventeen.

Measures for maintaining and improving the functions of the landscape

Protection of the forest

Legislation
The importance of the protective function of the forest for society was
recognised very early; regulations protecting the forests were established at
the end of the Middle Ages and in the mid-nineteenth century accumulated
earlier legal regulations were consolidated in the Forest Act of 1852. It was
not until 1975 that this Act was amended and extended in the Forest Act of
1975, Federal Law Gazette No. 44/1975. These regulations were supplemented at the provincial level by the Tyrolean assembly to introduce further
restrictions concerning the use of forest pastures and forest litter removal,
measures against forest fires, and measures to improve the forest (Tyrolean
Forest Regulation, Provincial Law Gazette, No. 29/1979).

Objectives
The forest regulations are designed to maintain and strengthen the economic
and ecological role of the forest. This applies in particular to the protective
function of the forest against natural hazards and adverse environmental
effects as well as to the maintenance of the soil structure. In a mountainous
area such as the Tyrol with a high population density in the lowlands this
function is of great value to society, and is fulfilled mainly by protected and
protective forests.

Protected forests are those needing protection on locations subject to
erosion. Timber felling in these areas is subject to special regulations. The
protected forest in the Tyrol covers 235,000 ha, of which only 36 per cent is
used for timber.

As protective forests protect persons and objects from dangers such as
avalanches and falling rocks, the regulations governing their use are even
stricter than for protected forests. The area covered by protective forests
totals some 3,420 ha at the present time. The Forest Act also seeks to

preserve the positive influence of the forest on the environment, especially its role in moderating the climate and the water balance, in cleaning and reactivating air and water and in reducing the noise level.

Implementation of policy objectives

The forestry laws contain provision for the protection of the forest from excessive felling, damage from haulage, forest fires, forest insects, air pollution, torrents, avalanches, etc., as well as injunctions against clear cutting and re-afforestation. These laws require a relatively large administrative body for their execution. The authorities implement forestry regulations through enforcing the law, counselling, promotional campaigns, and monitoring of the state of the forest. The administrative powers of the agencies concerned are far-reaching; if required, they can have the necessary measures implemented themselves. In the Tyrol the forest administration set up the Tyrolean Landscape Service. This service has been quite successful in repairing damage to the landscape and in enhancing valuable features of the scenery. The activities of the landscape service are very diverse: footpaths and cycling paths have been built, lakes and ponds beautified, children's playgrounds and educational trails created, etc. (Pevetz, 1986). In future, activities will concentrate on creating new ponds, protecting moors, planting vegetation around lakes and ponds and improving areas near settlements by planting trees and shrubs.

Costs and benefits

The protection of forests through restrictions on forest management limits the property rights of the owners. Thus forest owners carry part of the cost of protecting the forest in the form of higher felling and haulage costs, lower land prices, etc. Public funds are also applied to protecting the forest according to the principle of common responsibility. The costs incurred by the federal government consist of personnel costs, expenditure on promotional programmes, and expenditure on flood and avalanche control. During 1985 the federal government spent Sch 16·5 million on forestry in the Tyrol, which is 18 per cent of all funds it spent on forestry. The province of Tyrol spent Sch 20·4 million, with most of the funds going to the construction of forest roads (table 8.5). The construction of forest roads has been criticised in recent years by environmentalists, because it strongly affects the landscape and can result in erosion and damage to the water table. Such roads are, however, required for the economic use of the forest as well as for its protection (insect control, fighting forest fires).

The protective function of the forest is enhanced by building barriers against mountain floods and avalanches. In 1985 the federal government spent Sch. 127 million on the control of mountain floods and Sch 62 million on the control of avalanches. Thus the Tyrol absorbs almost 60 per cent of federal funds for avalanche control and a 20 per cent share for flood control. The province also contributed towards the control of floods and avalanches.

These measures may also have negative effects. River beds made of concrete and structures for controlling avalanches have impaired the scenery; furthermore, the faster flow of streams has increased the danger of floods in

Table 8.5 Assistance to Tyrolean forestry by federal, provincial and communal
governments, 1985 (Sch million)

	Federal	Provincial	Communal
Re-afforestation[a]	4·0	2·0	0·7
New afforestation[b]	1·3	0·4	0·7
Ameliorations, forest manuring	1·8	0·6	0·4
Thinning	2·0	0·6	0·2
Growing better timber	0·2	–	0·2
Wood protection: game, grazing			
livestock	0·6	0·4	0·5
Forest roads	6·6	16.4	38·3
Total	16·5	20·4	45·0

(a) Protection forests, damaged areas.
(b) Wasteland, pasture, welfare afforestation.
Source: Bundesministerium für Land- und Forstwirtschaft, *Jahresbericht über die Forstwirtschaft,
1985*, Vienna, 1986.

the lower courses of the rivers. In most cases, flood control using a larger area
would serve the purpose better as it is desirable to create large flood ways
whose exact form is shaped by the natural growth and the flow of the stream.
Only the edges of such flood ways would have to be secured. These flood
ways would not only be valuable from an environmental point of view but
could also serve as recreational areas. This kind of flood control would, of
course, involve a great loss of cultivated areas which are used mainly for cattle
raising, but such losses should also be viewed in the light of agricultural
surpluses in dairy products and beef.

Damages due to emissions originating outside the sphere of forestry have
already been pointed out. To a certain degree, the effectiveness of measures
to protect the forests in the Tyrol can be evaluated on the basis of the
condition of the forest. Thus the biological production potential and the
growing stock of timber is assessed via a systematic sampling procedure as
part of the Federal Ministry of Agriculture and Forestry's regular forest
inventory. The area covered by forests in the Tyrol has increased slightly in
recent years. During the last ten years some 5,700 ha were re-forested with
government aid. In addition, the self-seeding of abandoned alpine pastures
has also greatly increased the forest area. On the other hand, during the same
period, 1,720 ha of forest were cleared, most of it for ski runs, roads, power
plants, electric power lines or agricultural use.

According to the forest inventory the rate of felling has been substantially
below the rate of increase for some time, so that the stock of timber has been
steadily increasing. This increase is, however, not without problems for the
ecology, because it goes in step with aging of the forest in inaccessible
locations. Most of the forest in these inaccessible locations is protective
forest, whose resistance to avalanches and landslides has declined markedly
in recent years. The forest inventory also records tree trunk damage in the
forest. During the inventory period 1981/85, 38 per cent of trunks were

damaged, with damage due to haulage occurring most often. Damage caused by the construction of roads will gradually level off, as the objectives of road construction are attained. Damage caused by felling, and in particular by logging, can largely be avoided through better planning and the deployment of more suitable equipment and procedures as well as the better training of forest workers (Kastner, 1985).

Damage due to grazing livestock, mainly gnawing damage, is less important today but still substantial, and can be prevented only through the separation of forest and pastures.

A much more serious ecological and economic problem in the Tyrolean forest is damage caused by game. The selective gnawing of rare tree species such as fir, beech and maple which stabilise the environment is devastating in many areas. For example, the number of fir trees in the Auberfern has been reduced to one third (Grief and Schwackhöfer, 1979). It is noteworthy that in fenced areas firs do very well. In addition to bark peeling and rubbing damage, bark peeling results in infestation with fungi which, if trees are exposed to the full forces of storms or snow, can cause large areas of trees to fall over.

With the introduction of winter feeding of game about 100 years ago and the Tyrol Hunting Act, which tends to promote the overfeeding of game, the stock of big game has increased, despite a deterioration in their living conditions over time. The game population is now excessive and cannot be supported by the environment. An indication of the explosive increase in the stock of big game is given by the number of animals taken. In 1985 it was 5·5 times as high as in 1950.

The consequences of improper game management are particularly devastating for the protective forest. In some districts forest damage is so extensive that it amounts to devastation and some settlements and roads are subject to great risk. The high density of game in the protective forest derives from economic considerations. The timber in the protective forest can be exploited only at high cost, so that in many areas the income from the lease of hunting grounds is higher than the income from cutting trees. Furthermore, efforts to improve the state of the forest does not benefit the present owner but, at best, only his grandchildren. Thus the leaseholders are more interested in maintaining a high stock of game and achieve this goal by feeding in the winter and the summer and by shooting only a few females. Despite some promising initiatives to solve the big game problem, damage caused by game is still on the increase. This is clearly a case where measures to protect the forest have failed. A reduction in game damage through a reduction in the density of big game to a level that is tolerable from the point of view of forest management and the preservation of the landscape is crucial to the maintenance of the forest and its protective function. Ideally, no more than 10 per cent of new tree growth should be gnawed and no significant bark peeling should occur. A fundamental solution of the conflict between forestry and game management, however, has so far been foiled mainly because of the high social standing of hunters in Austria. But there is also another factor at work: many of the representatives of the hunters (e.g. the hunting club chairman) are officials of the forestry administration, that is, the forest supervising agency.

Land-use planning

Legislation
In Austria, land-use planning is the responsibility of the provinces. The Tyrolean Land Use Planning Act (Provincial Law Gazette No. 22/1972) applies to planning at the level of the commune and above.

Objectives
Under this Act the preservation and maintenance of the environment, especially the avoidance of harmful changes to the landscape and the eco-system, are listed as prime objectives. Other objectives are:

1 Preservation and appropriate development of sufficient areas for agriculture and forestry.
2 Provision of suitable areas for tourism and recreational use.
3 Provision of facilities for transport, communication and supply systems, and also of sites for industrial plants, training facilities, etc.

The local planning authorities of the communes are charged with responsibility for designating areas for construction, for settlements, for industrial plants, for mixed use, for open areas and for special uses.

Implementation of policy objectives
Regional planning, above the level of the commune, is mainly concerned with inventories of the features of the landscape, development programmes for the whole province, or parts of the province, co-ordination between the communes and counselling the communes. But specific land-use planning, crucial to the appearance of the landscape, is in the hands of the communes. Zoning plans, as approved by the communes, are subject to approval by the provincial government. With the exception of the city of Innsbruck, all Tyrolean communes now have approved zoning plans. The main activity of local planning now involves changes in zoning plans. Most of the applications concern the conversion of agricultural land to construction sites. In most cases it is not the urgent need for housing but rather the need to raise capital to pay back debts that prompts such re-zoning applications (Tyrolean Environmental Protection Report, 1985).

Costs and benefits
Most of the cost of these planning activities is born by the communes and province as part of general administrative costs. Costs arise for the landowner only so far as a refusal of the planning agency to re-zone land from agricultural use to construction represents a lost capital gain. Appropriate compensation is provided in the case of re-zoning from building sites to agricultural use.

Efficiency
The Land Use Planning Act could be an efficient instrument for shaping the landscape in accordance with environmental requirements. Its effective implementation depends, however, to a large degree on the distribution of

power within the commune. Public relations activity, such as information bulletins, exhibitions, lectures and discussions, can make the citizens and their representatives more interested in programmes that serve to improve the landscape. The government of the Tyrol also has the option of guiding the development of the rural landscape in the communes by designating areas as agricultural zones. This designation is binding on the communes in their development plans (Pevetz, 1986).

Nature conservation

Legislation
In Austria nature conservation is the responsibility of the provinces. The Tyrol's Nature Conservation Act (Provincial Law Gazette No. 15/1975) protects both wilderness areas and the man-made landscape.

Objectives
The Nature Conservation Act aims to 'preserve the eco-system, the diversity and beauty of nature as much as possible' by designing areas suitable to mankind and serving especially his health and recreation need, to be preserved, restored or improved. This also includes the landscape as shaped by man.

Policy implementation
On the basis of this Act certain zones can be designated as national parks, nature reserves, protected landscapes, rest areas, local nature reserves or protected areas. In the protected zones certain regulations regarding the height of buildings, industrial plant and changes in the landscape and in the ecological system apply. The Nature Conservation Act also protects rare or endangered species of plants and animals. The Act also refers to natural monuments such as rare, old trees, springs, watercourses, ponds, lakes, rock formations, characteristic ground formations, etc.

Landowners on whose property these protected sites are located are obliged to safeguard their existence and can be reimbursed if this entails substantial earnings losses or impaired agricultural management. At the end of 1985 the total area protected in the Tyrol as either a nature reserve, a rest area, protected landscape or a protected area was about 213,000 ha or 17 per cent of the total (Tyrolean Environmental Protection Report, 1986). There are 203 protected natural monuments in the form of moors, lakes, water courses, stands of timber, single trees, and tree avenues). Special protection is also afforded to all watercourses, including river banks. For the construction of electric plants and the control of river banks environmental permits are required. The administrative authority may require that the projects be carried out in conformity with nature and landscape conservation goals.

The construction of winter sports facilities such as ski lifts and ski runs also requires assessment or approval by the nature conservation authority.

Another important element of nature conservation is the protection of wetlands, areas which not only constitute the last-remaining elements of the original vegetation and fauna in the rural cultivated landscape but which are also of great importance for the water system. The province of Tyrol provides some of the funds necessary for tending and preserving wetlands. Other attractive features of the landscape, such as meadows with larches, are also protected, some funds also being provided by the provincial government.

In administrative proceedings regarding the Tyrolean Land Use Planning Act the nature conservation agency is accorded expert status by the agriculture agency conducting them. In proceedings regarding the consolidation of agricultural holdings an attempt is made to preserve wetlands or to create new ones, to purchase sites for the province, or to offer the landowners other sites in exchange.

Costs and benefits

In general, the costs of nature conservation are included in general administrative costs. Payments to landowners for conservation measures, for impaired agricultural use and for the acquisition of land are relatively small. The local population benefits from the rise in the recreational value of the scenery, either directly or indirectly through an increase in earnings from tourism.

The fund allotted to the conservation of nature in the provincial budget are relatively small. Nature conservation goals are mostly pursued by experts of the provincial government and conservation counsellors in all districts. The counsellors participate in land consolidation action and in many technical projects which have an impact on the landscape. It has not been possible, however, to create a national park in the Tyrol.

Conservation of alpine agriculture and improving agricultural income

Conservation of peasant farming in the alpine area has the objective of preserving the recreational and protective functions of farming.

Legislation

In general, matters concerning agriculture are the responsibility of the provinces, but certain important matters are within the competence of the Federal Republic. The Agriculture Act of 1976 (Federal law Gazette No. 261/1984) is the basis of the development programme administered by the Federal Republic for improving the production basis of agriculture and the structure of farm holdings. This Act also governs the manufacture and marketing of agricultural products. The Agricultural Marketing Act of 1985 (Federal Law Gazette No. 183/1986) and the Livestock Act of 1983 (Federal Law Gazette No. 621/1983) regulate the markets in milk, grain and fattened cattle. Tyrolean agriculture depends vitally on the price support system in the dairy market.

The Tyrolean Agriculture Act (Provincial Law Gazette No. 3/1975) empowers the government of the Tyrol to regulate agriculture in a similar fashion. The Tyrolean Land Use Planning Act, 1978 (Provincial Law Gazette

No. 54/1978), attempts to improve the performance of agriculture by regulating the ownership, utilisation and management of farm holdings. According to the Real Estate Transaction Act 1970 (Provincial Law Gazette No. 4/1971) the sale of agricultural and forestry land requires a permit which is designed to safeguard the interests of farmers.

Objectives
The basic objectives of these Acts are the preservation of healthy and productive farm holdings, the self-efficiency of the population with regard to food, and the preservation of the cultivated landscape. Mountain farms are deemed worthy of special protection because living and production conditions are made difficult by adverse climate, inaccessibility and steep slopes. The objectives of these agricultural programmes are:

1   The improvement of the production base through research, counselling, improvement of productivity in crop and cattle farming, technical rationalisation, levelling of farm land and the correction of watercourses for purposes of agriculture.
2   Improvement of the structure and management of farms through regional programmes such as the special programme for mountain farmers, building of roads, telephone lines, and electric power lines, agricultural campaigns, improvement of settlements and consolidation of land holdings.
3   Improvements in marketing by changes in the structure of the market, advertising, and exploration of new markets.
4   Lending policies.

Policy implementation
These objectives are attained by policy measures with regard to market regulation, price support, import tariffs and trade policies aimed at promoting the sale of Austrian agricultural products, and also programmes which reduce production costs.

Measures to set the price of and tariffs on agricultural products as well as trade policies are almost exclusively the responsibility of the federal government. Programmes concerning market regulations and promotional programmes, however, are also funded by the provincial government.

Compared with other provinces, agriculture in the Tyrol benefits more than proportionately from the regulations concerning the cattle sector. The dairy regulations guarantee the sale of all milk, and for a large portion of the output a guaranteed price is set. According to the transport cost regulation the mountain farmers, with their less accessible farms, receive the same price as dairy farmers close to the consumer. Mountain farmers also benefit from another provision of the regulations: milk, butter and cheese produced on mountain pastures are not subject to a levy used for promoting sales, and sales direct from the farm are not subject to a lump-sum payment as is the case otherwise.

The price of fattening cattle is maintained through a system of export subsidies and storage programmes for cattle and meat. Exports of breeding cattle are subsidised by the federal government and the province of the Tyrol, with higher payments for cattle, for example, from mountain farms, graduated by zones. With a share of 38·5 per cent of federal funding, the Tyrol benefits more than proportionately from these payments. Calf fattening is also of some importance and in this case the Tyrol's share of federal funds is 34 per cent. Subsidies for sheep rearing, however, continue to be small. Nevertheless, the marketing of sheep is promoted by transport and export subsidies, and direct payments are given for grazing sheep on mountain pastures.

The mountain farms, being net purchasers of coarse grain, are burdened by high prices fixed by the regulations governing the coarse grain market. These additional costs are partially offset by programmes reimbursing the transport costs of domestic coarse grain and programmes offering coarse grain at lower prices to mountain farmers.

In order to protect agricultural areas river beds are lined, streams regulated and avalanche dams built. These measures are implemented through direct payments, the granting of loans, and subsidised loans. For the development of agriculture in the Tyrol programmes regarding cattle rearing are the most important. A great number of Tyrolean breeders participate in a programme that monitors the milk production of dairy cows; this performance test is most important for decisions with regard to breeding. In the Tyrol 68·4 per cent of the stock of cattle were monitored in 1985 (Austria: 31·5 per cent). The promotion of ranching through a programme designed to take pressure off the dairy market whilst simultaneously slowing the decline in mountain cattle numbers and utilising the alpine grassland, however, has had little success in the Tyrol. Only 10 per cent of the cattle under this programme came from the Tyrol. Agricultural investment loans with a 50 per cent interest rate subsidy are available to mountain farmers and farms in specially designated regions.

Production costs may also be lowered through government-subsidised machinery pools. But in general, agricultural work in the mountain areas can be mechanised only to a very limited degree. Furthermore, because of the changeable weather, hay harvesting is concentrated in a very short period during which all farmers in the machinery pool need access to the harvesting equipment at the same time. The lack of good roads and the long distances between farms render the operation of the machinery pools quite difficult (Androschin, 1971).

In 1972, for the first time, a special programme for mountain farmers was introduced by the federal government and is considered to be of crucial importance for safeguarding the existence of mountain farms. It aims to preserve areas that are in danger of being abandoned by revitalising whole villages and regions over several years. The cost of these measures, so the reasoning goes, ought to be borne by the whole Austrian population, because of 'an economically sound alpine area which is culturally and socially viable and which conserves the environment as much as possible in its unadulterated form' makes an important contribution to the general welfare of the country (Federal Ministry of Agriculture and Forestry, 1973). This special programme improves the production basis of mountain farms and raises family income through direct payments. Direct payments account for 37 per cent of the funds dispensed so far, construction of transport routes accounts for 34 per cent, agricultural assistance programmes for certain regions (e.g. East Tyrol) for 19 per cent. The remainder is used for land levelling, forest management, and the erection of telephone and electric power lines.

Costs and benefits
The costs and benefits of the agricultural system can be evaluated in many ways. The costs are borne by the consumer in the form of higher food prices and by the taxpayer through price subsidies and export subsidies. Add to this the funds for the various special programmes, which also have to be borne by the taxpayer. In 1984 Tyrolean agriculture received Sch 109·5 million from the federal government and Sch 149·7 million from the province in the form of direct payments (table 8.6). The agricultural market system and the various assistance programmes raised farm income. On the other hand, the consumer benefits from a secure supply of foodstuff and from the external effects of agricultural management (such as the rural cultivated landscape).

Table 8.6   Assistance to Tyrolean agriculture by the federal and provincial governments, 1984/85 (Sch million)

Type of assistance	Federal	Provincial
Road building	58·2	25·8
Road building, special programmes	–	55·0
Cable laying	–	0·9
Electric power	0·1	0·2
Telephone lines	0·9	0·5
Consolidation of land holding	13·4	8·8
Settlement	0·8	0·5
Subsidies to areas losing population	–	10·5
Agricultural campaigns	4·6	24·7
Improvements of alpine pastures	5·7	7·0
Levelling farm land	0·9	0·8
Regional programme 'change-over'	11·4	2·3
East Tyrol special programme	11·6	8·1
Agricultural river regulation	1·9	7·6
Total	109·5	149·7

Source: Amt der Tiroles Landesregierung, *Bericht über die Lage der Tiroler Land- und Forstwirtschaft, 1984/85*, Innsbruck, 1986.

The agricultural marketing system and the promotional programmes have not succeeded in securing an adequate level of income for the mountain farmers. In 1984 the agricultural income per family worker in the Tyrol was 37 per cent lower than that of an industrial worker. The agricultural income of mountain farmers in zone 3, for example, was 55 per cent below that of an industrial worker. Many of the subsidised measures also had a negative impact on the visual aesthetic appeal of the cultivated landscape and the water system. For example, land levelling for agricultural purposes destroyed many of the smaller features of the landscape, although in some cases this practice created some green spots in barren areas. Programmes encouraging land levelling were phased out by the federal government at the end of 1985.

The on-going programme of encouraging the drainage of wetlands must also be viewed critically. Conflicts also occur again and again between environmentalists and proponents of agricultural use of mountain areas with regard to revitalising alpine pastures, road construction which impairs the landscape; the more intensive use of commercial fertiliser and wetland drainage (Schnitzer, 1973).

Conservation of alpine agriculture through direct payments for tending the landscape

Federal assistance to mountain farmers
As mentioned above, the federal government in 1972, initiated a programme designed to raise the family income of mountain farmers by lump-sum payments, in addition to encouraging production increases in these areas according to the general guidelines. This policy change was prompted by the view that measures to increase agricultural output were not sufficient to raise the agricultural income of mountain farms to an adequate level. The direct payments to mountain farmers can be viewed as a lump-sum payment for services rendered in maintaining the recreational value of the rural cultivated landscape. The payment scale depends on the 'assessed' value used for tax purposes and is well below the market value of a farm and, also, its categorisation according to zones of adversity. Thus a social component is also taken into account. At least two hectares must be cultivated or three hay-fed livestock units must be kept, while the assessed value of a farm must not exceed Sch 300,000. Since 1985 there have been four zones of adversity and the programme applies only to farms in zones 2 to 4. Farms in the highest zone of adversity and in the lowest level of assessed values (up to Sch 50,000) receive the highest payments; farms in zone 2 and in the highest level of assessed values (Sch 201,000 to 300,000) receive the lowest payments. In 1985 the highest payment amounted to Sch 15,000 per farm on an annual basis, the lowest Sch 3,000. In 1985 federal payments under this programme totalled Sch 430 million, with 20·4 per cent going to Tyrolean mountain farmers (table 8.7). A more than proportional number of Tyrolean mountain farmers were in the highest payment categories and the average amount paid was Sch 8,922, substantially above the average for Austria as a whole (Sch 7,538).

The graduation of the direct payments according to the assessed value and the exclusion of farms above the upper limit of Sch 300,000 raises some problems if these payments are viewed as compensation for preserving the landscape. Large mountain farms or farms whose owner runs the farm in the summer on a part-time basis are discriminated against, even though they,

Table 8.7   Federal assistance payments to mountain farmers[a] in Austria and the Tyrol, 1985 (Sch million)

Zone	Payment gradation	Austria	Tyrol	%
4	1	26·9	12·8	47·6
4	2	18·3	9·3	50·8
4	3	17·6	9·2	52·3
4	4	8·4	3·3	39·3
3	5	62·3	12·8	20·5
3	6	83·2	12·3	14·8
3	7	68·9	11·2	16·3
3	8	46·0	5·5	12·0
2	9	17·7	3·1	17·5
2	10	33·0	3·4	10·3
2	11	28·7	3·3	11·5
2	12	19·4	1·6	8·2
Total		430·3	87·8	20·4

(a) Preliminary.

Source: Bundesministerium für Land- und Forstwirtschaft, *Bericht über die Lage der österreichischen Landwirtschaft*, 1985, Wien 1986.

too, help to conserve the recreational role of the landscape. In the Tyrol some 30 per cent of mountain farmers received no assistance under this programme because their assessed value was too high.

Direct payments for agricultural use of the land
Direct payments are given by the province of Tyrol to farmers to encourage them to maintain farms in the medium and upper mountain zones for tending areas in the mountainous regions. The payments are graduated according to the degree of adversity in production of each farm and the number of hay-fed livestock units.

In 1984 8,976 farms received Sch 39·2 million under this payment scheme, which should be compared to Sch 78·6 million received by 9,980 farms under the federal grant programme. The average amount received by a farm for tending the landscape was Sch 4,393 under the provincial programme and under the federal programme it was Sch 7,875.

The provincial government grant programme for the conservation of alpine pastures of the Tyrol
An alpine pasture grant was introduced in 1977 to halt the abandonment or conversion of alpine dairy pastures to non-dairy use. The deterioration of the pastures in the absence of grazing animals and of people, and the decay of hay barns, were considered to be impairing the recreational value of the pastures which are an essential factor for the tourist industry.

The alpine payment is granted upon application and is based on the number of grazing animals, the accessibility and the elevation of the pasture.

In 1984 grant payments totalled Sch 5·4 million for 1,889 alpine pastures. On average, Sch 2,871 were paid per alpine pasture.

Over the last five years the agricultural use of alpine pastures has increased substantially. This upturn is due not only to the alpine pasture grant but also to the practice of excluding the milk produced on alpine pastures from the quota allotted to each farm as a means of offsetting the high costs of grazing cattle on alpine pastures.

Compensation for preserving the smaller features of the landscape

The Tyrol has a number of sensitive areas of local but not national importance and when this only involves the interests of the local tourist industry solutions are found at the local level. For example, in some Tyrolean valleys the farmers are no longer interested in taking care of larch meadows, because their management cannot be mechanised, and if these larch meadows, which are very attractive, are not mown a dense spruce–pine–larch forest develops. The owners of the larch meadows are paid Sch 1,000 per hectare annually by the Tyrolean government for maintaining these features of the landscape. The areas must be cleared of broken-off branches and spruce and pine trees must be uprooted. Moreover, larch may be cut only after giving notice to the forest supervisor and they must then be replanted. The total sum under this programme amounted to almost Sch 1 million. In one commune an additional amount of Sch 2,000 per hectare is paid for grazing the larch meadows and Sch 3,000 per hectare for mowing. The province of Tyrol also gives grants for renovating and maintaining hay barns in the larch meadows (Tyrolean Environment Report, 1985).

Keeping grassland open is considered of the utmost importance by some communes. Skiing, in particular, requires the management of grassland, and for this farmers are compensated. For example, in the Berwang valley (district of Reutte) extensive damage was caused by the erosion of abandoned farm land. So, to prevent further damage and to keep the meadows suitable for skiing, the commune now pays the farmers to manage the ski slopes (Pevetz, 1986). Similarly, the resort town of Fis offers grants for mowing or grazing certain areas. These payments are funded by the commune, the agricultural community, the tourist association, the ski lift company and the skiing school.

Similar means of financing the preservation of the landscape can be found in many other Tyrolean communes (Schermer, 1983). Many communes offer alpine pasture grants whose size depends on the number of cattle on the alpine meadow. Tourist associations, ski lift companies, skiing schools and communes also pay for the use of *pistes* and cross-country tracks. In 194 of the 251 Tyrolean communes (77 per cent) ski runs exist in agricultural areas and similarly cross-country tracks are on farmland in 72 per cent of the Tyrol's communes. For ski runs the payment depends on the quality of the area concerned. For example, payments for meadows which are cut several times a year range from Sch 14,00 to 40,000 per hectare, for meadows allowing only one cut from Sch 1,500 to 20,00 per hectare. For forest areas, payments vary between Sch 1,500 and 15,000 per hectare. The lowest payments are granted for the use of alpine meadows, between Sch 1,000 and

15,000 per hectare. Altogether, some 7,000 ha of agricultural and forestry areas were managed on a contractual basis.

General assessment and prospects

This section evaluates the various programmes which seek to preserve and restore the welfare function of the rural cultivated landscape, in particular the protective and recreational functions of agricultural management.

The protective function

Whether agriculture and forestry serve to protect the environment can be measured by the number of natural disasters such as floods caused by torrents, avalanches, landslides, etc. Since 1945 the number of natural disasters has been on the increase, and the extent of flood damage and the number of avalanche victims has been rising (Aulitzky, 1975). This upward trend is, however, mainly due to the fact that people have increasingly used danger zones for new settlements, roads and tourist facilities, and have started touring and deep snow skiing in avalanche areas.

The total area of forest in the Tyrol, which has the greatest protective value, has hardly changed over the last fifty years, although in Austria as a whole it has increased slightly. The state of the Tyrolean forest, however, has deteriorated markedly since World War II. This applies in particular to the protective forest. A substantial increase in damage due to erosion and avalanches can be expected in the future. Even though the 'dying of the forest' is mainly ascribed to air pollution, the use of forest for grazing, forestry (timber haulage) and hunting (gnawing, bark peeling damages) on steep slopes has also contributed to the present situation.

If the cultivated landscape is to protect the environment, the danger zones have to be afforested and protective forest revitalised. This requires a substantial reduction in the stock of game (Donaubauer, 1979). As the state of the forest improves, the capacity to store water also increases. The restoration of wetlands might also serve to improve the water system locally.

The recreational function

The recreational value of a landscape depends very much on its diversity. Statistics on land use do not provide much information on changes in the pattern of land use, as abandoned agricultural areas, in particular, are not covered adequately by the surveys (Haefner and Günter, 1984). Aerial surveys might depict these changes in the cultivated landscape more accurately. But no such surveys are available for the whole region, and the following assessment must rely on the land-use inventory (table 8.8).

In the Tyrol the area of farmland is on the decline, but the rate of decrease is slightly lower than in the rest of Austria, where forests and unproductive

Table 8.8 Land use in Austria and in the Tyrol 1937–85 (ha 000)

Year	Farm land		Forest		Unproductive area	
	Austria	Tyrol	Austria	Tyrol	Austria	Tyrol
1937	4,356	520	3,135	426	896	318
1950	4,176	513	3,057	412	1,039	324
1960	4,052	507	3,142	423	1,112	337
1970	3,896	482	3,206	426	1,206	359
1975	3,789	459	3,250	425	1,259	381
1980[a]	3,741	458	3,282	423	1,298	393
1985[b]	3,549	447	3,221	426	11,300[b]	400[b]
			Index (1937 = 100)			
1970	89	93	102	100	135	113
1985	81	86	103	100	145	126

(a) Limit of inclusion risen form 0·5 ha to 1·0 ha.
(b) Wasteland estimated.

Source: Österreichisches Statistisches Zentralamt, *Ergebnisse der landwirtschaftlichen statistik.* Wien, relevant years.

areas are expanding at a faster rate than in the Tyrol. In general, it can be said that the distribution of land as between agriculture, forestry and unproductive areas has changed only slightly over the last fifty years. The extent of arable land, however, declined substantially, amounting now to no more than one third of the area before World War II. Arable land has been converted mainly to permanent grassland. This conversion, it is true, has reduced the variety of features to be found in the cultivated landscape, but it also enhanced the protective function, because grassland on steep slopes is less prone to erosion than arable. Alpine pastures and hay meadows, land especially valuable for recreation, have been better preserved in the Tyrol than in other parts of Austria. It must be pointed out, however, that large areas indicated as cultivated land in the land-use inventory may no longer be in agricultural use.

The diversity of domestic animals kept in the open also contributes to the recreational value of the rural landscape. The livestock statistics record a more or less constant number of head of cattle. This is mainly due to the price support system in the dairy market; milk production is the most reliable and profitable source of income for the mountain farmer. The number of horses dropped by half from 1938 to 1970, but has remained unchanged since then. Work horses are increasingly replaced by riding horses. By 1970 the number of sheep had declined to one third of the level before the war, but then doubled by 1985. This development has been very favourable to the conservation of the landscape and was facilitated by various market support programmes (Hoppichler and Groier, 1986); payments for grazing were also a contributing factor. Sheep grazing is less labour-intensive than cattle grazing.

Goat raising is also on the increase again, but in 1985 the stock of goats was no more than one fifth of the level in 1938.

The peasant farmer, his working conditions, customs and buildings are an essential factor in the recreational value of the rural cultivated landscape (Pevetz, 1978). Many of the tourists seeking recreation appreciate the opportunity to observe the farmer at work, or even to help in the farm work, to participate in the customs, to talk to the farmers, or to be put up on the farm. In this way the sheer number of agricultural holdings is also relevant to the recreational value. From 1930 to 1980 the number of farms in the Tyrol declined by only 14 per cent (Austria 29 per cent). This stability must be ascribed to a great extent to the ample sources of non-agricultural income in the tourist industry, which have secured the existence of farms in the higher regions of the Alps. The family income of mountain farmers has also been supplemented by social transfer payments and by direct payments at the federal, provincial and commune level in exchange for conserving the environment. The price support system and production cost subsidies alone were not sufficient to raise agricultural income to a level adequate to secure the existence of the farms.

There are several methods of evaluating the natural scenery on an objective basis (Kastner, 1981), but in the final analysis the aesthetic appeal of the scenery remains a subjective matter. Just as action to improve the protective function of the cultivated landscape can be measured by the number of natural disasters, success in conserving the recreational value of the landscape might be measured by the development of tourism.

Summer tourism has decreased slightly during the last few years in the Tyrol. This development does not, however, necessarily imply a deterioration in the recreational value of the cultivated landscape. The main factors contributing to the loss in market share were economic, such as the loss of price competitiveness by the tourist industry in the Tyrol, the fall in the cost of air travel and changes in travel preferences, especially on the part of younger tourists. The tourist attraction of 'Alps' simply as scenery (passive experience of nature) is no longer enough (Tschurtschenthaler, 1985). Tourists must be offered ample opportunities to engage in their preferred vacation activities, mainly sports. A package combining sporting activities and the 'experience of the mountains and nature' is an area where alpine tourism still excels and may draw new customers.

Winter tourism, more harmful to the environment and less dependent on the aesthetic appeal of the scenery, is still on the increase. Here, a successful combination of recreation and sport has been achieved. Untouched nature, with all damage to the environment masked by the snow cover, takes care of the recreation aspect: the thrill is provided by alpine or nordic skiing, ski tours and hikes, and, last but not least, *après-ski* is provided every night (Tschurtschenthaler, 1985).

In the future, the landscape is likely to be characterised by tourism, more than now because golf courses, horse trails, ski runs and other facilities require open space. In planning these facilities the protective functions of the environment and the aesthetic appeal of the scenery should be taken into account. As these changes occur, man is likely to accept increasingly a cultivated landscape dominated by tourism, just as he got used to a landscape characterised by mountain farms without arable fields.

The question arises whether the conservation of the landscape in its present state should be left to the farmers alone. Studies in Switzerland have shown that taking care of the landscape by way of agricultural management is the least-cost alternative, even though the deployment of a government landscape service would guarantee nature conservation to a higher degree. But programmes carefully designed to encourage local farmers to take environmental concerns into account will also achieve these goals. This applies not only to areas with especially attractive scenery but also to recreational areas near cities, because the continued existence of the 'world of the peasant farmer' is essential to the recreational value of a landscape.

Summary and conclusions

The Tyrol is an easily accessible alpine area, with a highly developed tourist industry. The main products of agriculture and forestry are milk, cattle and coniferous logs; however, the share of agriculture and forestry in gross value added of the Tyrolean economy is only 2·2 per cent. While the economic importance of these two industries is on the decline, the public's interest in the positive environmental effects of agriculture and forestry is on the increase. The welfare effects consist in the protection of settlements and traffic routes against avalanches, landslides and floods, and in maintenance of the recreational value of the landscape.

The Austrian federal government, the provincial government of the Tyrol and the communes endeavour to safeguard these benefits of the rural cultivated landscape by a variety of measures. The protective function on steep slopes is best performed by the forest. Relative to its climatic and morphological conditions, the Tyrol's forest endowment is too small; the protective forests are also in bad shape. Forest damage is still on the increase, despite rigorous regulations and extensive public subsidies. Much of the damage is due to emissions outside the sphere of forestry, and inappropriate game and forest management.

The recreational functions of the alpine landscape consists in its tranquillity, clean air, sports facilities and visual aesthetic appeal. Measures to maintain the recreational function concern the establishment of recreational areas and preserving the existence of the world of peasant farmers. Land-use zoning plans of the communes provide for recreational zones. The provincial government has the option of guiding land-use planning in the communes by designating areas as agricultural zones and by withholding approval of the zoning plans.

Under the Nature Conservation Act certain areas can be designated as national parks, nature reserves, protected landscapes, rest areas, local nature reserves or protected areas. At the end of 1985 the total area protected in the Tyrol was about 213,000 ha or 17 per cent of the total area. The Tyrolean Landscape Service, which is organised by the forest administration, is quite successful in repairing damage to the landscape and in enhancing valuable features of the scenery. Nature conservation counsellors participate in land consolidation action and in technical projects which have an impact on the landscape.

The importance of agriculture to the recreational function consists not only in tending and conserving the landscape by keeping certain areas open (through mowing and grazing) and in maintaining farm buildings, alpine pasture huts and the road network, but also in preserving the existence of the peasant farmers, their working conditions and customs. The federal and provincial governments and many communes make a financial contribution to safeguard the recreation function of the world of peasant farmers.

By raising farm income and improving the infrastructure farmers are encouraged to stay on their farms. Dairy farming and cattle breeding are supported by guaranteed prices and subsidies. The region also benefits from various investment grants and special loans that help to modernise agriculture. But the agricultural marketing system and the promotional programmes have not succeeded in securing an adequate level of income for mountain farmers who work under very harsh conditions. In these areas net agricultural incomes are still less than half industrial wages. Given the present surplus, a general increase in the prices of milk and cattle will only aggravate the agricultural surplus problems in Austria, and the more intensive utilisation of the most suitable areas is not desirable from an ecological point of view. Moreover, further development such as land levelling and drainage of wetlands would impair the natural scenery. Promotional programmes have therefore shifted to direct payments and measures to improve the infrastructure by facilitating the farmers' participation in tourism and other economic activities. Direct federal payments (up to Sch 15,000 per farm) are intended to safeguard the existence of farms in inhospitable regions, regardless of the volume of landscape tending services rendered.

There are also direct income payments for conserving the landscape and maintaining traditional farming practices. Special management payments of the province of Tyrol, for example, are disbursed for mowing and grazing of steep slopes. These payments are graduated by the degree of adversity and the number of cattle. Payments are also granted for managing dairy alpine pastures. Similar grants are paid by communes. Other maintenance services, such as mowing and grazing larch meadows, ski slopes and other aesthetically appealing meadows are carried out or financed by the provinces, the communes, local tourist boards, lift companies and hotel owners.

Over the last thirty years the Tyrolean landscape has lost much of its visual attractiveness: arable land and meadows in the valley bottoms have been built over; new road, electric power lines, ski slopes, structures for controlling avalanches and torrents are now highly visible landmarks. Agriculture itself has reduced the diversity of the landscape by converting half the arable into permanent grassland and by abandoning or afforesting grassland in unfavourable regions. The number of farm enterprises, however, has declined by only 13 per cent since 1960 while in the rest of Austria it has declined by 25 per cent. The existence of mountain farms has been secured mainly by integrating them directly or indirectly into the tourist industry.

The economic situation of the Tyrolean mountain farmers would not be affected very much by the recent pledge by OECD Ministers to 'a progressive reduction of assistance to end protection of agriculture' across all commodities. The farmers in the higher mountain region benefit only from the price support payments for milk; a mere sixth of their total income depends on dairy prices. In the future, direct payments to farmers for conservation measures should be increased and the participation of farmers in the tourist industry should be promoted. More funds should be applied to strengthening the infrastructure of the tourist sector or should be converted into direct payments for conservation services. In this way the farmer is encouraged to make the transition from the primary to the tertiary sector (Lichtenberger, 1979). Environmental concern should also play a larger role in the design of support programmes. The participation of conservation counsellors, mandatory now only in land consolidation, should be extended to other programmes.

The incorporation of peasant farming into the service sector is taking place more or less automatically in the highly developed tourist centres. The beneficiaries pay indirectly, through local fees and lift tickets, for the management of ski runs and recreational areas. But landscape conservation in valleys not easily accessible to tourists cannot be financed by the beneficiaries alone. In some quarters, 'soft tourism' has been recommended as a route to an environmentally favourable regional economic development (Keith, and Dörr, 1985). 'Soft tourism' appeals to certain groups (Spiegler, 1984), but the direct earnings generated will certainly be too small to safeguard the existence of mountain farms. If society wishes to preserve these isolated valleys in their present state and to protect them from mass tourism, the incomes of mountain farmers and thus the preservation of the landscape must be financed by general tax revenue. This implies a redistribution of income towards those who prefer museum-like alpine scenery, just as expenditure on museums and other cultural attractions has certain distributional effects.

References

Amt der Tiroler Landesregierung, Abteilung Umweltschutz (1986), *Tiroler Umweltschutzbericht 1985*, Innsbruck.
Androschin, M. (1971), *Die landwirtschaftlichen Maschinenringe in Nord- und Osttirol*, Beiträge zur alpenländischen Wirtschafts und Sozialforschung, 120, Innsbruck.
Aulitzky, H. (1975), 'Zur Veränderung der Landschaft in den österreichischen Alpen', in W. Danz, *Die Zukunft der Alpen*, I, Schriftenreihe des Alpen-Instituts, 4, Munich.
Bätzing, W. (1984), *Die Alpen*, Frankfurt.
Bätzing, W. (1985), *Bad Hofgastein. Gemeindeentwicklung zwischen Ökologie und Tourismus*, Institut für Stadt- und Regionalplanung der TU Berlin, discussion paper 20, Berlin.
Bericht über die Lage der Tiroler Land- und Forstwirtschaft, 1984–1985.

Brugger, E., Furrer, G., Messerli, B., and Messerli, P. (1984), 'Welche Politik für das Berggebiet?, in E. Brugger *et al.*, *Umbruch im Berggebiet*, Berne and Stuttgart.

Brunner, G. (1977), 'Mäglichkeiten zur Sicherstellung der Flächennutzung in landwirtschaftlichen Problemgebieten', dissertation, ETH, Zürich.

Bundesministerium für Land- und Forstwirtschaft (1972, 1985), *Bericht über die Lage der österreichischen Landwirtschaft.*

Bundesministerium für Land- und Forstwirtschaft (1988), *Tätigkeitsbericht.*

Creditanstalt, Tirol (1986), 'Die Österreichischen Bundesländer', Extra Series, 3, Vienna.

Donaubauer, E. (1979), 'Die Notwendigkeit standortgerechter Wildstandsbewirtschaftung aufgrund der Forstinventurergebnisse 1971/1975', in R. Kreisl, *Standortgerechte Wildbewirtschaftung*, Schriftenreihe des agrarwirtschaftlichen. Instituts des Bundesministeriums für Land- und Forstwirtschaft, 30, Vienna.

Greif, F., and Schwackhöfer, W. (1979), *Die Sozialbrache im Hochgebirge am Beispiel des Außerferns*, Schriftenreihe des agrarwirtschaftlichen Instituts des Bundesministeriums für Land- und Forstwirtschaft, 31, Vienna.

Haefner, H., and Günter, Th. (1984), 'Landnutzungswandel und ökologische Veränderungen im Schweizer Berggebiet', in E. Brugger, *et al.*, *Umbruch im Berggebiet*, Berne and Stuttgart.

Hoppichler, J., and Groier, M. (1986), 'Schafhaltung in Österreich', *Der Förderungsdienst*, 11.

Kastner, A. (1985), 'Waldschäden in Österreich', *Der Förderungsdienst*, 6.

Kastner, M. (1981), 'Methodische Ansätze in der Landschaftsbildbewertung', in M. Kastner, *Stellenwert des Landschaftsbildes im Natur- und Landschaftsschutz und seine Bewertung*, Seminarbericht des Instituts für Landschaftsplanung und Gartenkunst, Technische Universität Wien, Vienna.

Keith, W.J., and Dörr, H. (1985), 'Sanfter Tourismus'—Impuls für die regionale Entwicklung?', *Boku Raumplanung*, 'Extrakt', series 11, Vienna.

Kreisl, R. (1982), *Regionale Waldausstattung in Österreich*, Schriftenreihe des agrarwirtschaftlichen Instituts des Bundesministeriums für Land- und Forstwirtschaft, 38, Vienna.

Krysmanski, R. (1971), *Die Nüstzlichkeit der Landschaft*, Düsseldorf.

Lichtenberger, E., 'Die Sukzession von der Agrar- zur Freizeitgesellschaft in den Hochgebirgen Europas', in P. Haimayer *et al.*, *Fragen geographischer Forschung*, Innsbrucker geographische Studien, 5, Innsbruck.

Meleghy, T., Preglau, M., and Walther, U. (1982), *Die Entwicklung Obergurgls vom Bergbauerndorf zum Tourismuszentrum, sozialhistorische Analyse und evolutionstheoretische Rekonstruktion*, Forschungsbericht des Instituts für Soziologie der Universität Innsbruck 18, Innsbruck.

Pevetz, W. (1978), *Struktur und Motive von Urlaubern auf österreichischen Bauernhöfen und Einstellung zur Landwirtschaft,*Schriftenreihe des agrarwirtschaftlichen Instituts des Bundesministeriums für Land- und Forstwirtschaft, 27, Vienna.

Pevetz, W. (1986a), *Landwirtschaft, Agrarpolitik und Umweltschutz in Österreich*, Bericht der Bundesanstalt für Agrarwirtschaft, Vienna.

Pevetz, W. (1986b), *Entschädigungen der Landwirtschaft für Einschränkungen und Auflagen im Interesse des Natur-, Landschafts- und Wasserschutzes*, Monatsberichte über die österreichische Landwirtscahft, 7.

Schemel, H.-J., and Ruhl G. (1980), *Umweltverträgliche Planung im Alpenraum*, Munich and Berlin.

Schermer, M. (1983), 'Die Förderung der Landwirtschaft durch die Gemeinde in Tirol', dissertation, Universität für Bodenkultur, Vienna.

Schnitzer, R. (1973), 'Naturschutz gegen Almsanierung', *Der Alm- und Bergbauer*.

Spiegler, A. (1984), 'Sanfter Tourismus'—Chance für 'typische österreichische Ferien', *Umweltschutz*, 5.

Tschurtschenthaler, O. (1985), 'Probleme des Sommerfremdenverkehrs in den Alpen', in W. Hämmerle, *Problem Sommerfremdenverkehr in den Alpen*, Institut für Verkehr und Tourismus, Schiftenreihe B, Innsbruck.

Tyrolean Environmental Protection Report (1985).

9. Changing landscapes and land-use patterns and the quality of the rural environment in the United Kingdom

J. Bowers and T. O'Riordan

The aim of this chapter is to show, with reference to two case study areas:

1 How in terms of agricultural practices the needs of the environment can conflict with those of farmers seeking to maintain and improve incomes from their farms.
2 How the mix of agricultural and other policies bearing on farming activities is being and has been modified to reduce or eliminate the conflict.
3 The problems that arise during the phase of transition to integrated policies which harmonise farm income support with a sound environment.

In order to do this it is necessary to explain:

1 How and why attitudes are changing towards the incorporation of environmental values and action within agricultural management practices that are themselves undergoing an important transition.
2 The influence of recent developments in the provision of information and specific advice geared to integrating environmental considerations with good agricultural practice.
3 The role of particular national or local policies geared specifically to improving this relationship, notably policy measures of inducement (income assistance, including tax relief, management agreements, compensation, advice) and those of compulsion (designation, planning controls), in relation to different patterns of land ownership.

Definition

The words 'environment', 'conservation', 'natural history', 'wildlife', 'landscape', 'scenic beauty' tend to be muddled up and at times used interchangeably. This looks like sloppiness, but there can be good reasons for it. Relative simple definitions will suffice.

Landscape refers to the visual character of the land, including its intrinsic beauty, historical features and evidence of past and present land management. *Conservation* applies to the sustained management of natural resources, notably wildlife, soil and water systems so that the land is capable of being replenished for continuous use and enjoyment. *Environment* is essentially the whole concept of conservation, beauty, history and economy. It is usually applied to the values of sustained enjoyment of the land and its natural resources.

Background to the case studies

'There has never been an age, no matter how rude and uncultivated, in which the love of the landscape has not been in some way manifested'. So wrote John Constable, probably the greatest English landscape painter, in the early nineteenth century. He was speaking for his fellow cultivated Englishman, but not for Englishmen of all time. The love of landscape is a cultural phenomenon some 300 years' standing. Prior to that there was no English word for either 'landscape' (introduced in the seventeenth century by the Dutch from their *landskip* or 'scenery'. The landscape movement came of age in the early eighteenth century when men of vision created the park estate, and the agricultural revolution gave capital and prestige to land ownership.

It is therefore part of British folklore that farmers are stewards of the land. In 1942 an official inquiry into the future of rural land use under the chairmanship of Lord Justice Scott concluded, ' . . . the cheapest, and indeed the only way, of preserving the countryside in anything like its traditional aspect would be to farm it'. In a 1981 parliamentary debate an Agriculture Minister proclaimed, ' . . . the countryside is a living, ever-changing entity, and I know of no greater conservationists than those people who live in it and who eke a living from their work within it'.

There is much truth in this statement. Prior to 1870, when Britain was in competition with grain-producing colonies, vast areas were converted from pasture to arable production. In the century before that, enclosure transformed the face of the countryside, much to the dismay of the lovers of a pre-enclosure landscape.

Historians, however, recount that hedge-fringed fields were by no means a feature of the early nineteenth century. Many Tudor estates of the fifteenth and sixteenth centuries removed the ancient Anglo-Saxon strip fields to pasture their stock. And after 1870, when agriculture passed through a prolonged depression, neglect was almost universal. Much of what the British now regard as important semi-natural habitats—moorland, heath, marsh, bog and downland—flourished during this period. Large numbers of hedgerows grew rank, but wildlife and wild flowers proliferated. Parsimonious manuring practices maintained the special ecological conditions necessary for the survival of flower-rich meadows, which thrive on very low-nutrient regimes.

In many respects, therefore, neither agriculture nor the rural scene has stood still. There is no such thing as a 'traditional' landscape: these are nostalgically imagined landscapes of childhood and yesteryear. The fascination with landscape and natural history, which in some European eyes amounts almost to an obsession, is an outcome of British intellectual, economic and social history. This is why the agriculture–conservation link is such a 'political' issue in contemporary Britain. It has deep roots, it arouses cross-party support, and it generates powerful emotions.

Changing policies towards agriculture and the environment

After 1945 the fortunes of British agriculture changed rapidly. Food production was regarded as a strategic and economic necessity. While a comprehensive system of land-use planning was introduced in 1947 to regulate the pattern of development, few equivalent public controls extended to agriculture or to forestry.

This distinction is vitally important. The planning system is widely acclaimed as one of the most important determinants of built land-use patterns in the contemporary rural scene. No building used for commercial purposes and no new houses can be constructed without permission from the local planning authority. In general no compensation is paid if permission is not granted. There is an elaborate but well established quasi-judicial system of appeal where planning permission has been refused, granted subject to conditions or no decision has been taken (see 3.4.1 in the United Kingdom's country information paper).

Such was the support for and faith in the benign hand of farming in 1947 that most agricultural activities were exempt from these planning controls, with a few notable exceptions that will be discussed below. Alterations to fields, field boundaries and natural habitats could take place without notice or reference to a publicly accountable authority. More substantive operations on agricultural land, such as building construction or engineering works, constitute development but some are permitted under the General Development Order. In 1950 the Landscape Areas Special Development Order (LASDO) extended additional discretionary controls to the design and external appearance of buildings in some national parks. Compensation was payable only where permitted development rights are removed, but the LASDO has usually been used only to attach conditions to development permission, so compensation was not normally involved. The LASDO controls were discretionary, however, and did not apply to roads. In 1986 the LASDO controls were extended to cover the design, external appearance and siting of farm and forestry buildings and roads in all national parks in England and Wales under the Agricultural and Forestry Development in National Parks, etc., Special Development Order (AFSDO). However, application of these powers to a particular farm would require compensation for any denial of income involved. Nor is there any compulsion in its use. AFSDO is a voluntary arrangement, at the mercy of negotiation, goodwill and compromise.

Farmers were essentially trusted to care for their land and the landscape. Self-regulation based on a voluntary code remains overwhelmingly the British philosophy in the regulation of land use, so is given some prominence in this report. It is a matter of party political controversy today.

Post-war governments subsidised product prices so as to encourage the growth of output and implemented a number of policies, including subsidies on capital investment, farm management advice, the promotion of agricultural technology and research and development designed to raise agricultural productivity. Initially farmers were guaranteed prices and markets for most major agricultural commodities except horticultural produce, where seasonal tariffs were the normal protectionist device. From 1973, when Britain joined the European Community, policy for British agriculture became part of the Common Agricultural Policy. From then on, an important dichotomy developed.

Agricultural policy was largely dictated by Community policy. Price support, farm development grants, and special subsidies for upland farms were, for the most part, determined by negotiations among the Community Agricultural Ministers. Admittedly there was a United Kingdom financial element to these policies, but most of the United Kingdom programmes conformed to Community regulations.

Environment policy, on the other hand, is largely determined by member states. In Britain it is the responsibility of the Department of the Environment, for England, together with its Welsh, Scottish and Northern Ireland counterparts. Working within and alongside the Environment Department are the statutory agencies responsible for the countryside, nature conservation and archaeological and historical monuments. In England, where the two case studies are situated, these agencies, respectively, are the Countryside Commission, the Nature Conservancy Council and the Historic Buildings and Monuments Commission.

This brief history provides three lessons for the case studies that follow:

1 Landowners, especially owner-occupiers, have long been interested in the landscape heritage of their farms, even as agricultural practices have dramatically altered. That ethos is now reasserting itself, and is a key factor in the case studies that follow.
2 Agriculture is essential to the peculiar combinations of wildlife and landscape that the British so love. Admittedly there are some semi-natural and wild habitats where agriculture has no place (the foreshore, bog, ancient woodland). But the majority of nature conservation sites *depend* on certain kinds of agricultural management for their survival (grassland, moorland, heathland, marshland, woodland).
3 Land-use change within agriculture has taken place, until recently, almost entirely on the initiative of the landowner or occupier. Although policies are changing, for nearly 90 per cent of the countryside this is still the case. Substantial public accountability over what a farmer does to the land is confined to highly protected Sites of Special Scientific Interest (SSSIs) covering some 8 per cent of the land. This means that land designation, ownership of the land, conditions and money attached to

management agreements, and the influence of advice and encouragement, are important influences when public intervention in private agricultural practice is contemplated.

Changing policies towards agriculture and conservation

Three factors have served to change policies towards agriculture and conservation since 1975:

1 The Common Agricultural Policy proved to be too successful in promoting production. Food surpluses dominate the political scene, and the open-ended nature of price guarantees has caused severe budgetary difficulties for the Community and for certain member States. Overproduction and over-protection in agriculture have become politically unpopular—at a time when farm incomes are falling and indebtedness is rising.
2 Agriculture was blamed as largely responsible for the remarkable postwar alterations to habitats and landscapes depicted in figure 9.1. Other factors, such as urban development, neglect and pollution have all contributed to this transformation. The official landscape and wildlife agencies, together with the politically well connected and active pressure groups, are lobbying very hard for fundamental changes in the policy linkage.
3 The Wildlife and Countryside Act, passed in 1981, while meritorious in many respects, is regarded by many as inadequate as regards the matters of compensation and site safeguard. The compensation principles applying to management agreements through which an owner or occupier agrees to regulate land use in the interests of wildlife and/or scenic features have been criticised because:
   (a) They apply to 'profit forgone' which essentially provides landowners and occupiers with risk-free incomes over the life of the agreement, based on some subsidised inputs and some guaranteed prices. This is regarded by many as iniquitous even though it is equitable. The official policy line is that no landowner or occupier should be disadvantaged in relation to neighbours when being encouraged to meet the public interest.
   (b) They are, for the most part, voluntary, so the onus is on the wildlife or landscape conservation agency to settle, and for the recipient of the agreement to drive a hard bargain. the final terms are reviewed by the district valuer, and there is no hard evidence of serious abuse. But concern over certain highly publicised agreements remains. When information, advice and a conservation ethos are lacking, management agreements can be awkward and possibly expensive to complete.
   (c) The payments come from the purse of the 'environmental' agencies, which are, in effect, offsetting agricultural expenditures. This tends to make the cost of conservation to the nation appear much higher than it actually is. So long as demand-led agricultural subsidies

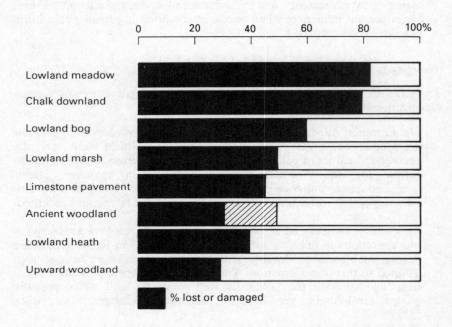

Figure 9.1 The loss of wildlife habitat in England and Wales since 1945. Public opinion in western Europe favours the retention of ecological diversity and areas of landscape value even if it means greater public expenditure. The redirection of agricultural support payments would be a preferable option

exceed budget-restricted conservation agency payments, management agreements on a 'profit forgone' basis are regarded by critics as a second-best solution.

(d) There are no restrictions on how the recipient can use the 'conservation' money. It is possible for some or all of it to be diverted to environmentally damaging operations on other land under the farmer's control

These safeguarding measures are regarded as inadequate because:

1   In practice they apply to certain designated areas, notably sites of Special Scientific Interest, and national parks. While it is possible for local authorities voluntarily to enter into similar agreements anywhere on the wider countryside, very few have done so because of budget limitations and a dislike of the guidelines underlying the payments.
2   There is very limited 'last resort' compulsion. Indeed, only for certain cases in a relatively small number of SSSI's is real force possible.
3   Compulsory purchase powers, in the event of irresolvable disagreement, are, in effect, unavailable except for a few internationally important

national nature reserves. In practice these powers are not used because of the commitment to the voluntary principle.

4   While, prior to 1980, the Agricultural Development and Advisory Service of the Ministry of Agriculture had to be notified in advance of an impending farm development project (e.g. drainage) before the award of grant aid, between 1980 and 1985 this prior notification no longer applied outside SSSIs and National Parks. Since 1985 two new schemes for agricultural development—the Agricultural Improvement Scheme of 1985, which became the Farm Conservation Grant Scheme of 1988—both require advanced notification of intent. At present this scheme is extending only slowly across the farming community. This is because the Farm Conservation Grant Scheme is only a grant aid. Farmers who have a limited income, or who are burdened by debt, cannot afford the high cost of support payments for capital expenditure (on such items as slurry and silage effluent control), restoration of farm buildings in traditional materials and the protection of upland woodland by stock-proof fencing or walls).

All this controversy has led to some important policy changes. By way of summary, the main aspects are the introduction of environmentally sensitive areas (ESAs), the extension of conservation advice on farm management and further changes in the provision of statutory controls.

All these policy changes combine in a significant shift in the pattern of financial and advisory signals guiding agriculture. Not only has the Common Agricultural Policy changed, so too has the total amount of grant aid available for farm improvement and the activities that are eligible. The money and the authority for farmers to drain, convert land or construct buildings unsuitable for the local landscape or habitat are now much reduced. True, this has not stopped such activity, but it has caused much more of it to be paid for by the landowner or occupier. During a period of falling incomes these harsher conditions have contributed to what seems to be a slowdown in the pace of alteration depicted in figure 9.1. Detailed official information on the changing pattern of alteration is still to be published. Preliminary indications from the Nature Conservancy Council suggest that the rate of loss is still very considerable, and that farm-induced damage to the 'protected' SSSIs is still serious.

At the same time, policies are changing towards afforestation, the felling of trees, the planting and management of broadleaved woodland, tourism, recreation, and non-agricultural income generation in farms. Forestry agencies and private companies are under pressure to balance better the relevant interests of timber management and wildlife–landscape considerations. Similarly tourism is regarded as an activity that benefits from conservation investment in scenic areas. Over a fifth of farmers' income in many of the national parks comes from tourism. Increasingly, too, farmers see income opportunities in opening up their land to public access, through management agreements with the local authorities, backed by grant aid from the Countryside Commission, or via special commercial arrangements. There is now serious interest in reorganising the pattern of age-old rights of way so that farmers can encourage the public to come into areas specifically managed for recreation, notably walking, shooting, picnicking and educational enjoyment. Such arrangements would not diminish effective rights of access and should enhance total recreational pleasure. Present grant aid and advisory programmes are not quite ready for what could be an explosion in public access on to private land.

Public access is an intrinsic feature of contemporary British environmental policy in the countryside. Many grant aid schemes offered by the conservation agencies are contingent on public access being provided. Public access in its own right is also eligible for grant aid. Scenic beauty and wildlife or archaeological/historical interest combine to increase public enjoyment, and are the focus of information and educational programmes on many farms.

The choice of case studies

The two case studies selected for the analysis that follows are: (1) East Anglia, with special reference to the Broads marshes; (2) the northern Pennines, with special reference to the Pennine dales Environmentally Sensitive Areas. These two areas are complementary in a number of respects. They both contain a variety of land designations, including SSSIs, National Park or National Park equivalent, and Environmentally Sensitive Areas, as well as important non-ESA-designated areas. So comparisons can be made across the designation boundary.

While East Anglia is a rich, productive lowland, with a predominance of arable output, the north Pennines are far less productive uplands, including large areas of rough grazing, where livestock rearing predominates.

Agricultural incomes, turnover and indebtedness per farm are generally much higher in East Anglia, profit margins generally much lower in the north Pennines. In both areas farm incomes are falling and borrowing is increasing, though indebtedness depends much more on ownership and past investment decisions than on geography.

As a very broad generalisation, East Anglian farms are likely to continue to rely primarily on agricultural output as their main source of income and activity, while north Pennine farms will probably diversify more, where possible into tourism, plus, in some areas, conservation, and in some cases providing local services, such as farm-based education.

The emphasis on borrowing and scope for income supplementation is deliberate. Conservation investment is less likely on an impoverished farm or where future returns look very shaky—unless forthcoming national policy measures connect income support with conservation integration.

The East Anglian region and the Norfolk and Suffolk Broads

East Anglia comprises the counties of Norfolk, Suffolk, Essex, Cambridgeshire and parts of Lincolnshire. This is lowland eastern England, characterised by low rainfall, rich soils (for the most part) and intensive arable production. The average farm size, at 142·6 ha, is a little above the national average of 125 ha. Real net farm income is £30,000 per year at current prices, though for farms over 500 ha income exceeds £140,000. For owner-occupiers, net worth was calculated at around £5,500 per hectare, with total external liabilities £500 per hectare (1985 prices). The shift towards cereals has been dramatic in recent years. In 1952 47 per cent of land was under cereal, compared with 69 per cent today. While 34 per cent was in fallow or grass in 1952, only 9 per cent is so today. Productivity is consistently high, with winter wheat yields averaging 6·88 tonnes per hectare, up by 2·47 tonnes per hectare since 1948.

There are no regional statistics for alterations or losses to 'valued' landscapes and natural history features in the East Anglian region. Recent evidence suggests that the rate of hedgerow loss, estimated at one time to be 2 per cent per year, is now less than 1 per cent annually and still falling. Some two-thirds of small woodlands are in a state of neglect, though this is common elsewhere in England. About a quarter of broadleaved woodland is used for sport, mostly pheasant rearing. Though this does not guarantee special attention for wildlife, the landscape value of such woodland is significant.

The major cause of change has been drainage, both improvements to on-farm drainage and major arterial drainage, financed by internal drainage boards. Some of the on-farm investment has been aimed at improving grass yields, mostly for silage production. Most of the arterial schemes, which were eligible for Community grant aid (until 1980) and Ministry of Agriculture grant aid (though rates have fallen steadily since 1983), were geared to the conversion of grass to arable production (about 60 per cent of the whole) and to intensive silage production for dairy units (about 30 per cent of the total).

The Norfolk and Suffolk Broads case is selected primarily because of the arterial drainage issue. 'The Broads' is a regional term applying to the river valleys of a number of east Norfolk and north Suffolk catchments draining out to the North Sea at Great Yarmouth. These valleys consist of three kinds of 'scapes'. First there are the waterscapes of the rivers, the broads (flooded medieval peat cuttings) and the drainage dikes (ditches). The whole area lies below high tide, so both the coast and the river margins have to be protected by floodwalls. Drainage is therefore vital for any agricultural activity. Most of the drainage is now operated through electrically driven pumps, though gravity drainage still takes place in some upper valleys. Formerly drainage was provided by wind, the disused windpump being of considerable historical interest, and a valued landscape feature.

Second, there are the marshscapes of fen and wet meadow and fen woodland (carr). Here the water table is at or near the surface, so no agricultural activity takes place. The fen vegetation, dominated by reed and sedge, needs to be cut and protected from invasion by scrub but is of great importance for wildlife, and for the commercial thatch industry. The carr woodland is the climax vegetation and is, for the most part, unmanaged. Very little wet meadow remains but what there is can be of great value for overwintering bird populations.

The third 'scape' is the drained marsh landscapes of high water table grazing meadow and deep-drained arable and intense farm production. These two landscape types dominate the region, accounting for 60 per cent of the 20,000 ha of river valley. Between 1970 and 1984 drainage improvement removed some 800 ha annually from the high water table marshes for intensive cropping. The steady but geographically unpredictable conversion of high water table meadow gave rise to great local and national concern during the period 1976–84 and led to a shift in both Community and national policy.

The Broads form an incomparable area of wetland habitat of both national and international importance. The fen is noted for bittern, marsh harrier, hen harrier and bearded tit, all national rarities. The grazing marshes receive important overwintering populations of bean geese, widgeon, Bewick swan and other arctic migrants. Nesting species include redshank, greenshank, yellow wagtail and snipe. Floristically the Broads grazing marsh dikes (drainage ditches) harbour about 30 per cent of all the freshwater aquatic species in the country. There are some national rarities. Amongst the invertebrates, the swallowtail butterfly and the Norfolk (Aeshna) dragonfly are very special to the region. But the assemblage of higher aquatic plants is hardly paralleled in Britain.

The north Pennine dales

The northern Pennines are carboniferous limestone hills, sometimes with a capping of gritstone, covering parts of the counties of North Yorkshire and Durham and, in the north-west, Cumbria. Their tops range in altitude from 300 m to 730 m and consist of grass and heather moorland utilised mainly as rough grazing for sheep, with the production on the heather of red grouse (*Lagopos 1. scoticus*). These hills are intersected by a number of river valleys known as dales, and it is these that form the subject of the case study. The principal dales in North Yorkshire are Wharfedale, Arkengarthdale, Swaledale, Langstrothdale, Waldendale and Wensleydale, this latter, having somewhat different agricultural characteristics, being excluded from the study area. In Durham are Teesdale and Weardale and in Cumbria Dentdale and Deepdale. The Yorkshire Dales National Park covers much of the dales in North Yorkshire. The more northerly ones, together with the Durham dales, form the recently designated North Pennine Area of outstanding Natural Beauty (AONB). These two designated areas encompass both the dales and the adjacent uplands. The newly designated North Pennine Dales Environmentally Sensitive Area comprises eight blocks drawn from the dales mentioned but excluding the surrounding uplands. The case study focuses on the ESA and thus does not consider the hills as such. Most of the information that it draws on relates to the National Park or the AONB. Statistical information on the ESA as a unit is largely, for the present, lacking.

The dales that constitute the ESA are all limestone valleys with characteristic scenery. They have an enclosed landscape of stone walls and many stone field barns, with scattered farms and predominantly small villages. The walls frequently extend beyond the sides of the dales on to the adjacent hills, enclosing areas of the moorland as allotments. The streams in the lower parts of the dales are frequently lined with deciduous trees and there is a scattering of trees around buildings. But the landscape, particularly in the higher parts of the dales, is largely devoid of timber. A network of stone walls, appearing often startlingly white, dominates the scene. Small relics of woodland often serving as woodland pasture and thus bare of under-storey occur in parts of the area. Small blocks of dense coniferous plantations are found on the valley sides or can be seen on the skyline of the upland, but typically the dales are devoid of timber and complement the bare uplands in which they are located.

The principal activity in the dales is farming, and the landscapes and landscape features that the various administrative designations exist to conserve are essentially artefacts of the agricultural system. The farming system is generally pastoral and the dales consist of a mosaic of different types of grassland: rough grazing on the allotments, with better pasture, hay meadows and occasional wet areas in the valley bottoms. It is the grassland that the ESA exists to conserve, together with the dominant features of stone walls and field barns. Of particular value are the hay meadows which constitute in total the major examples of a scarce, fragile and rapidly disappearing habitat type. These meadows are normally small, of at best only a few hectares in extent. They constitute an important landscape feature, providing in season an abundance of flowers, but their real importance is probably a scientific one as a highly diverse series of species-rich neutral grasslands whose diversity derives from subtle differences in grazing regimes and nutrient sources. Because of this, many of the meadows are scheduled as Sites of Special Scientific Interest.

The ecological interest of these hay meadows and the attendant pasture, while primarily botanical, is not confined to the flora. The variation in species diversity is associated with variation in density and variety of invertebrate fauna. The hay meadows and the wetter areas also provide breeding grounds for a number of wading birds, principally curlew (*numenius aquata*), redshank (*tringa totanus*), lapwing (*venellus vanellus*) and snipe (*gallinago gallinago*). Along the streams additionally are found common sandpiper (*tringa hypoleucos*) and occasionally oystercatcher (*haematopus ostralegos*). Passerines, however, are few, principally wheatears (*oenanthe oenanthe*) in the stone walls, meadow pipits (*anthus pratensiis*) and yellow wagtail (*motacilla flava flavissima*) in the pasture, whinchats (*saxicola rubetra*) on the valley sides and dippers (*cinclus cinclus*) on the streams. Where the allotment meets the moorland ring ousels (*turdus torquatus*) are found. The dales additionally provide hunting for the declining population of merlins (*falco columbarious*) that breed on the moors. In parts where relic pasture woodland is on the valley sides black grouse (*lyrurus tetrix*) occur.

The other feature of the area for which conservation is desired are its archaeological remains. These are principally of two types, prehistoric agricultural and ceremonial remains (field systems, burial mounds, standing stones) and the remains of past industrial activity. The principal industrial activity was mineral extraction, particularly lead mining, which continued in parts of the dales well into living memory. Spoil heaps from lead mining are also interesting habitats with a unique flora, including plants that are otherwise confined to salt marshes. Some ares of these are also SSSIs.

The East Anglian case study

In East Anglia the scope for integration of environmental and agricultural policies rests on two themes:

A land designation approach: the wholesale protection of high water table grazing meadow in order to safeguard a regionally distinctive landscape type of national significance, and to introduce more wet meadow marshes in selected locations.

A whole farm approach: the incorporation of environmental management within farm planning so that:

1 Existing habitats and landscape features are retained and appropriately managed.
2 New habitats and landscape features are created in areas appropriate to the running of the farm.
3 New income-creating opportunities are seized, notably from grant aid, taxation relief, woodland output, recreation, and training for conservation-minded farmers and farm workers elsewhere.

The first relies on national policies that protect whole areas of land so that large numbers of adjacent farmers can be enabled to manage their lands to meet a national purpose. The second depends upon a much more farm-specific mix of policies encompassing incentives, advice and controls, so that farm planning can proceed with environmental conservation practices pro-grammed in from the outset. Although in this case national policies are still very important, the focus of attention turns to advice and income support rather than land-use planning and control. The East Anglian case study allows a comparison to be made between these two approaches to policy integration.

This section is divided into three parts, namely, the Broads, United Framlingham Farms, and two large estates (Mannington and Thornham) where whole farm conservation plans are being prepared by 'in house' advisers. The Broads provide an illustration of the application of national policies to a specific geographical unit. The other two examples signify the provision of advice, taxation relief and specialised conservation planning and marketing.

The Broads marshes

The peculiar features of the Broads have already been introduced. The first section also outlined the intent and criticism of agricultural and environmental policies that have had such a bearing on the Broads grazing marsh history. A little more detail is now necessary before policies can be evaluated.

The Broads have long frustrated those who believe in holistic land and environmental planning. The area is a complex mix of land, water and settlement that cannot effectively be administered by separate agencies. The links between water pollution, causing increased sedimentation, and, indirectly resulting in the removal of protective reed vegetation from the river banks, require that navigation, water quality management and conservation promotion are organised by a single responsible agency. Otherwise co-ordination between these essentially complementary functions will be awkward and, possibly, unnecessarily costly. At present management responsibilities are split between land and water; between the built environment and agriculture; between landscape beauty and nature conservation; and between short-term expediency and long-term cost-saving.

The Broads have always been administered by a number of different agencies, each with specialised responsibilities and functions, operating particular policies. In 1978 a partial solution to this problem was launched. This was the establishment of the Broads Authority, an amalgam of local authorities, the Countryside Commission (which provided about 40 per cent of the funding) and the navigation and water authorities. The rivers are navigable for both coastal craft and pleasure boats and are at present run by an independent statutory body, the Great Yarmouth Port and Haven Commissioners. The water functions of quantity, quality, recreation, drainage and flood protection are run by Anglian Water, which received its guidance, and grant aid, though two different departments, Environment (for water management) and Agriculture (for land drainage and flood protection).

The Broad Authority was charged with an overriding concern for protecting and enhancing wildlife and scenery. To this end it published a map depicting traditional or characteristic landscape units (figure 9.2). These consisted of sizable areas of each of the three 'scapes' already referred to, with the exception of intensive arable and grass production. The aim was to identify locations of scenic intactness, representing typical landscape assemblages deemed 'traditional'.

The open grazing marshes of the lower valleys were regarded as especially 'traditional'. They are regionally distinctive and have no parallel elsewhere in the UK. They evoke very special sensory pleasures—a vast, flat landscape; a big, expansive sky; horizons dotted with carr; church towers and windpumps; the criss-crossing of innumerable dikes; and the presence of grazing livestock on tussocky, tousled grass.

These open grazing marshes are important almost exclusively because of their landscape value. The wildlife interest was confined to a few marginal dikes, where spring-fed water encouraged the proliferation of diverse and numerous aquatic plants. In the more enclosed marshes of the middle valleys, the spring-fed dikes on the valley edges are of immense nature conservation interest. They harbour some 120 aquatic plants plus a great variety of invertebrates, including two rare species of dragonfly. Throughout the Broads marshes the grass swards are not ecologically interesting, no comparison at all with the north Pennine hay meadows. Bird life is also impoverished, because the marshes contain little standing water, but overwintering geese and ducks do graze some of the grasslands. All the important sites are either SSSIs or reserves managed by the Royal Society for the Protection of Birds or

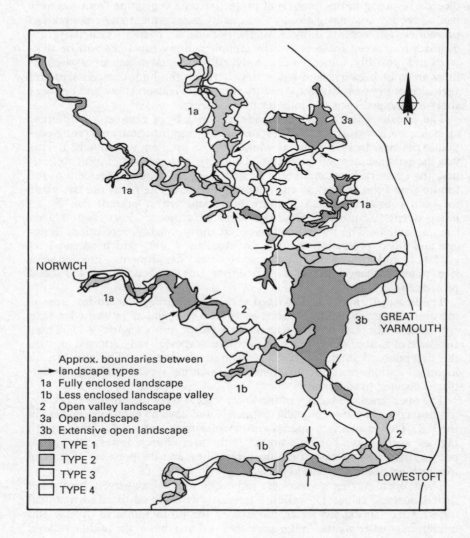

Figure 9.2 The landscape and types of vegetation in the region of the Broads. Dark shading applies to most representative areas

the Norfolk Naturalists' Trust. But landscape values in these very expansive marshes make very demanding requirements. Vast areas of contiguous land need to look the same and be managed under similar regimes. The stamp of the public interest has to be placed on the operations of large numbers of neighbouring farmers, irrespective of their personal predilections. The key issues for the Broads marshes are therefore land-use planning, controls over conversion, and how much and by whom compensation is to be paid for management agreements or restrictions on the use of agricultural land.

The pressure on farmers to convert their Broads marsh holdings were very great indeed during the period 1976–84. The European Community provided 25 per cent of the cost of arterial draining schemes through its farm structure policies (until 1980). The Ministry of Agriculture financed another 50 per cent of such schemes (until 1984). The high cereal support prices (calculated to be subsidised by at least 35 per cent) and reasonably lucrative livestock 'outgoer' schemes of the Common Agricultural policy meant that net farm profit could be increased by anything between £300 and £700 per hectare, depending on whether the farmer had an existing arable holding on the adjacent non-valley lands. Farmers with the capital, the skill and the equipment could have paid off all loans within two years. Additional subsidies of between 5 per cent and 10 per cent extended to seed, to fertiliser, to machinery and to other inputs. Even a traditional livestock farmer, without arable production but with enough land, could have made a return within five years. Almost all the advice going to farmers during the early part of this period was in favour of at least some conversion. The Agricultural Development and Advisory Service of the Ministry of Agriculture said so, the seed and fertiliser representatives said so, and the agricultural lobbies said so. Only the poor, the small-scale, the absentee landlords and the traditionalists chose not to convert.

Other factors influencing the rate of conversion during this period were the limited amount of access for heavy vehicles, the high water table and the tradition of letting out grazing marshes on annual tenancies. Many local farmers lease part of the marsh for a year to allow young stock to fatten up or to become ready for full dairy production. About 40 per cent of the open grazing marshes are let on annual tenancies, at prices ranging from £90 to £200 per hectare per year, depending upon access and the quality of the meadow. Some of these tenancies are run by large landowners. Many were let, through agents, by small-scale absentee landowners. This somewhat fragmented pattern of grazing ownership and management has contributed to the charm of the area. It also delayed the onset of major arterial drainage.

Before 1984 there was no agricultural policy that deliberately assisted farmers to retain low-productivity regimes. Almost all agricultural money worked in favour of intensification. The Wildlife and Countryside Act of 1981 complemented rather than challenged this position. Where a landowner or occupier was required to alter a predetermined management in favour of a pattern of operations that was deemed to be in the public interest, compensation was to be paid based on net income otherwise made. In 1981–83 about 3,000 ha of the open grazing marshes were threatened with major arterial drainage schemes and on-farm underdrainage. The only way to control this and encourage a landowner to retain 'old-fashioned' grass was for the Broads Authority to offer a Section 39 management agreement under the Act. The terms of this agreement were flexible, the payment (until 1984) being shared equally between the Broads Authority and the Countryside Commission. Since 1984 the Countryside Commission finances 75 per cent of such payments.

If a landowner refused to accept an agreement there was nothing the Boards Authority could do. However, Article 4 of the GDO (made under the 1971 Town and Countryside Planning Act) enables a local planning authority to issue a direction removing permitted development rights so that specific planning permission would be required. In the case of agricultural works, such a direction would take away the right to undertake engineering operations on agricultural land: it could not seek to control activities which constitute use of agricultural land, such as ploughing, since the use of land for agriculture is not covered by the planning system. Such directions need the approval of the Secretary of State before they can take effect: or the Secretary of State may make such a direction himself, as happened in one instance in the Broads. The direction cannot compel a landowner to adopt any specific farming practices: it merely requires that the kind of developments covered by it should be subject to prior planning permission. Should the local authority refuse to grant permission, the applicant can appeal to the Secretary of State; he may claim compensation if permission is then refused. In the one instance referred to above, a management agreement was eventually negotiated and the Article 4 direction was then lifted.

In 1985 the Broads were declared in effect a National Park under the provisions of Section 41 (3) of the 1981 Act. This arrangement means that all applicants for agricultural grant aid under the Community structural directives must first consult the Broads Authority before such claims are submitted. Prior to this designation no prior notification needed to be given.

In at least two cases, that arrangement has stopped land underdrainage that would almost certainly otherwise have taken place. Advice at a critical point in an investment decision can be of immense importance in harmonising agricultural and environmental policies. but that advice has to be backed up by suitable financial inducements.

The inappropriateness of the Act's provisions for safeguarding and compensation, over large and contiguous areas, led to the search for a new approach. But the Act also provided an interim solution. Under Section 40 the Countryside Commission can conduct experimental schemes that might have wider application. The commission proposed to launch such a scheme out of its own funds for the Broads open grazing marshes, but the Ministry of Agriculture found a 'catch-all' clause that enabled it to share the costs of the scheme with the commission for an experimental period.

On 1 April 1985 the Broads Grazing Marsh Conservation Scheme was launched. A total of £1·7 million was allocated over three years to pay any farmer on the designed areas (figure 9.3) £123 per hectare to retain 'traditional' livestock practices. The precise conditions are:

1   To maintain permanent grassland with grazing livestock for three years.

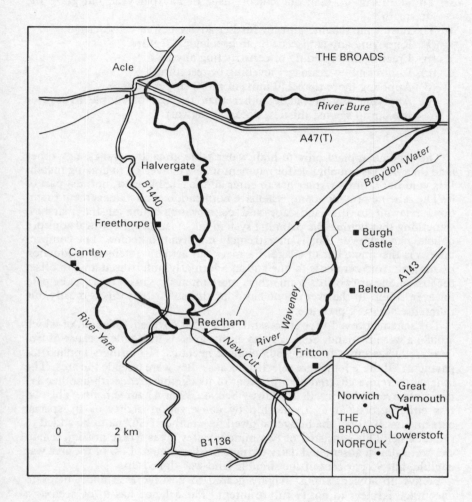

Figure 9.3   The area of the original Broads Grazing Marsh Conservation Scheme.
Source: based on the Ordnance Survey map, Crown copyright

2   To keep an average stocking density of between 1·25 and 3·75 livestock
     units per hectare during the period May–November.
3   To take no more than one cut of silage or hay per year and graze the
     aftermath.
4   To consult the scheme unit for further advice before considering:
     (a) Removing any landscape or archaeological feature.
     (b) Erecting any building or constructing any road.
     (c) Improving by drainage, levelling or reseeding.
     (d) Applying more than 250 units of nitrogen per hectare.
     (e) Applying any herbicide, other than the use of simple MCPA or
          mecoprop against thistles, docks or ragwort.

The scheme applied only to high water table grazing marshes: any other
land on a farm is not eligible for payment unless converted to grazing marsh.
It is voluntary, but a farmer has to enter all the eligible land, not just part of
it. The scheme provides comprehensive notification of all management intentions relevant to the landscape and conservation value of the marshes,
something that neither the planning system nor existing agricultural advisory
policies can achieve. This is an extremely important outcome. The compromise was the single cut of silage. Farmers can actually intensify production
by moving from zero cuts to one cut on a formerly unfertilised area to offset
the loss of one or two cuts from other, more intensive grass areas, yet be paid
for everything. In the event none has done this so far, but this is mostly due
to intense advisory pressure.

The scheme proved very successful. In 1985 it covered 4,874 ha, of which
4,008 ha was grassland. Some 200 ha of this grass is ineligible because of two
or more silage cuts. In the 3,803 eligible hectares 116 farmers applied for
payment. Of these four were rejected because they were outside the area. The
scheme therefore covered 89·5 per cent of its eligible area, with another 1·5
per cent covered by Broads Authority Section 39 or 41 agreements. The rest
was either protected by ownership by conservation bodies or by special
agreements. By 1986 the budget allowed for another 1,367 ha to be added to
the scheme. This consisted of two middle valley areas where grazing tenancies were almost absent and dairy farmers predominated. Less of the area was
eligible, but 95 per cent of the suitable land was signed up.

Follow-up advice, almost wholly geared to positive assistance, helps to
encourage farmers to notify future intent. The scheme has to be seen as a
source of counsel, not as a police measure.

Voluntary agreement adopted by the vast majority can expose a potentially
non-compliant minority to effective peer-group pressure. Publicity also
helps. No farmer ploughed any of the eligible marshes in 1985 of 1986. Only
minor infringements were reported on less than 7 ha of land. Two landowners seriously threatened to plough but one was bought out by a conservation organisation, the other settled for a special Broads Authority payment
of £50 per hectare to retain grass (but three cuts of silage) in a highly visual
area. The rest seem to be prepared to 'toe the line' even on a voluntary basis.
Much of this can be attributed to all the factors cited above. If compulsion is

required, it need only be used as a last resort. The proposed landscape conservation order operating within the scheme should prove adequate.

Some more land will be reconverted to grass as a result of the ESA scheme. This will include cereal land that is too wet for a food crop and some arable land that is suffering from clay deflocculation. This is a condition of soil instability created by deep drainage which results in dissociation of the clay fraction. This clogs up the drainage pipes and forms impermeable layers in the upper horizons of the soil. Soil deflocculation affects marine-laid soils in eastern England, notably in East Anglia and north Kent. About 100 ha would be involved over the next two or three years. In both instances the Ministry of Agriculture grant aid paid in the late 1970s helped to finance such conversion.

The provisions of the Wildlife and Countryside Act still apply within the scheme area. But to date no farmer has sought or obtained a Section 41 agreement. On one occasion this was hinted at, but the ADAS part of the scheme intervened and suggested that no conversion should take place because of deflocculating soils. This suggests that notification to the scheme unit prior to the consideration of any agricultural grant aid can act as a guide and controller of higher-level payments.

The Broads ESA

In August 1986 the Agriculture Minister announced that the whole of the Broads area and some parts of the river valleys extending westward from the Broads would be designated an environmentally sensitive area. The formal process was completed in January 1987, with inauguration on 1 April 1987.

The Broads ESA is a quite different proposition from the Grazing Marsh Scheme. It is wholly paid for by the Ministry of Agriculture and will be managed through an office in the Agricultural Development and Advisory Service (ADAS) to become the Farm and Countryside Services (FCS). It covers a much larger area than the scheme, taking in water meadows and areas of fen and carr. The kinds of agricultural enterprise that would be eligible are far more varied, and the potential damage or alteration of conservation value is drawn from a long list of possible action.

The management objectives are:

1   To maintain and support traditional livestock grazing on permanent grassland.
2   To conserve the landscape, wildlife, archaeological and other features which are responsible for the character of the traditional grazing marshes and wet meadows.
3   To encourage the continuation of grassland farming, especially in the upper river valleys, where the threat of abandonment is greatest.
4   To extend the area covered by traditional grazing marshes and wet meadows by encouraging the adoption of traditional farming practices on land now managed more intensely.

It is obviously premature to forecast the success or otherwise of the ESA compared with the Grazing Marsh Scheme. In any case, at the time of writing a number of important details remain to be settled, as follows.

Adequate administration of an ESA requires the attention of at least two full-time officials plus continuing advice from official and voluntary conservation bodies. It seems that the Farm and Countryside Service will be allowed to select only one individual for the task. This may mean that in the critical six months leading up to the onset of ESA its objectives and opportunities may not be adequately explained.

Two tiers of payments are being suggested, one equivalent to the Grazing Marsh Scheme (£125 per hectare per year) and one involving more specific management (£200 per hectare per year). The first payment covers maintaining the skeleton and fabric of the land—the dikes, woods, field gates and grassy swards. The higher payment is offered for particular conservation measures. These would include maintaining dikes to retain and restore aquatic vegetation, shallow flooding of marshes near river meadows for roosting and nesting birds, managing reed margins or woodland to protect landscape and wildlife features. Depending on what requirements are demanded, the relationship between these two sets of payments could be crucial, with many farmers seeking only the higher-tier payments. Here is where appropriate advice, in advance and on a continual basis, will be of the utmost importance.

Farmers may be able to exclude some of their eligible land within the payment regime. If this is the case, then it is possible that they will only voluntarily include certain areas, and intensify or otherwise alter the rest. This could destroy the very purpose of designation.

The cost of managing woodland and other specialised habitats may be greater than even the second-tier payments would cover. Yet the benefits would not be 'agricultural'. It is therefore possible that in some instances a third tier of payment, financed by the conservation purse, might be made available. This might well apply to SSSIs, but could take place on other valued features.

Meanwhile the provisions of the Wildlife and Countryside Act still operate. There is nothing to stop a landowner from seeking a Section 28 agreement (on an SSSI) or a Section 39 agreement (elsewhere) for all or part of the land. The major differences would be proof of intent, payment by a conservation agency or local authority, and a basis of payment that would be tuned to the farmer's requirements. Depending on the character of advice available within the local agricultural department, this need not be a problem.

By mid-1990 12,400 ha of marshland were covered by ESA agreements, including 90 per cent of the eligible grazing marsh. About a third is covered by Tier 2 agreements. An additional 10 per cent will be included following a scheme, financed by the Ministry of Agriculture, to raise water tables in a block of otherwise well-drained marsh. Over 80 per cent of the farmers have participated. Some 400 ha of arable land has been converted to grassland under the Tier 2 scheme. The total cost of the scheme if £1·7 million annually, with an administration cost of some £200,000.

Observations on the ESA experience

Farmers seem to prefer reasonably equivalent treatment, so long as it can be broadly justified. This is particularly important, as the scheme provides an element of justice where little exists under the Wildlife and Countryside Act. Under the scheme all farms are eligible for payment, irrespective of whether they can or cannot carry out potentially damaging operations, so long as they meet the conditions of the scheme. Under the Act only those who can prove they have the capacity and intent to take action receive management agreements. The poor, the traditionalists or those who simply find the inertia of doing nothing attractive do not get compensation. Yet in many cases they are the very people who are serving the public interest. The area-based scheme is more equitable and more successful at protecting large areas of open landscape.

The payment needs to be carefully fixed so that it makes agricultural as well as conservation sense. The £124 per hectare neatly approximated the higher costs of servicing drainage charges and falling grazing rentals. In effect it held livestock returns more steady during a period of falling real incomes, though even on the Broads marshes income fell by 20–30 per cent during 1985–89. Inertia is a powerful brake. However, it is possible that returns will be so low for marginal farmers, especially on the poorer-quality land, that some of it may be abandoned. This would result in vegetation quite out of keeping with the scheme's objectives. Reeds would flourish in the wetter areas, and thistles and ragwort in the drier. The future pattern of payments during a period of uncertainty over livestock returns and encouragement to farmers to get out of livestock production could be very important for the whole philosophy of the scheme.

Surveillance needs to be comprehensive but unobtrusive. Farmers need to know that they can get caught if they break the rules, but authority should not appear threatening or heavy-handed. Aerial photographic surveillance will help, and is required under the monitoring part of the programme. The crucial relationship is between the farmers, the project officer and the conservation community. The ESA scheme has created an atmosphere of much greater trust where self-policing has become a vital ingredient. Nevertheless the scheme operates only on a five-year payment regime. It will have to be very carefully renegotiated in 1992 if it is to continue to meet its objectives.

United Framlingham Farmers

United Framlingham Farmers was formed twenty-six years ago as a non-profit-making co-operative aimed at reaping the economic advantages of bulk buying. The co-operative consists of 350 members covering 65,000 ha, mostly in Suffolk. The co-operative is divided into a number of crop production syndicates. A syndicate is a group of twenty-five to thirty farmers who 'buy' a full-time adviser whose job it is to increase cereal yields. He suggests the seeds, the chemicals and the management. His income is paid for out of the profits: he earns his money by increasing the income of the syndicate.

In 1985 a small group of the UFF decided to hire a conservation adviser. They levied a rate of £1 per hectare on every participating farm, and topped up the £20,000 per year budget with grant aid from the Countryside Commission and a donation from a local agro-chemical supplier. This individual covers some forty-two farms and 10,000 ha. His job is to:

1 Walk the whole farm and indicate a conservation management strategy.
2 Search out all grant aid options, discuss them with the farmer, and complete the necessary forms.
3 Buy in bulk various conservation inputs, e.g. seedlings, wild flower seeds, equipment.
4 Train farm workers how to manage hedges, ponds and headlands.
5 Organise farm holiday visits for tourists, in league with a local hotel chain.
6 Sell his specialised service to other interested parties, including conservation organisations.

The UFF experiment is innovative and imaginative. It is essentially a self-financing operation aimed at marketing conservation output as well as integrating conservation within farming. Its aim was to create a multi-farm conservation enterprise with bias towards woodland marketing and tourist provision. About a third of the conservation syndicate also receive advice from other agencies, notably the farming and wildlife advisory group and the local authority. They tend to use this advice to supplement the suggestions of the conservation specialists, and frequently add their own imprint to the final management plan.

The UFF scheme is one model for the farm-based approach to integrating conservation and agriculture, benefiting from economies of scale. The disadvantages and dangers of this approach are as follows.

It depends on voluntary co-operation of a critical mass of farmers. There is a sub-threshold stage where there are insufficient funds to pay for the adviser and a super-threshold stage where there is too much work for one adviser to undertake but not enough money to pay for an assistant. The participating farms are also not adjacent to each other. This restricts the scope for neighbouring farm collaboration, which could bring important conservation and income benefits, and extend the operational effectiveness of the conservation adviser.

Farmers' expectations are high but the adviser may not be able to meet their demands. They look for a comprehensive plan, continuing input, income from their investment, and diversity of conservation output.

Woodland management presents a particular burden. Many of the woodlands are ill managed and need specialised advice. Most are small (less than 25 ha) and scattered. It is difficult to organise a joint marketing arrangement or even to finance a woodchip burner to create on-farm fuel utilising the brash and thinning wood. Ideally the whole woodland operation requires a specialised assistant.

Inevitably the conservation adviser tends to offer similar suggestions for quite different conditions. Conservation investment should not be homogeneous: it should reflect local ecology, topography, drainage, management history and landscape characteristics. To give really appropriate advice takes time and requires a lot of local knowledge. The danger therefore is that this arrangement creates a similarity of conservation effort artificially and inappropriately superimposed on a living and evolving landscape. Ideally this can be alleviated if the adviser co-ordinates with other advisers. This is not always possible, and in any case other advisers are very hard pressed.

Grant aid for conservation investment is often too little to entice farmers 'out of the field corners'. The adviser is often forced to concentrate on the readily applied and marginal activities, such as ponds, tree planting and field margin protection plus some hedge maintenance. The more fundamental and truly integrative activities are often delayed.

The UFF idea is exciting but tenuous. To be truly effective it requires more cash, a second adviser plus a part-time administrator, and more grant aid, taxation relief or other financial inducements to complement the efforts at integrating conservation within changing farm practice. External policies and monies have to be more flexible and generous if this co-operative idea is genuinely to succeed. Clearly, changes in policy that provide more cash for conservation investment, either as grant aid or tax relief, would make an enormous difference.

In the event none of these conditions were met. In 1989, incomes fell for a third year running and many participants pulled out. No new external funds were forthcoming. The syndicate tried to raise revenue by selling its landscape management services to the industrial market. That further weakened the effectiveness of the small administrative unit. In 1990 the scheme failed and the project officer left the district. The real economics of agriculture took over and conservation resumed its secondary place.

Mannington Hall and Thornham estate

These are two large properties, both exceeding 800 ha, where the owners have employed, with grant aid (50 per cent) from the Countryside Commission, a full time 'in house' conservation adviser. The commission is involved because both estates contain high-quality landscape with the potential for much landscape restoration and improved public access, but which could fall into disrepair without management and public-sector funds. The holdings are very different but there are certain similarities.

The prime aim is to extend management of important scenic and wildlife features so as to increase the total conservation value of the property within a viable regime. About half the effort will be devoted to creating new habitats (hedges, ponds, bog, meadow, grassland) and some of the existing agricultural enterprise will be deliberately geared to fit in with conservation requirements.

Public access is an essential consideration. The farm plans include provision for opening up a number of new rights of way, including facilities for the handicapped, and for reorganising existing (concessionary) footpaths. Reorganisation will only take place with the agreement of rambling groups, the local authority and other interested parties.

Education through interpretation and specialised courses in conservation management for local schools and farmers are a long-term objective. The 'in house' advisers hope to be able to sell their services to neighbouring properties. After three years the advisers are expected to be self-financing, earning their income through the marketing of specialised services and the output of conservation investment (tourism, educational visits, woodland products).

Tax advantages to the landowner, through relief on capital transfer tax, is also a consideration.

Additional labour will be provided through government-funded youth employment schemes. Given good organisation and management, this can become a very effective labour force.

Both these projects are passing through regular bouts of cash shortages. Their success will depend upon generating sufficient income to pay for themselves. Even though both estates have some financial leeway, neither can afford to subsidise the scheme. Marketing and management are the key variables for success, together with a firm element of 'public interest' support funding. The greatest danger is the loss of the country-wide ranger–educator who has not only become 'field wise' but who can also build up a constructive *rapport* with the visiting public. Because incomes are low, and career paths uncertain, the very best are often forced to leave just when they are most valuable and experienced.

North Pennines case study

The key to conservation of the landscape, field archaeological and habit features of the north Pennine dales is the continuation of current farming practices. Four agencies have varying responsibilities:

1 Within the National Park, the National Park Authority, with powers of physical planning control but also under Section 39 of the Wildlife and Countryside Act, 1981, powers to negotiate management agreements with the farmers for the continuation of desired agricultural practices and the preservation of artefacts such as barns and walls. No similar body exists to provide such agreements within the AONB although some country councils have similar powers. The National Park authorities can also offer grant aid for capital works of various types provided that they further the objectives of the National Park.

2 Outside the National Park, the Countryside Commission alone can provide grants towards capital expenditure which assist in the conservation of landscapes and their attendant communities or assist in the use and enjoyment of the countryside. These are available equally in the AONB and indeed within the National Park the county council plays essentially a supporting role.

3   The Nature Conservancy Council negotiates under Section 15 of the 1981 Act management agreements, with annual or lump-sum payments for the conservation of SSSIs. The financial guidelines that accompany the Act require that compensation be based on profits forgone as a result of the imposition of restrictions on agricultural activities.

4   The Ministry of Agriculture, Fisheries and Food (MAFF) supports farm income through the EC Less Favoured Areas regulation and provides capital grants which aid farm development or are for certain specified agricultural investments.

The objectives of all these bodies are different and their interaction in the conservation of the north Pennine dales is the subject of the case study. The ESA scheme offering a fixed payment per hectare to participate in return for farming in traditional ways, accepting restrictions on grassland management and stocking practices, and conserving landscape features (and, where applicable, managing small deciduous woodlands), a new initiative from MAFF, is superimposed on this already complex system. The case study investigates the problems by concentrating on the problem of the hay meadows, and the maintenance of the characteristic stone walls and barns, the features which led to the ESA designation.

Farming in the ESA

The typical dale farm will consist entirely of livestock enterprises. The farm will consist of a relatively small area, probably less than 100 ha of grassland on the floor of the dale, the in-bye. Here will be located the farmhouse and the farm buildings, including the field barns already referred to. The in-bye will be divided partly into hay fields and partly into permanent pasture. In addition to in-bye the farmer may own or rent an allotment—walled rough pasture—on the edge of the moor proper, of low agricultural value but capable of sustaining stocking densities greater than can be maintained on the moor. The farmer will also have grazing rights on the moor itself. These may be specified as the right to graze a certain number of animals, known as sheep gaites or stints, or could be the exclusive right to a specified area of moor, perhaps with a restriction on the number of animals that may be kept there.

On the moor he will 'heif' a sheep flock, i.e. the flock will be kept entirely on the moor except for lambing and perhaps for tupping when it is brought to the in-bye. As well as the heifed sheep flock he will keep additional ewes on the in-bye as well as some cattle. The cattle are normally beef cattle and the enterprise is single suckling autumn-born calves which are housed during the winter and weaned at a date which avoids digestive problems when they are turned out on the in-bye in the spring. In some cases there may be a small dairy herd instead of single suckling beef, although dairying has been declining in the upper dales. In the past, certain dale farms specialised in the breeding of dairy herd replacements, and this still continues to some extent. Indeed, on one farm there is a management agreement to maintain the practice. The imposition of dairy quotas cannot have assisted this enterprise.

In this simple agricultural system the hay meadows play a pivotal role. They provide winter keep for the animals housed over the winter and for the heifed sheep in severe conditions. In addition the lambing is conducted in the hay fields in the period before the fields are closed for hay production, usually around mid-May. Once the hay is cut and removed lambs and/or cattle are grazed on the aftermath and the meadows may subsequently be used also for tupping. The capacity of the hay fields is a limiting factor on the ability of the farmer to over winter stock. Substitution of bought-in feed is expensive and, given typical low margins, cuts unacceptably into profits. The late date on which the meadows are closed for hay production necessitates a late date for haymaking: hay is rarely made until July, and late August haymaking is not unusual. This factor increases the risk entailed in the enterprise, since the hay has to lie in the fields to dry before baling. It appears that some loss or reduction in quality of the hay crop is experienced about one year in five.

The use of hay fields for aftermath grazing together with their late closure limits also the possibility of making silage. Silage-making reduces the risk in the process, since the crop is cut at an earlier stage, before the seed is developed, and requires that the cut crop lies for less time before gathering. Silage yields typically average 2·9 times those of hay crops for equal levels of fertiliser application. However, much of this difference in weight is water. This means that the food value of a single silage crop, taken from a given area, would for identical fertiliser inputs be only 70–80 per cent of that of a hay crop. On the other hand silage has a higher crude digestible protein content. This makes it more suitable for more intensive livestock enterprises, where it can to a degree be substituted for feed compounds. But this is of no great interest to the typical farm in the ESA. Silage production would be a superior product to hay if a second cut could be taken or if fertiliser input and hence yields with either crop were substantially increased. The question of fertiliser use is taken up in the next section; two-cut silage is not possible without fairly drastic changes in the farming system to provide substitutes for the grazing value of the hay field aftermath.

Thus, despite the fact that silage has 70–80 per cent lower average production costs than hay, it is not much practised. An intermediate system (but a silage in essence) known as big-bag silage is used occasionally in bad years, but as an expedient to get some crop from the fields when the weather makes haymaking impossible. This was done extensively in 1985. This system entails expenditure of about £15 per tonne to contractors.

Hay production is a major constraint in the existing system. The meadows cannot be extended at the expense of pasture, since that is the other limiting factor on operations. Increasing the output of hay from the existing area is incompatible with the retention of the conservation interest. This is examined further in a subsequent section but first a statistical analysis of the farming system is necessary.

The viability and development of ESA farming
Between 1967 and 1983 the number of agricultural holdings in the AONB declined by 31 per cent. The decline in numbers was brought about by the amalgamation of the smaller holdings so that the average size of in-bye

increased by 42 per cent (tables 9.1 and 9.2). These changes were probably replicated in those parts of the area within the National Park. For the National Park as a whole between 1974 and 1987 holdings of 20 ha in-bye decreased by 20 per cent in number, 2–40 ha by 31 per cent, 40–60 ha by 49 per cent and those over 60 ha increased by 55 per cent, with total holdings declining by 15 per cent.

Table 9.1 Estimated distribution of holdings in the Area of Outstanding Natural Beauty by size, and percentage change, 1967–83

Size category[a]		Year					
Ha	Acres	1967	1971	1974	1978	1983	% change
0–19	0–49	726	504	443	391	364	−50
20–39	50–99	511	452	404	348	306	−40
40–200	100–500	477	449	469	484	481	+1
200+	500+	9	9	11	15	18	+100
Sub-total		1,723	1,414	1,327	1,238	1,169	−31
Rough grazing only		21	24	23	32	42	+100
Total		1,744	1,438	1,350	1,270	1,211	−31

(a) Crops and grass only.

Source: M. Whitby (ed) (1985) *Agriculture in the North Pennines*, University of Newcastle on Tyne.

Table 9.2 Estimated average size of the holdings in the Area of Outstanding Natural Beauty (hectares) and percentage change, 1967–83

Type of holding	Year					
	1967	1971	1974	1978	1983	% change
Crops and grass	33·1	38·6	41·8	45·2	46·9	+42
All[a]	71·5	85·2	92·1	97·0	102·9	+44

(a) Including rough grazing.

Source: Whitby op. cit.

These declines represent a reduction in agricultural employment. Between 1971 and 1983 in the AONB full-time farmers declined by 29 per cent in number, part-time farmers by 26 per cent and family and hired workers by 43 per cent. The changes were accompanied by some fairly small increases in stocking density. The main component of this was an increase in sheep numbers. Between 1967 and 1983 the AONB increase was 38 per cent in sheep and only 6·0 per cent in cattle. The small change in cattle numbers masks a substantial change in composition. Thus over the same period a decline of 6,400 dairy cows and heifers was matched by an increase of 8,400 in beef cows and heifers. The increase in stocking rates appears to have been brought about by increased stocking of the upland rough grazings rather than by increased intensity of use of the in-bye. With effective limitations on the capacity of in-by, the upland commons provide the only source of forage that can be exploited. Since they are suitable only for heifed sheep, it is to be expected that the increases in stock numbers will be composed largely of sheep. The role of the in-bye, particularly the hay fields, in the sheep cycle prevents any substantial shift to cattle on the in-bye land.

The economics of farming in the north Pennines ESA

The best measure of the economic performance of farming is net farm income (NFI). This represents the payment for the own-account labour of the farmer and spouse plus the management and investment income (MII), the return on capital and enterprise. It is thus a measure of the income to the farm. It is not an appropriate measure of value added, the contribution of the farm to GNP, since it includes subsidies and excludes the wages of hired labour. In 1983/84 NFI for Less Favoured Area (LFA) farms in England was £3,412 for small farms and £10,795 for medium (which in ESA context are large) farms. For the small group this was 80 per cent of the NFI of dairy farms, but over twice that of lowland cattle and sheep farms, which are typically much smaller. Furthermore LFA cattle and sheep farms have not performed badly in recent years. Nevertheless incomes are declining, possibly by over 30 per cent in real terms since the 1960s.

While public expenditure at both the national and the EC level is a major element in the evolution of farm incomes, the LFAs receive additional subsidies. The headage payments, by their nature, directly affect net farm incomes; additional capital grants depend on take-up and in certain circumstances could lead to the reduction of NFI. Without the Hill Livestock Compensation Act (HLCA), and the comparable national schemes that it replaced, the plight of hill farming would have been grim. In 1983/84 HLCA/ha payments in the AONB exceeded NFI/ha. Without them, farmers would have made losses. Over the previous three years HLCA payments averaged 65 per cent of NFI/ha. An analysis of headage payments in sustaining the viability of ESA farms suggest that without them many holdings would not even have yielded an adequate wage for the manual labour provided (table 9.3). It is not far from the truth, then, to say that the farmer is being paid simply for manual labour in working the farm, some of that payment being presented for accounting purposes as a return on capital and enterprise, as entrepreneurial rent. In the circumstances there would seem no

Table 9.3 Importance of Less Favoured Area subsidies to viability of upland and hill farms

	1980/81	1981/82	1982/83	1983/84	1984/85
Own-account labour[a]					
Upland		4,253	4,599	5,978	6,899
Hill		5,801	6,787	6,624	7,254
Management and investment income (MII)					
Upland	1,946	3,865	8,512	−1,451	2,441
Hill	3,926	8,157	8,336	1,489	3,274
Net farm income					
Upland		8,118	13,111	4,527	9,340
Hill		13,938	15,123	8,113	10,528
LFA subsidies					
Upland	3,514	4,388	5,267	6,359	6,989
Hill	4,653	7,439	6,825	8,492	9,464
Subsidies as % NFI					
Upland		54·1	40·2	140·5	74·8
Hill		53·4	45·1	104·7	89·9
Margin of subsidies over MII					
Upland	1,568	523	−3,245	7,810	4,548
Hill	727	−718	−1,511	7,003	6,190

(a) Estimated.

Source: Askham Bryan College of Agriculture and Horticulture, *Farming in Yorkshire* annual.

moral case against placing constraints on farming activities in the interest of the community.

The conservation of hay meadows

The problems
The dale hay meadows are fragile habitats, of considerable diversity. Their species composition depends on two broad factors: the grazing regimes to which they are subjected and the rate of application of fertiliser.

The species diversity of the meadows results from these factors limiting the dominance of several common grass species which constitute the bulk of the sward in improved grassland in the area. The late closure of the hay crop with grazing until mid-May is thought to give an interval for these less common and less palatable species to get well established before they are shaded out by the faster-growing dominant species. The lateness of the haymaking season is thought to be an important factor in explaining the unusual species composition of the meadows. Thus it is likely that changes in the timing of grazing use and particularly earlier closure could lead to a decline in species diversity and the scientific interest (since earlier closure is widespread) of the habitats.

But if this is likely, their decline with the application of artificial fertiliser is well established. The common grass species respond rapidly and vigorously to increased nutrients and crowd out the rarer species. The conservation of these meadows requires in most cases an absence of the use of artificial fertilisers. Conservation of the biological interest often requires restriction also on the application of farmyard manure and limitations on this factor feature also in the conditions of Section 15 management agreements.

There are probably between 3,000 and 4,000 hay meadows in the Yorkshire dales, not all of them by any means within the ESA, and an unknown number are to be found in the rest of the area, with Teesdale being particularly rich in them. However, the vast majority have been improved to some extent. Some 590 could be herb-rich, amounting to about 4·7 per cent of all enclosed fields and 2 per cent of the area of grassland. Other scientific surveys suggest that the floristically rich are much rarer than this. Possibly as few as fifty to eighty high-quality meadows can be identified in the study area.

The application of artificial fertiliser to the hay meadow is the key to improvement on the farm. So there is a direct conflict between the conservation of hay meadow habitats and raising the agricultural return from the holding. Application of compound fertiliser at a 'normal' rate of 250 kg/ha will typically double the yield of hay. Where silage production is planned, application rates 50 per cent greater than this would be used. The extra winter keep will permit an increase in the stock that the in-bye can support. From the evidence of the next section this will normally mean more single suckling beef production and perhaps reduced agisting of in-bye sheep. An application of nitrogen to the meadows after the hay crop is taken aids the recovery of the aftermath and allows some fattening of the in-bye lambs, raising their value in the autumn sales.

Interaction of policy instruments: Section 15 agreement, HLCA and
ESA payments
The Nature Conservancy Council provided information on management agreements in existence and being negotiated under the provisions of Section 28 of the Wildlife and Countryside Act relating to north Pennine dale hay meadows. Data, some of them incomplete, concerning twenty-five agreements on twenty-four sites were analysed and some results are given in table 9.4. Thirteen of the sites were within ESA, the rest were adjacent and in any case concerned similar problems in closely similar areas. In all cases a primary purpose of the control was to restrict the application of artificial fertiliser on the meadows, in many but not all cases eliminating its use. Additional restrictions concerned liming, basic slag and the frequency and intensity of the application of farmyard manure. The continuation of traditional grazing practices—lambing on the meadows before closure for hay production and grazing the aftermath by cattle or sheep—was considered necessary (by the NCC) for maintaining the ecological value of the hay fields. In a few instances restrictions on stocking rates on the aftermath and on out-

feeding in the winter (to avoid poaching of the grassland) were also thought necessary.

While the conservation of hay meadows as a major purpose of the scheduling of the sites, in some cases the SSSI boundaries extended beyond the meadows to encompass pasture. Fertiliser use was the main restriction sought here, as well of course as more drastic improvements such as drainage and reseeding. Procedures on SSSIs can be triggered in two ways:

1 As a result of the occupier's objection to the restrictions placed on his husbandry when the site is notified or renotified.
2 In consequence of notice by the occupier of intention to carry out agricultural improvements to the site.

The financial guidelines require compensation to be based on the annual profits forgone as a result of the restrictions imposed on the SSSI. In case 1 above, compensation is calculated either as the market value of the hay that would otherwise have been produced, less the fertiliser costs saved, or as the net margin on the livestock enterprises that the restriction renders inviable. With case 2 a more detailed appraisal is required entailing *inter alia* an evaluation of the practicability of the farmers' improvement plan as well as a scrutiny of its costings. In this exercise the advice of local MAFF officials is sought.

The distinction between the two types is not altogether clear, since notification under case 1 sometimes reveals plans that the farmer had for improving his land. Nonetheless it appears that the payments in type 2 cases are typically higher than under type 1, as indeed from the intentions of the Act they should be. The analysis in table 9.4 uses the proposed farming system in type 2 cases as the 'without restrictions' benchmark, and the actual system in type 1 cases where no improvement intentions have been revealed. Since management agreements are reached by a process of bargaining, there is a tendency for farmers to overstate their losses from SSSI restrictions, and compensation claimed is normally above compensation paid. Where expert appraisal was available, type 2 cases were adjusted in the light of it. Where that stage in the process had not been reached, judgement was exercised, resulting in some cases being excluded from the data set. It remains likely, however, that the effects of the SSSI restrictions on stocking rates are somewhat overstated in this analysis.

All the holdings analysed were devoted entirely to livestock enterprises. All possessed a sheep flock heifed on the hill. The extent of hill grazing rights was not known in many cases, although both sheep gaites on common grazing and specified areas of upland rights were mentioned. The lack of information means that the conceptual problems of measuring hill grazing are avoided. The analysis is confined to in-bye. Out-bye, including allotments, is ignored. Depending on the quality of allotment pasture, this could in one or two cases mean that the intensity of the holding is overstated.

On most of the farms involved in management agreements there are just two enterprises: sheep—both in-bye and heifed—and single suckling beef. One holding had a small dairy herd and one produced dairy replacements as a

Table 9.4 Analysis of Section 15 payments for Pennine dales hay meadows

Type of payment	Mean	SD	Range	No. of observations	Coefficient of variation
In-bye (ha)	70·3	26·9	29·2–108	14	0·38
Total stock (LU)	75·1	43·2	30·7–177·6	16	0·58
In-bye stocking rate (LU/ha)	1·15	0·54	0·56–2·05	14	0·47
Area of restriction (ha)	11·7	13·1	0·6–58·5	25	1·12
Restriction (% of in-bye)	24·1	15·7	5·1–60·3	14	0·65
Payment (£/ha)[a]	176·4	121·9	40–575	19	0·69
	(139·4)	(52·3)	(40–235)	(17)	0·38
Stock reduction from restriction					
(% of LU)	28·6	17·5	7·5–47·6	14	0·61
Sheep	11·6	11·5	0–34·7	14	0·99
Cattle	31·1	23·8	0–86·1	14	0·77
HLCA without restriction (£)	4,731	2,682	813–1,107	16	0·57
HLCA reduction (£)	836	589	0–2,286	13	0·70
HLCA/in-bye (£/ha)[b]	74·1	27·9	42–130	16	0·38
HLCA reduction/in-bye (£/ha)[b]	13·4	6·7	0–23·5	12	0·50
HLCA reduction/restriction (£/ha)[b]	64·2	44·4	0–193·5	14	0·69

(a) Excluding two observations.
(b) Excluding dairy farms.
LU livestock units.

Source: Author's estimates from data supplied by the Nature Conservancy Council.

minor enterprise. Some farms had a few other stock cattle. Some of the proposed improvements entailed a shift to multiple suckling of beef calves.

The typical improvement that was restricted involved fertiliser application at the rate of 250 kg/ha on hay meadows. This would normally give yields of 4·5–5 tonnes/ha, about twice that which could be achieved without fertiliser. A hundred and eighty kilograms of 33 per cent nitrogen on the aftermath would allow additional fattening of lambs for autumn sales. Where restrictions involved pasture, the normal improvement practice was to apply 125 kg/ha of fertiliser. There were also one or two proposals (none in the ESA) to change from hay to two-cut silage production with heavy fertiliser application and to reseed pasture. 'Big bag' silage was seen as an occasional expedient, when the prospects for a hay crop were bad, rather than a normal practice.

The main effects of restrictions, therefore, were reduced stocking rates, less feed and hence fewer suckling cows, lower lambing percentages and lower weight gains for in-bye lambs. Alternatively, or in some combinations additionally, it meant more purchased feed, only partially offset by lower bills for fertiliser.

In terms of in-bye, the farms affected were small to medium-size, with a range of 30–110 ha. Livestock numbers and stocking rates per unit of in-bye (unrestricted as defined) showed greater variation than farm size, partly because of differences in out-bye but partly properly reflecting differences in intensity which would follow from proposed improvements. The areas subject to restriction averaged 12 ha and ranged from 1 ha to almost 60 ha. The larger areas related to cases where the SSSI included areas of pasture, the small ones were all small hay fields. But if the absolute areas were small they involved a significant area of the in-bye land of the farm—24 per cent on average, with a range from 5 per cent to 60 per cent. Because of the pivotal role the meadows play in the farming system the restrictions resulted in a larger percentage reduction in livestock numbers expressed as grazing livestock units than in area of in-bye, 29 per cent against 24 per cent.

The bulk of the reductions in livestock were borne by cattle rather than sheep enterprises, the percentage reduction for the latter averaging almost three times that of the former. In one case the farmer observed that reducing cattle rather than sheep was technically necessary, given the nature of NCC restrictions on the land, but in general this was clearly not so. Rather, in terms of financial performance, beef is the marginal enterprise on these farms. This is not altogether surprising. The complex structure of support of beef production confers an advantage on intensive systems and particularly those such as grass silage beef, where much of the fodder is produced on-farm. Single suckling is not a very profitable enterprise; more intensive systems require substantial pasture improvement. No such advantage confers to intensive sheep production. Rather, comparative advantage lies with extensive systems as practised in the hills. Conservation of hay fields therefore in the Pennines seems likely to shift the balance of livestock towards sheep. Its role in this regard is, of course, a small one, since the trend towards sheep, as noted earlier, is of long standing.

In type 2 cases where special restrictions were imposed against the likelihood of undesirable improvements intensive grass production under farm improvement plans necessitated annual payments under management agreements of £400–£500/ha. For the rest, which are thought more typical, payments averaged £139/ha with a range of £40–£235. There is no relationship between payment per hectare and the proportion of in-bye subject to restriction. While further analysis is needed, the probable explanation is that the payment reflects the potential for improvement, while the survival of large proportions of traditional unimproved grassland on these farms is explained by the limited potential for improvement. These farms may well be the most valuable from an ecological viewpoint, but the financial guidelines under the Wildlife and Countryside Act take no account of this dimension of site value.

Despite the small number of observations, there is a strong positive correlation between compensation payments per hectare and loss of Hill Livestock Compensation payments (HLCA) per hectare of restricted land, HLCA being defined to include the suckler cow premium. This implies that the proportion of management agreement payments that can be seen as direct compensation for loss of HLCA declines as the size of the payments increases and hence the proposed intensity of exploitation of the grassland increases. Thus small payments per hectare by the NCC represent simply the substitution of one form of upland subsidy for another, but the larger payments are not so clearly categorised.

The point can be put another way. HCLA payments provide a substantial incentive for agriculturally minor but environmentally significant changes in farm practices (principally increases in the use of nitrogen) which raise the intensity of grassland exploitation. With agriculturally more significant changes the incentive effect is less important. However, another factor comes into play here. The larger changes in intensity resulting in higher per-hectare compensation require programmes of investment in drainage, reseeding and the provision of livestock housing. These elements could all qualify for grant aid under the Agricultural Improvement Scheme (AIS) at a rate of 30 per cent for Less Favoured Areas, twice the standard rate. The 15 per cent grant differential seems to have added substantially to the profitability of the projects but was not in itself sufficient to tip the balance between viability and non-viability. Positive payments would still have been required by the NCC in these cases even if neither the AIS differential nor HCLA applied, but they would of course have been significantly smaller.

A final observation from the data concerns the interrelationship between Section 15 payments and payments under the ESA. The management agreement payments as operated by the NCC, as has been seen, are confined to specific SSSIs and normally relate only to parts of the in-bye of a farm. The ESA proposals could relate to whole farms and although set at a fixed rate of £100/ha, which is lower than the average for the NCC data that we have examined, may appear more advantageous to the farmer because of their greater extent. At the mean levels of payment and proportion of the farm affected, management agreement payments amount to only a third of the level of ESA payments on a whole farm basis, and the break-even level of the management agreement, to make it of equal value to the ESA, would be £415/ha. Provided that the restrictions are the same with the two payments, only farmers with a potential for substantial intensification would decline to enter the ESA. It might be thought then that, with the introduction of ESA payments, the NCC would be required to negotiate agreements only with those farmers who saw a prospect of a substantial increase in intensity on their land. The mean payment would rise very substantially, possibly increasing threefold, but the number of agreements and the total amount paid to protect Pennine dale hay meadows would fall substantially. This reasoning, however, oversimplifies the problem, since its premises are false.

The ESA requirement to maintain traditional farming practices as a condition of payment may be construed as placing limits, for example, on application rates of artificial fertilisers on all fields of the in-bye. Section 15 agreements, *per contra*, are confined to the SSSI and the possibility exists of the sort of 'slippage' seen in set-aside programmes in the USA; the farmer compensates for his reduced yields on the restricted areas by increasing his exploitation of the rest of his land.

It is possible that the sort of restrictions on agricultural practice imposed under the ESA scheme, while sufficient to maintain the visual or landscape impact of the hay fields, are insufficient to ensure the conservation of their ecological value. Thus the NCC might wish to maintain management agreements over some meadows on farms in receipt of the ESA payments. The acceptance of an ESA agreement will of course alter the level of payments necessary under Section 15, since the possibilities of improvement and hence the profits forgone are reduced by the ESA agreement. But this will influence the attitude of the farmer towards acceptance of a Section 15 agreement. The relatively small additional payment may well not compensate for the inconvenience of differential restrictions on his activities. The NCC could then find itself in a position where it has to offer a greater incentive to the farmer so that the total cost (ESA + Section 15) of protecting SSSIs could exceed the previous cost under Section 15 alone. The extreme case might be where the farmer is unwilling to have both and will accept a Section 15 agreement only if it yields sufficient payment to compensate for forgoing the option of an ESA agreement. In this case the cost of protecting SSSIs rises drastically with, from society's viewpoint, a landscape gain in areas outside SSSIs.

Conclusions

Conservation of the landscape and habitat features of the north Pennine dales requires a continuation of traditional unintensive husbandry with low stocking densities and an absence of artificial fertiliser application. If these traditional farming practices are to continue, then they must be profitable. Given the general lack of opportunities for alternative activities, the in-bye, with its attendant out-bye and upland grazing, must yield a sufficient income to keep the farmers on the farms. But if this is sufficient for conservation in the short term it is unlikely to be adequate for long-term survival of the meadows. The main explanation of the survival of unimproved meadows was that the farmers who managed them were old; they had learned their farming practices in a different era and were resistant to change. A reason given in several instances for proposed farm improvements which resulted in Section 15 agreements was succession to the holding of a young, ambitious occupant wishing to demonstrate farming skills.

Thus a requirement for long-term maintenance of the habitat is the creation of some criteria of success for the farmer: a means of demonstrating efficiency in some other way than the traditional one of increased intensity and rising output. Such a conservation ethic is not impossible: it exists, presumably, on nature reserves, where the young warden has an ability to display skills and has some yardstick for gauging success. But it does not exist at present for the farmer, whose success is to be gauged in agricultural terms. The development of improved enterprises, increased yields from grassland, and the pioneering of new enterprises and more intensive production modes are commonly accepted criteria here. In this context conservation is seen as a constraint on enterprise, an obstacle to success and not a source of satisfaction in itself. Seen in this light, the Wildlife and Countryside Act is wholly negative.

The development of a conservation ethic may entail to some extent a process of re-education for the farmers. But this must be accompanied by management plans which allow scope for achievement—some possibility of development in landscape and habitat terms, of improving the farm environment and not merely preventing further deterioration. The ESA scheme, which should lead to total farm conservation plans, is a step in this direction, although it carries with it the seeds of conflict with the stricter criteria of wildlife conservation. ESA shares with the specific management agreements the conflict with HLCA payments, which provide a direct incentive to increased stocking rates. The implicit message is that agricultural improvement is desirable but may be constrained by the (political) pressure for conservation. In a changed world that is the wrong message. Changing it, after forty years of continual reinforcement, is a mammoth task.

It is premature to evaluate any of the initiatives analysed in this report. Speculation would triumph over measured assessment. The evaluative criteria that should be applied are:

1 The scope for genuine integration policies.
2 The scope for improving administrative efficiency nationally, regionally, locally.
3 The scope for increasing the value for money invested.
4 The scope for establishing a learning capability.
5 The scope for retaining flexibility on the ground to take into account individualistic aspects.
6 The scope for improving biological diversity and retaining historic landscapes.

Total policy integration between agriculture and the environment is still being developed. The United Kingdom's country information paper has highlighted the efforts to improve co-ordination. But so long as agricultural policy is controlled primarily through the European Community and environmental policy effectively managed via the member States, truly effective co-ordination remains unlikely. A particular difficulty is the awkwardness of trying to transfer surplus agricultural monies to environmental budgets. The ESA is a key ingredient in establishing a firmer footing.

The relationship between encouragement and control is bedevilled by the inadequate co-ordination of policy at the European Community level. The inadequacy, for the most part, of statutory controls over agricultural alteration of habitats and landscape features outside selected 'high profile' areas is a serious drawback, as is the lack of sufficient resources within local authorities for financing management agreements.

Proposals for reducing the cost of surplus agricultural production in the European Community include price restraint, quotas, voluntary or compulsory retirement of land, and diversification of farm income. None of these to date specifically and consciously includes an environmental component. Price restraint may increase neglect and/or increase intensification. Quotas could freeze unsuitable practices. Retirement of land offers scope but there is a long way to go before this idea becomes instituted as a conservation measure, as opposed to a production limitation device. So long as real agricultural incomes fall the scope for conservation investment remains limited, and dependent upon conscience.

The best hope for more effective policy integration is for member countries to take a stronger national initiative and experiment even more ambitiously with schemes such as an extended ESA. Carefully monitored experimental schemes seem the most promising way forward. Future schemes should include a strong bias towards diversification of farm income.

Administrative efficiency seems best achieved if conservation payments are made on a clearly defined and relatively simple basis where rough, but reasonably equal, justice prevails. It is also crucial that the various payments available to farmers do not conflict, as they do at present. This is inevitable as a system develops piecemeal, with separate Ministries and agencies pursuing contradictory objectives. Efficiency would also be improved if neighbouring farms could rent or buy an adviser charged with the task of 'opening the gates' to advice and back-up funds. Such people could be essentially 'farm enterprise advisers' with a remit to diversify farm incomes and to link farmers to a variety of specialist advisory services. Ideally they should be within the Farm and Conservation Service, but many could be self- employed or employees of local authorities.

Value for money invested is a difficult item to measure. Much of conservation value is a public good, usually a long-term public good, for which there is no definable market. So returns should be visualised in resource costs saved, in improved property values, in public enjoyment and in the marketing of conservation output. Experimental schemes should be monitored to assess how important each of these items is.

The learning curve is all-important. Aspects of the scheme depicted here are on the low point of a learning curve. But part of their purpose is actively to pass on the benefits of their experience elsewhere. Adequate monitoring and evaluation are therefore desirable.

Flexibility for idiosyncracy remains a stumbling block. This can be at cross-purposes with efficiency and value for money invested. In addition, the advisory services are stretched and can rarely 'fine-tune' the final product. This will remain an awkward problem for large-scale schemes (e.g. ESAs) and for 'in house' advisory arrangements unless adequate labour with sufficient experience is provided. One must never forget the 'folk knowledge' of local naturalists and farm workers. That is an under utilised resource.

Land use designation is an advantage for the integration of policy so long as:

1 It does not confuse by adding layers of designation on top of existing designations. This is a potentially serious problem that must eventually lead to the simplification of land-use nomenclature
2 It is backed up by adequate 'stop' powers.
3 It is also supported by sympathetic policy adjustments in the agricultural sector aimed at diverting more of existing budgets geared to increasing agricultural surplus to a variety of conservation investments.

Planning controls should be 'last resort'. But there should be improvements to the indicative planning of local authorities, the forestry community and agricultural interests, so that the areas of land or particular features suitable or unsuitable for particular forms of development are clearly identified. The principle has long been established in the structure plan approach adopted by county councils. This needs to be superseded by more adventurous land-use indicative plans, also prepared by the local authorities. Such plans should be based on the capacity of the land to support particular uses (e.g. forestry, agriculture, wildlife and recreation) as modified by amenity and historical land-use considerations.

Prior notification of intent to alter protected environmental interests is an essential requirement in designated areas, and in those zones of importance depicted on indicative maps. This should be encouraged if whole farm plans become more fashionable, triggering improved advice and financial assistance.

While the provision of an adequate income from environmentally friendly farming and adequate incentives for improving the farmed environment are plainly necessary, conservation has to 'be sold' to its practical champions.

Income assistance is the best way of portraying the value of conservation investment. Few farmers like the idea of museumisation or keeping public pleasure places. Conservation has to be promoted as part of good agricultural practice, a means of maintaining property values and an avenue to income supplementation via grant aid, tax relief, on-farm business activity and product marketing. This package is still in the process of being sold to British farmers. The task awaits better integration of policy, a higher point on the learning curve, more diversion of money from the over-provided agricultural budgets, much more attention and resources devoted to advice, and improved education and training in marketing and management skills.

As agricultural incomes fall and quotas or price cuts are put into effect, the British farm of tomorrow may well prove to be quite a different enterprise from that of today. Variations will abound, but most likely is a pattern of income that derives only partly from food production. For some farm enterprises as much income may come from woodland products, tourism, recreation, education and interpretation services, and off-farm businesses.

Already the agriculture departments are providing advice and grant aid for supplementary income generation, including the more adventurous use of redundant farm buildings. The development agencies are providing similar advice and some grant aid to non-farm enterprises in rural communities. Co-ordination between these two socio-economic development bodies needs to be improved. Both programmes are very modest, with insufficient advisers on the ground. It is likely that this initiative will be extended into many other realms, involving conservation agencies, local authorities and the tourist boards. Surrounding the future farmer will be an array of advisory and grant-aiding bodies, including private advisory consultants, whose job it will be to encourage the development of a 'whole farm enterprise plan' in which agriculture, conservation, non-farm business activity and providing local community service will combine in various ways, as suggested in fig. 9.4.

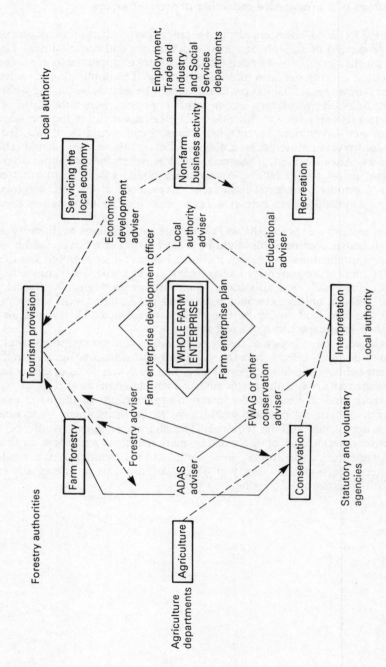

Figure 9.4 Advice, farm plans and community initiatives in a healthy rural economy

Implications of a progressive reduction in product prices

In May 1987 OECD Ministers pledged a progressive reduction of assistance to and protection of agriculture across all countries and commodities. The reasons for their position are understandable, as price support costs are rising almost out of control, yet farm incomes are falling. The implication of such a move for landscape and habitat protection would be mixed, but on the whole negative, unless compensatory income support policies were substituted, at least up to a certain level. On the plus side, price falls would reduce pressure for major new investment on marginal farms. Land would be abandoned. Some might revert to interesting ecological climaxes though this would take generations. Abandonment, if planned, could be beneficial for habitat diversity. Land prices would fall, so conservation bodies might be in a better position to purchase marginal farm areas. Experience in the UK suggests, however, that landowners extract a high price for the best conservation estates.

On the minus side, price falls will encourage some farmers to intensify in order to maintain income. This might be a short-term move but it could have devastating implications for sensitive wildlife habitats whose interest lies in a particular form of low-intensity management, for example, the Pennine hay meadows. Price falls will also cause some farmers to neglect ecologically valuable habitats, so this too would be undesirable. As farms went bankrupt, some of the sale land would be bought and intensified. This would also reduce their landscape and ecological value.

On balance, then, a policy of steady reduction in commodity prices would result in undesirable effects from the viewpoint of maintaining agriculturally dependent habitats and landscapes. It would be necessary to compensate for such income reduction with specific management agreements or through the sale of spare land to *bona fide* conservation organisation. In the latter case some form of farming practice would have to be maintained, so income compensation would still be required. The alternative option would be a progressive deterioration of the existing marginal areas and some natural habitats which are agriculturally dependent, and the beginning of a whole new pattern of eco-systems, many of which would not be so biologically or intrinsically healthy or interesting.

Part V
The impact on agriculture of pollution from other sources

10. The economic impact of air pollution on agriculture: an assessment and review

R.M. Adams and T.D. Crocker

The modern industrial version of alchemy, which transforms otherwise harmless natural elements into a pervasive toxic burden, harms agriculture. Its air, soil and water pollution manifestations frequently reduce agricultural yields, increase the prices that consumers of agricultural products must pay and alter the returns accruing to owners of agricultural inputs.

Cameron (1974) has discussed some of the plausible economic consequences; however, research in the area has been heavily dominated by biological dose-response research directed solely towards understanding the behaviour of various facets of plant systems subjected to pollution-induced stresses. The relationship of these facets to an economic parameter, such as yield, has infrequently been a question of interest to biological researchers. Prior to 1980, whatever yield response information was available for economic assessment was dully and unimaginatively used. Yield changes for a particular crop multiplied by an invariant market price served as the measure of 'economic' benefits. Economically relevant changes in output and input prices, producer and consumer risks, input uses, and cropping and location patterns were overlooked.

Many parts of any society regularly confront the problem of 'optimal' eco-system management. A farmer's choice of crop to grow, a householder's choice of yard plantings, and multiple-use management of a forest can all be cast as problems of eco-system management. The eco-system can then be seen as a process that produces desired outputs and services, which may be facilitated by some human inputs such as fertilisation and hindered by others such as atmospheric pollution.

Just as a manufacturing firm requires a description of its production process in order to allocate its resources efficiently, the eco-system manager must have in mind a description of the process by which the outputs he desires are produced, and of what effects any factors that he may control have on those outputs. A model of the internal workings of the eco-system is fundamental to economic analysis of management strategies. Questions about the optimal combination of inputs with which to produce one or more types of eco-system output under assorted environmental conditions have therefore traditionally been attractive research topics for plant scientists and economists.

The plant science and the economic perspectives

Although he always draws upon known biological principles, the plant scientist, particularly the agronomist, is a type of engineer who focuses on the design and operation of processes that can produce desired outputs with a minimum expenditure of valuable material inputs such as fertilisers. The input prices that determine economic costs are taken as given. The agronomist thus seeks to solve a constrained optimisation problem involving detailed knowledge of both the biological possibilities and the expected costs of material inputs. If all non-material inputs, such as labour, were free, these agronomic solutions would alone, in economic terms, decide the optimal design process. However, non-material inputs are also costly, and both material and non-material costs can change. Agronomic results thus identify the technically efficient set of material combinations, within a larger set of ultimate biological, environmental and technological limits, that will produce a given amount and type of desired output (Adam, Crocker and Katz, 1985).

Economists who study agriculture also focus on cost minimisation, but they assume that the agronomist has already solved his constrained optimisation problem. The economic theory of cost and production describes the effects of changing variable input and output prices on cost-minimising combinations of material and non-material inputs and, also, objective-maximising quantities and types of outputs. From the set of cost-minimising combinations the theory provides rules for identifying the lowest combination of material and non-material costs; it then specifies how this 'minimum of the minima' alters as relative prices change. For given output quantities and types, it identifies economically efficient input combinations. Economics, then, portrays the results of agronomic re-optimisation in terms of the effects of changes in biological, environmental and technological possibilities, and of changes in the relative prices of inputs and of outputs, on the cost-minimising combination of inputs and the objective-maximising amounts and types of outputs. When dealing with pollution, it describes how differences in pollution impacts alter the economic consequences of the alternative physical and biological ways a farmer has to meet his production objectives.

The agronomist viewed as engineer considers all input combinations consistent with known biological and physical principles, since his objective is to develop and ultimately implement a detailed plan for a particular production process. Subsequent efforts to improve on this plan centre on any input or sub-set of inputs that might allow substantial cost reduction. In contrast, most basic scientific studies of plant damage from pollution have concentrated on how a single pollutant affects a single input combination in some production process. Many have also studied such effects independently of any cost-minimising production process. Hence, at best, the focus has been on a small set of the feasible input combinations. These highly detailed studies typically assume that other input types and qualities remain constant. They therefore frequently fail to provide the information the economist requires in order to estimate the changes in cost-minimising input combinations that pollution induces. Much, perhaps most, plant science research on pollution inadequately reflects commercial or field conditions and neglects other sources of stress upon the plant system. When applied to commercial and field conditions, common plant science research practices create other sources of incomplete and even misleading response information.

A basic science, piece-by-piece approach to the study of how pollution affects vegetation makes truly awesome the task of covering and synthesising the feasible or even the technically efficient input combinations. In agriculture alone, environmental and edaphic factors create thousands of different types of inputs. Each input type, in turn, is embodied in one or more production processes or crops, which may appear in a variety of cultivars and which can be put to many distinct uses. Moreover, environmental co-factors, such as moisture and temperature, may act in concert with pollution to alter its impact. Natural science studies of how pollution affects plants have neither received nor provided much guidance about which of these many input combinations have economic significance (Adams and Crocker, 1982).

One key problem, then is to reconcile the natural science and the economic approaches to plant damage by developing criteria for deciding how much plant science information to generate and retain in any particular pollution problem setting. When one of the objectives of pollution damage studies is to provide useful information for estimating economic consequences, a basic question is whether more or less natural science detail will alter the economic estimates in a non-trivial way. Both the plant scientist and the economist must refine their knowledge so that the relationship between pollution and vegetation damage is defined in dimensions which are simultaneously coherent, consistent with agreed biological and chemical principles, and which can be described in terminology understood by farmers (Adam and Crocker, 1985).

Beginning in 1979, the United States Environmental Protection Agency attempted to improve data on crop response to air pollutants by funding the National Crop Loss Assessment Network (NCLAN). NCLAN is a multi-site, multi-disciplinary programme under which field experiments have been conducted on the yield response of major crops to ozone. The experiments have been designed in collaboration with economists to support economic assessments. As these policy-oriented NCLAN experiments progressed, questions arose regarding programme-research priorities (see Adams and Crocker, 1985). Specifically, in order to allocate scarce research dollars, administrators wanted to know the sensitivity of economic benefit assessments to improve biological yield-response data. Several studies have now demonstrated that there are some important classes of pollution impact upon agriculture for which rather imprecise yield-response information is sufficient to distinguish among the economic consequences of alternative ambient pollution levels.

In their Bayesian enquiry into the robustness of economic estimates of crop damage from ozone, Adams, Crocker and Katz (1984), showed that the policy value of additional plant-yield response information declines rapidly. For example, forty or fewer yield response observations per crop, similar to those found in Heck et al. (1982), adequately discriminated among the differences in the economic surplus for corn, cotton, soya beans and wheat production at ambient ozone levels of sixty-six, fifty-three, forty-eight and forty parts per billion. In an entirely different setting Feinerman and Yaron (1983) demonstrated that the value of information to Israeli growers about potato yield responses to soil salinity was maximised with as few as fifteen and not more than twenty-seven observations. Adams and McCarl (1985), while studying the economic benefits of ambient ozone control for corn, soya beans and wheat grown in the United States corn belt, found that changes in key physical parameters had to be substantial if they were to alter benefit estimates significantly. For given pollution levels, yield responses across alternative cultivar selections and across temporal and spatial variations tended to be quite homogeneous among experiments, implying that additional yield-response observations of the NCLAN type do not greatly affect economic benefit estimates. Even where some disparate cultivar, spatial and temporal responses were observed, their effect on benefit estimates was not as great as the effect of using the NCLAN data rather than yield-response data generated without economic assessments in mind. (Adams and McCarl, 1985, p. 274, concluded, '. . . that even a limited set of crop response data, when generated in accordance with the needs of those doing the assessments, appear adequate to measure the general benefits of pollution control'.

Adams and McCarl (1985) add two strong qualifications to this conclusion. First, they note that Tingey et al. (1982) hypothesise that climatic and soil factors are important in modelling the real-world effect of ozone on crop yields. The NCLAN data confirm the biological importance of the interaction between moisture stress and ozone, generally showing that water stress tends to suppress negative yield effects. When response functions reflecting these moisture stress–ozone interactions were included in their economic analysis the changes in benefit estimates were as great as 50 per cent. This result could be due to the fact that commercial crops within the United States corn belt typically suffer from moisture stress, while the NCLAN response experiments were usually conducted under conditions where moisture availability did not limit production.

Adams and McCarl (1985) also warn that their results are only the latest in a succession of studies showing that yield response and economic benefit estimates are sensitive to the type of yield-response function used. The Adams and McCarl (1985) benefit estimates differed by as much as 60 per cent according to whether a linear or a non-linear (Weibull, 1951) yield response function was employed. This result is similar to that found between the linear and quadratic forms by Adams, Crocker and Katz (1984). Note, however, that while some plant scientists suggest the Weibull (1951) distribution for studying yield responses to pollution (Rawlings and Cure, 1985), most non-linear forms, including the Box–Tidwell or other flexible forms, seem to give very similar results.

Plant responses to pollution or any other form of stress are a complex, possibly non-normal function of absorption, distribution, metabolism, detoxification and excretion parameters. Nevertheless, whatever the correct form of yield-response function, most efforts to fit observed yields statistically to air pollution exposure have employed classical least-square techniques to account for random variation. The importance of testing the goodness-of-fit of crop yield probability density functions other than the normal and log-normal was first recognised by Day (1965). Atkinson *et al.* (1985) argue that the possible non-normality of yield response date justifies the use of maximum likelihood estimators rather than the classical least-squares approach. They assert that the maximum likelihood approach is more efficient and that it more readily incorporates censoring, as the cessation of the productive life of the plant results in a truncated probability density function which, if unaccounted for, leads to biased response estimates. Examining the NCLAN data on the yield responses of soya beans to ozone, they found that maximum likelihood estimates for the Weibull distribution are about 25 per cent greater than the least-square estimates obtained by Heck *et al.* (1984) for the same distribution and similar data. To date, however, no economic analyses of pollution impact upon agriculture have been performed in which the embodied yield-response functions were estimated by maximum likelihood rather than classical least-square techniques.

The economic assessment problem

Assessments of the economic consequences of pollution upon agriculture try to estimate differences in income equivalents, defined as differences in the sums of buyer 'surpluses' and seller 'quasi-rents' over two or more policy-relevant pollution levels.

Figure 10.1 traces the set of factors that must be accounted for if a complete assessment of the economic consequences of pollution-induced vegetation damage is to be performed. Assume that a grower is trying to decide what kind and how much of a crop to grow. One or more of these prospective crops is susceptible to pollution damage. For given pollution exposures, the portion of the figure lying to the left of the left-most dashed line represents the problem of finding the least costly way of producing a given crop. The question to be answered is: given the alternative ways I have of producing this crop, which ways allow me to employ the minimum cost combinations of the inputs biologically and physically necessary for production? By affecting the productivity and the availability of inputs, variations in pollution levels will alter these minimum necessary combinations. Identification of these combinations for each pollution level enables the farmer to maximise the output of the crop obtained from a given expenditure on inputs. However, the resolution of this problem cannot decide which of the alternative crops to grow: it can only show the least costly plan for growing a particular crop, and the manner in which this plan will vary as pollution varies.

The portion of fig. 10.1 lying between the dashed lines depicts the problem of choosing among the alternative crops, given prior identification of the least costly means of producing each crop. In this portion the producer is allowed to substitute crops that are more or less prone to yield and/or quality reductions from expected pollution levels. The problem of this middle portion can be contrasted with the problem of the first portion, where the producer could do not more than manipulate the process used to produce a

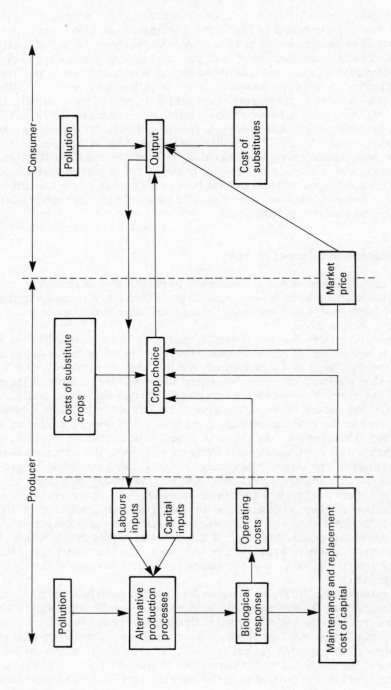

Figure 10.1 The sources of economic consequences

crop. Basically the producer has a set of alternative crops dictated by biological, physical, technical and institutional constraints. For each known and allowable process for each crop, the producer estimates the least-cost process, and, for given output prices and pollution levels, selects the combination of crops expected to generate the maximum quasi-rent.

In the right-most portion of fig. 10.1 resides the consumer. There are two routes whereby pollution-induced vegetable damage can influence consumer behaviour, thus altering the surplus that the consumer obtains from the output in question. First, an increase in vegetation damage (leading to a decrease in yield) will increase costs and therefore increase the minimum price the producer must receive in order to be willing to commit himself to supply any given quantity of the output (and vice versa for a decrease in damage). In addition, altered levels of pollution may affect the quality attributes of the output, thus changing the consumer's willingness to pay and the surplus he acquires from any quantity of the output. The change in cost implies a shift in the seller's supply function, whereas a shift in the buyer's demand function represents a change in his willingness to pay. Finally, if the grower produces for home consumption as well as the market (Strauss, 1986), pollution, by affecting his nutritional intake as well as his market income, can influence his labour productivity. The nutrition-induced effect can include changes in both the productivity of a given labour effort and a nutrition-induced change in the quantity of labour supplied.

Given that producers and consumers have already adapted in order to minimise their prospective losses or to maximise their prospective gains, fig. 10.2 depicts one example of the changes a pollution increase can have upon buyer surplus and seller quasi-rent for a single crop. The pollution increase reduces the desirable properties of the output, making the buyer's willingness to pay smaller and causing his demand function to shift from D_0 to D_1. It simultaneously increases the least cost of producing any particular output quantity, thereby causing an upward shift in the supply function from S_0 to S_1. Market price drops from P_0 to P_1, partly because of the greater relative magnitude of the shift in the demand function and partly because of its lesser relative slope. Buyer surplus was the triangle aP_0b; it is now the triangle dP_1e. Seller quasi-rent was the triangle fP_0b; it is now the triangle gP_1e. Total economic surplus from the production and use of the output in question is thus reduced by the area enclosed by $fghb$ plus the area enclosed by $adeh$.

Of course, the other relative shifts in demand and supply functions, reflecting different effects in the production and consumption sectors, will yield other results. For example, if the demand function shifts to D_2 rather than D_1, market price will rise to P_2. Thus, if the supply shift is small enough, producers could actually see an increase in their quasi-rents. Qualitative results are unchanged, however: alterations in producer quasi-rents and in consumer surpluses result from the pollution-induced changes in the two sectors. These two examples illustrate the issues of concern here: consumers and producers can bear very different economic gains or losses, depending on the relative shifts of the demand and supply functions. Moreover the distribution of these economic consequences can differ

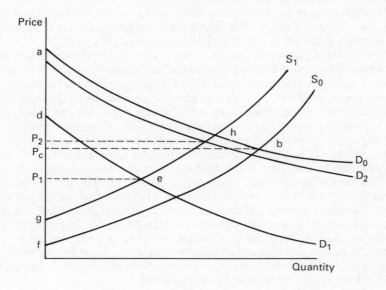

Figure 10.2   Changes in buyer surplus and producer quasi-rent

drastically, depending on how the slope of the demand function is related to the slope of the supply function.

The preceding observations describe, in rather broad brush strokes, the problem of calculating what economic benefits may flow from controlling pollution-related vegetation damage. In the following two sections we discuss technical procedures commonly used to assess these benefits, and the empirical results that have been obtained.

Methods

The discussion of the last section implies that a thorough assessment of pollution control benefits, the gainers' income equivalents, requires three kinds of information: (1) the differential changes that pollution control causes in each person's production and consumption opportunities; (2) the response of input and output market prices to these changes; and (3) the input and output changes that those affected can make to minimise losses or maximise gains from changes in production and consumption opportunities and in the prices of these opportunities. Natural science studies of biological dose-–response functions are the primary source of information for the first requirement. Evaluation of the latter two requirements represents the economics portion of any benefit assessment exercise. If pollution control causes substantial changes in outputs, price changes can occur which, in turn, lead to further market-induced output changes. Moreover, even if prices are constant, natural science information will still fail to provide accurate indications of output changes when individuals can alter production practices and the types of outputs produced. Thus accurate information on the economic consequences for agriculture of pollution can be achieved only if the reciprocal relationship between physical and biological changes and the responses of individuals and institutions are explicitly recognised.

There are two dominant methodologies used to assess the complete economic consequences of pollution impact upon agriculture—mathematical programming and dual profit functions. Mathematical programming and dual profit function methods are analytical representations of an identical problem: the effort by a decision-maker to meet his objectives. For similar pollution levels the two methodologies should produce identical empirical results. In agriculture a reasonable index of the grower's objectives in market economies is his net revenue. The mathematical programming method, by identifying those choices among producers' alternatives which maximise net revenue, estimates change in quasi-rents and surpluses; the profit function method by presupposing net revenue-maximising behaviour, allows the investigator to recover (infer) changes in these rents and surpluses. The mathematical programming approach simulates the behaviour of the agricultural sector by actually solving the producer's problem; while the dual profit function approach, by presuming that the producer has solved his problem, synthetically describes what the behaviour of the agricultural sector has been. That is, the mathematical programming approach estimates what is best for the producer to do under alternative pollution levels which may or may not have occurred historically. The dual profit function approach presumes that the producer has always done what is best for him. Historical data on the revenue he has acquired and the costs he has chosen to incur under alternative pollution regimes are then used to estimate the economic consequences of his behaviour.

Takayama and Judge (1971) provide a detailed treatment of the analytical foundations of mathematical programming applied to the problems of economic analysis; Silberberg (1978) derives the theory underlying the dual profit function approach. Whatever their advantages in giving the researcher the ability to evaluate a wide variety of historical and prospective data and event scenarios, mathematical programming methods possess vast appetites for research time and data. In principle, at least, the dual profit function method requires substantially less data and research time. This is because the method is based on the grower's actual solutions to his own decision problem. That is, the researcher is able to exploit the grower's solution; the grower performs the technical optimisation that is done by the researcher in the programming method. Fewer parameters and less complex interactions must be confronted by the researcher, and, because of this, the trade-off between analytical completeness and empirical tractability which any applied researcher inevitably meets becomes less severe. In fact, as Diervert (1974) proves, the dual cost function has a one-to-one correspondence with a real, unique production (dose-response) function and is the mirror image of the textbook economic supply function (Lau and Yotopoulus, 1979, pp. 21–2; Silberberg, 1978, pp. 309–13). Dose-response functions can thus be established indirectly by applying duality methods to observations on growers' behaviour rather than through direct observations of affected organisms. Alternatively, one can use duality methods to assess economic consequences without having direct knowledge of dose-response function (Garcia *et al.*, 1986).

Empirical results

Tables 10.1 and 2 respectively portray the conditions specified and the empirical results obtained for selected studies of the economic consequences of air pollution for North American, particularly United States, agriculture. The first two studies in table 10.1 and the initial three studies in table 10.2 are examples of the traditional naive approach to economic assessment in which no producer adaptations to altered pollution levels are allowed and in which the effects on consumers are disregarded. With the exception of the duality study of Mjelde *et al.* (1984) the remaining studies cited in these tables employed simulation, particularly mathematical programming approaches.

Caution should be exercised in trying to compare numerical estimates across these studies, since, as the tables make evident, crops, response information and assumed conditions differ considerably. Because the methods used range from the economically naive approach which does not allow for changes in producer and consumer responses to the complex simulation and multi-sector analyses which capture such changes, an individual estimate can provide little insight unless one delves deeply into the unique conditions of the study that produced it. However, even though numerical estimates and the conditions under which they were derived may differ, divergent studies can still exhibit common patterns of behavioural responses and sensitivities to imposed conditions and to data accuracy and precision. Subsequent researchers can then use these commonalities to restrict the dimensions of they analytical and empirical problems. The following discussion emphasises nine general conclusions which emerge from the studies in tables 10.1 and 10.2 and some other, more specialised studies.

1 *Increasing air pollution causes losses in the total economic surplus from the production and consumption of agricultural outputs to increase at an increasing rate.* Many studies illustrate this pattern but almost all deal only with ozone pollution. However, as indicated in tables 10.1 and 10.2 very few have dealt with more than two ambient pollution levels. Thus it is impossible to evaluate the likely rate of change in economic surplus as a result of changes in pollution levels. No thoroughly consistent pattern emerges as to the absolute magnitudes of this rate of change in total surplus.

2 *Growers can gain from increases in air pollution.* Adam, Callaway and McCarl (1986) found that substantial increases in acid deposition enhance the quasi-rents of soya bean growers. Shortle *et al.* (1986) estimated that these growers benefit similarly from ozone increases. Adam and McCarl (1985)

Table 10.1 Selected recent economic studies of air pollution impacts on US agriculture

Study	Pollutants	Ambient concentrations	Completeness				Crops	Results (1980 US dollars)		
			Price changes	Crop substitutions	Input substitutions	Quality changes		Consumer benefits	Producer benefits	Total benefits
Stanford Research Institute (1981)	Ozone SO$_2$	(a) universal reduction to 80 ppb (b) Universal reduction to 260 mg/m³	No	No	No	No	Corn, soybeans, alfalfa, and 13 other annual crops	None	$1800×10^6$	$1800×10^6$
Shriner et al. (1982)	Ozone	(c) universal reduction to 25 ppb	No	No	No	No	Corn, soybeans, wheat, peanuts	None	$3000×10^6$	$3000×10^6$
Kopp et al. (1985)	Ozone	(d) universal reduction from 53 ppb to 40 ppb	Yes	Yes	Yes	No	Corn, soybeans, wheat, cotton, peanuts	Not reported	Not reported	$1300×10^6$
Adams and Crocker (1984)	Ozone	(d) Universal reduction from 53 ppb to 40 ppb	Yes	Yes	Yes	No	Corn, soybeans, cotton	Not reported	Not reported	$2200×10^6$
Adams, Crocker and Katz (1984)	Ozone	(d) Universal reduction from 48 ppb to 40 ppb	Yes	No	No	No	Corn, soybeans, cotton, wheat	Not reported	Not reported	$2400×10^6$
Adams, Hamilton and McCarl (1986)	Ozone	(d) universal reduction from 53 ppb to 40 ppb	Yes	Yes	Yes	No	Corn, soybeans, cotton, wheat, soybeans, barley	$1160×10^6$	$550×10^6$	$1700×10^6$
Shortle, Dunn and Phillips (1986)	Ozone	(d) Universal reduction from 53 ppb to 40 ppb	Yes	No	No	Yes	Soybeans	$880×10^6$	$-90×10^6$	$790×10^6$
Adams, Callaway and McCarl (1986)	Wet acid deposition	(e) Universal decrease from 4·8 to 4·5	Yes	Yes	Yes	No	Soybeans	$-172×10^6$	$30×10^6$	$142×10^6$

(a) Averaging time of one hour; not to be exceeded more than once a year.
(b) Averaging time of 24 hours; not to be exceeded more than once a year.
(c) Annual geometric mean.
(d) Seven hour growing season geometric mean. Given a log-normal distribution of air pollution events, a 7-hour seasonal ozone level of 40 ppb is equal to an hourly standard of 80 ppb, not to be exceeded more than once a year [Heck et al., 1982].
(e) Annual arithmetic mean of rainfall pH.

Table 10.2 Selected regional studies of air pollution impacts on agriculture

Study	Pollutants	Ambient concentrations	Completeness				Crops and place	Total benefits (1980 US dollars)		
			Price changes	Crop substitutions	Input substitutions	Quality changes		Consumer benefits	Producer benefits	Total benefits
Forster (1984)	Wet acid deposition ozone	(e) Universal increase to 5·6 (a) Universal reduction to ppb	No	No	No	No	All Eastern Canada	None	$75×10^6	$105×10^6 $23×10^6
Crocker and Regans (1985)	Wet acid deposition ozone	(e) Universal increase to 5·6	No	No	No	No	All Eastern US	None	$130×10^6	$130×10^6
Page et al. (1982)	Ozone	(a) Universal reduction (b) Universal reduction to 260 mb/m^3	No	No	No	No	Corn, soybeans, and wheat Ohio River Basin, US	None	$400×10^6	$400×10^6
Adams, Crocker and Thanavibulchai	Ozone	(a) Universal reduction to 80 ppb	Yes	Yes	Yes	No	Cotton, sugar beets, and 12 annual vegetables Southern California	$14×10^6	$51×10^6	$65×10^6
Adams and McCarl (1985)	Ozone	(a) Universal reduction to 80 ppb	Yes	Yes	Yes	No	Corn, soybeans, and wheat Corn Belt, US	$207×10^6	$−11×10^6	$66×10^6
Howitt, Gossard and Adams (1984)	Ozone	(d) Universal reduction to 40 ppb	Yes	Yes	Yes	No	38 crops California	$17×10^6	$28×10^6	$45×10^6
Mjelde et al. (1984)	Ozone	(d) 10 per cent increase from 46·5 ppb	No	Yes	Yes	No	Corn and soybeans Illinois	None	$226×10^6	$226×10^6

(a) Averaging time of one hour; not to be exceeded more than once a year.
(b) Averaging time of 24 hours; not to be exceeded more than once a year.
(c) Annual geometric mean.
(d) Seven hour growing season geometric mean. Given a log-normal distribution of air pollution events, a 7-hour seasonal ozone level of 40 ppb is equal to an hourly standard of 80 ppb, not to be exceeded more than once a year [Heck et al., 1982].
(e) Annual arithmetic mean of rainfall pH.

obtained the same result for corn, soya bean and wheat growers, considered as a package. A sufficient condition for these findings is that the pollution-induced percentage reduction in output quantity should be less than the percentage increase in market price that the supply reduction causes. All other studies conclude that growers lose from increased air pollution. However, those studies, such as Crocker and Regans (1985) and Shriner *et al.* (1982), which disregard market price changes and which therefore guarantee producer losses with increased pollution cannot be taken seriously as evidence supporting the hypothesis of producer losses.

3   *Losses in buyer or consumer surpluses are a very significant portion of the total losses that air pollution causes agriculture.* Of the studies that accounted for air pollution-induced changes in the prices of agricultural outputs, consumer surplus losses as a percentage of total losses range from a low of 22 per cent (Adams, Crocker and Thanavibulchai, 1982) to a high of 100 per cent (Adams and McCarl, 1985). Given that producers can sometimes benefit from air pollution increases, methods which disregard consumer impact can, it seems, understate total losses in surplus and grossly misstate the distribution of these welfare effects.

4   *For air pollution increases, percentage losses in total economic surplus are no greater than and nearly always less than the associated percentage changes in biological yields caused by the increased pollution; similarly for air pollution decreases, percentage gains in producer surplus are no less than and nearly always greater than the associated percentage changes in biological yields caused by the decreased pollution.* Every study which accounted for price effects or crop and input substitutions obtained this result. The result is unsurprising to anyone familiar with micro-economic theory. Consider an improvement in air quality from the grower's perspective. The improvement removes a shackle from the grower, causing crops, cultivars, input combinations and growing locations that had previously been non-viable to become realistic alternatives: the set of production possibilities available to the grower has expanded. For a given product price, air quality improvements will certainly increase the net revenues obtained from a former cultivar, input combination and growing location. However, the greater the expansion in the grower's set of possibilities, the greater the odds that a new cropping pattern which produces greater net revenues will be adopted. Consumers will respond analogously to any price reductions; that is, they will substitute towards the input which has had a fall in relative price and be able to buy more of that input for the same expenditure. Depending upon price elasticities, their percentage gain in surplus could also exceed percentage gain in yields. The substitutions allow producers and consumers to magnify the gains they would otherwise acquire if they continued to adhere to their old pattern. Similarly, substitutions allow them to reduce the losses they would otherwise suffer from air quality declines.

5   *Air pollution increases can cause growers of relatively pollution-tolerant crops to suffer production declines greater than those triggered by the pollution increase.*

This phenomenon is explicitly noted by Howitt *et al.* (1984), where growers of pollution-tolerant crops like broccoli, cantaloupe, carrots and sugar beet were also estimated to experience production yield declines as air pollution increased. Adams, Crocker and Thanavibulchai (1982) in another California study, and several studies (e.g. Brown and Smith, 1984; Kopp *et al.*, 1985) of the maize–soya-bean–wheat economy in the US mid-west have produced the same finding. This counter-intuitive result occurs because of a 'crowding out' phenomenon. If, after a pollution increase, the pollution-intolerant crops continue to have higher unit values than the pollution-tolerant crops, and if inputs such as land and water are scare, growers of the pollution-intolerant crops will substitute land, water and similar inputs for air quality. The more profitable affected crops now require more land and water to produce the market equilibrium quantities.

6 *Changes in air pollution affect both the productivity and aggregate demand for factors of production.* Several studies have demonstrated that changes in air pollution will change the demand for specific inputs. Mjelde *et al*, (1984) estimate that a 10 per cent increase in ozone results in a 4 per cent decline in demand for variable inputs such as labour, water and fertiliser. Crocker and Horst (1981) found that the ambient ozone conditions prevailing in southern California during the mid-1970s reduced the economic productivity of selected agricultural workers by an arithmetic mean of 2·2 per cent. Howitt *et al.* (1984), Garcia *et al.* (1986), Kopp *et al.* (1985), and Brown and Smith (1984) have all shown that increased ozone increases the demand for land inputs.

These inputs demand effects result from two interdependent processes. First, changes in air pollution may change the marginal productivity of a given output. For example, if changes in air pollution are viewed as a neutral technological change, air pollution reductions increase the productivity of all inputs proportionally. The value of any input in the agricultural production process is thus increased, causing grower willingness to pay for the input to rise. The second process that results in increased demand for inputs refers to the 'crowding out' phenomenon noted above. That is, some crops and regions receive a greater relative increase in crop productivity from reductions in air pollution. Consequently, an expansion of individual crop acreage or total regional crop production may occur, resulting in an increase in aggregate input use in these more favoured crops and regions. Conversely, crops and regions that do not realise net relative productivity gains from pollution control may reduce individual or total crop acreage, thus causing a reduction in aggregate input use. The exact nature of these input demand changes will be a function of the input mixes used for each crop or region and the relative effects of pollution changes within and across regions.

7 *The contribution of additional biological information to the accuracy and precision of economic assessments declines as more information is collected.* How much plant response information is needed in order to perform credible economic assessments? Several studies have attempted to provide guidance on this issue: Crocker (1982), Adams, Crocker and Katz (1984); Howitt *et al.*

(1984); Adams and McCarl (1985); Kopp *et al.* (1985); Adams and Crocker (1985). Typically, these studies have focused on how the economic estimates change as the researcher modifies assumptions about the nature of the underlying plant response data, such as the functional form of the response model, the number of observations used to estimate such models and the number of cultivars over which to estimate 'representative' response models. The results have generally demonstrated that the gains in economic precision (as measured by changes in the confidence intervals of the estimates) tend to decline quite rapidly as more information is obtained on each issue. Such an observation is consistent with general findings in statistical and other literature on the trade-offs between additional model restrictions and predictive capabilities. These findings, however, do not necessarily mean that additional information on a particular model parameters is never warranted. Rather, the costs of acquiring this information need to be weighed against the consequences of being 'wrong' with regard to the use of the model predictions.

8 *Changes in air pollution have differential effects on the comparative advantage of agricultural production regions.* Numerous studies have focused on the consequences of air pollution across large geographical areas, such as the United States (see Adams and Crocker, 1982; Kopp *et al.* 1985; Adams, Hamilton and McCarl, 1986). In order to correctly represent aggregate supply response across a large geographical region, economists typically define the large regions as being composed of a collection of distinct sub-regions representing different characteristics in terms of crop production alternatives, resource and environmental conditions and other factors within each sub-region. Such economic representations are labelled 'spatial equilibrium' models to indicate that the subsequent analyses assure an equilibrium of supply and demand conditions across the spatial representations (sub-regions) found in the model. A useful feature of such models is that the aggregate economic consequences of the policy alternatives as well as the effects of each policy on each sub-region are measured. The literature demonstrates that the economic effects can vary sharply across regions. Specifically, studies by Adams, Hamilton and McCarl (1986), Kopp *et al.* (1985) and Adams, Callaway and McCarl (1986) indicate that reductions in air pollution resulted in gains to producers in areas simultaneously characterised by high ambient pollution levels and crops that are sensitive to air pollution. Such gains were frequently in the form of expanded crop acreage. Conversely, areas with the opposite set of characteristics experienced reductions in acreage of some crops and hence a loss of 'market share' for those crops. This implies that pollution-induced shifts in comparative production advantages can occur among sub-regions. Thus, while these analyses all found net total gains to society from reductions in air pollution, some sub-regions gained at the expense of others.

9 *Air pollution has economic transboundary effects that alter international trade flows, with attendant gains and losses to exporters and importers.* The economic gains from free and open trade are well recognised. International trade in agricultural commodities has profound effects on the welfare of consumers

and growers in both exporting and importing countries. Some of the recent studies of air pollution effects on United States agriculture have explicitly recognised the role of exports by incorporating a trade component into the model (see Adams, Hamilton and McCarl (1986); Shortle *et al.*, (1986). In the united States, increases in agricultural output have typically moved into the export market. The results of changes in air pollution have thus been to alter slightly the supply of commodities entering this market. Specifically, reduction in air pollution lead to increases in the production of agricultural commodities, with much of this increase moving into exports. The net effect is that foreign consumers of these commodities capture most of the consumer gains. Indeed, Adams, Hamilton and McCarl (1986) estimate that, of the total consumer gains from ambient ozone reductions, 60 per cent accrued to non-United States consumers. Conversely, increases in air pollution imply a reduction in the welfare of importing countries. As a result, national environmental policies can readily have economic transboundary implications even in the absence of a transboundary pollution phenomenon.

The above commonalities emerge from the substantial number of economic studies about air pollution and agriculture. Apparently economists and those who fund their work have believed that information on air pollution impacts is extraordinarily valuable. Assessment studies about pollution issues other than air have been much less numerous. Other than a very few soil salinity studies (see, Feinerman and Yaron, (1983), the literature seems to be devoid of assessment of water pollution effects on agriculture. Over the last 15 years, a few economic studies of sewage sludge spreading have appeared (see, Seitz, (1974); Ott and Forester, (1978); Young and Epp, (1980), and Diner and Yaron, 1986). These studies are all highly site-specific, however, and all focus exclusively upon the direct costs of applications and the gains to growers from enhanced nitrogen and phosphorus. None consider the trace metal and disease organism issues surveyed in Matthews (1984), Ott and Forester, (1978, p. 557), capture the attention level economists have devoted to these latter issues when they state: 'With proper handling, environmental problems caused by the heavy metals and disease-causing organisms can be minimised.' In general, economic assessments of pollution impacts other than air upon agriculture strike one as neither robust nor complete. Studies tend to be time-and site-specific and therefore not robust, and little, if any, attention is given to the price effects and producer and consumer adaptations that truly constitute the economic analysis portion of any assessment exercise.

Problems and data gaps in current economic assessments

Previous sections of this chapter have reviewed economic and biological issues involved in performing assessments of the consequence of air pollution and other pollutants on agriculture. Despite recent knowledge gains about pollution effects on managed eco-systems and concomitant improvements in economic modelling, much that probably has high policy pay-offs remains to be done. These unresolved issues in current economic assessment involve natural science data, aerometric data, and economic assumption or modelling shortcomings.

Natural science knowledge about the dynamic (sequential) responses of vegetation to pollution stresses is, at best, superficial. The dynamics of these responses have implications for both annual and perennial cropping systems. For example, for annuals, we know that the pool of genetic material available to growers tends to change over time. Hence analyses performed today may not be relevant to the gene pool that will be available in the future. More importantly, however, there are large categories of vegetation for which we know little or nothing about responses over time. Perennial plants such as forest species are one clear example where current knowledge is limited mostly to guesses as to long-range consequences.

A related natural science issue is the need to identify potential interactions among pollutants and between pollutants and other environmental stresses. The existing response literature consists mostly of studies of the effects of individual pollutants on individual crops. This crop by crop, pollutant by pollutant approach to crop response, however, is a severe abstraction from real-world conditions. Plants are simultaneously exposed to a range of environmental stresses. It is also likely that changes in pollution precursors associated with specific pollution control policies influence simultaneous pollution stresses. Thus, if one is to assess accurately the economic benefits of an air pollution control policy on crops, one needs to know: (1) the range of pollutant changes that will be associated with changes in pollution control strategies; and (2) what the specific effects of that range of pollutants will be for the crops of interest.

Another issue involving the natural science basis of current economic assessment involves the multitude of crops and crop varieties grown worldwide. Against this backdrop the typical crop by crop, pollutant by pollutant approach to generating response information cannot be expected to resolve a measurement problem of such scope (Adams and Crocker, (1982). If economic assessments are to address pollution impact over large areas, then at least some future plant science research must focus on acquiring a structured and robust understanding of the fundamental processes underlying crop responses to pollution; that is, timely policy responses to pollution problems of large geographical scope require the development of generic plant response information. This may be a less demanding task than first appears. As earlier noted, the literature dealing with the adequacy of limited data sets in economic assessments of pollution suggests that a basic understanding of the sign and general magnitude of biological responses is frequently sufficient to discriminate among the economic implications of policy alternatives.

The currently available response functions used to measure yield changes arising from pollution must be supported by corresponding and appropriate measures of actual crop exposure to air pollutants. Most empirical economic studies have focused on the United States, where a network of monitoring stations exists. The network provides a historical representation of ambient air pollution levels in many rural and agricultural areas. However, the reporting form for these data refers to the annual hourly levels on which urban ambient standards are based. It is difficult and costly to translate changes in these annual hourly average levels into growing season daylight levels appropriate for describing crop responses in agricultural areas. In spite of the fairly stringent assumptions the translation requires, all North American assessments have been forced to proceed with it. The situation in countries with less dense networks which have been operating for shorter time spans would appear to be even more frustrating. In essence, a very critical type of information required to implement the economic assessment methods described in this chapter is lacking in many areas of the world.

The economic portion of the assessment exercise also still faces a number of problems. Though existing studies address a wide range of the economic consequences of pollution, including distributional consequences across producers and consumers, importers and exporters, and production regions and individual crops, opportunities remain to improve the dimensionality of the models. One such opportunity deals with grower uncertainties induced by prospective changes in pollution types and levels. Specifically, if changes in pollution events increase the natural variability of crop yields, the risk-averse grower confronts additional costs: he will expend resources preparing for many situations of which no more than one will be realised at any given time. No assessment study to date has explicitly incorporated risk or the income variability associated with pollution events into its analysis. Procedures to do so are available, though they require information on yield probability distribution as well as data on the income variability associated with changes in pollution. Admittedly, however, large-scale stochastic models like this are substantially more difficult to solve than the static, non-stochastic and discrete models that are now found in the literature.

The conventional economic models universally assume that agricultural outputs are bought and sold in long-run, perfectly competitive market settings. This assumption provides a useful benchmark for analysing issues of efficiency in market-oriented economies. Perfect competition is generally an appropriate assumption for modelling agricultural production in North America, where large numbers of growers respond to market signals as if their individual behaviour has no effect on market prices. However, North American agriculture is typically not in an equilibrium situation but, because of assorted social goals for rural areas, is instead characterised by chronic oversupply. Governments try to combat this oversupply through a mix of commodity production quotas and price subsidies to participating growers. Pollution reductions which increase crop yields therefore conflict, as McGartland (1987), points out, with farm support programmes intended to restrict supply. If one accepts the validity of the social goals which the farm support programmes express, then correct assessments of the net economic consequences of pollution changes makes obligatory the inclusion in the analysis of the programme provisions. From the perspective of the programmes, pollution reductions which lead to supply increases are costly, if only because the increase government carrying costs. In addition, from a strictly market perspective, they expand the agricultural use of inputs such that their social costs exceed the value of their contribution to production.

Another issue now receiving little attention in the literature concerns potential changes in crop demand associated with pollution changes. With the single exception of Shortle *et al*. (1986), all analyses focus upon changes in supply and then integrate the supply changes against assumed static demand relationships for each crop. However, pollution may change the qualities of some commodities. For example, ambient ozone can affect the protein content of soya beans; similarly, sewage sludge spreading can increase the trace-metal burden of plants. These potential demand consequences have yet to be addressed for most crops.

Challenges

The preceding discussion of extant economic assessments, including common themes and common limitations, have been drawn almost exclusively from the North American experience. These studies provide a consistent set of signals concerning the costs that gaseous air pollutants impose on agriculture in the United States and some regions of Canada. In addition, the list of observations common to these North American studies provides a set of restrictions that may be useful in formulating a common approach to similar research efforts in other pollutant, sit and time settings. Specifically, economic enquiries into air pollution effects on agriculture are likely to have theoretical and empirical counterparts with water pollutants, sludge disposal and environmental changes. Thus information gleaned from North American results could serve as a template for other studies.

However, some pollution effects on agriculture are a worldwide phenomenon and may require a different policy perspective. For example, the current scientific and policy discussions concerning global climatic change involve a complex set of pollutants/climatic change interactions that, taken together, have truly global implications. Analysis and resolution of global pollutant issues, such as those inherent in the global climate change problem, creates a set of new challenges for researchers. The first such challenge pertains to the different orderings of economic processes, markets and institutions across countries. This diversity requires an assessment framework that exploits the common relationship between economic orderings while adequately representing the relevant individual characteristics of each country or region. The second challenge relates to a fundamental lack of biological data from which to predict crop responses across matrices of pollutant, site and time combinations. Innovative procedures are required to determine the 'transferability' of response parameters within this matrix. A crucial starting point in economic analysis is the correct representation of the choice problem facing the various decision-makers in each economic setting. In market-oriented economies the neoclassical paradigm of producer and consumer behaviour provides appropriate bases from which to analyse this problem. However, in many settings, individual supply and demand behaviours are essentially subsumed into planned, aggregate supply and demand decisions based on perceived social objectives. In addition to such planned, socialistic economies, a sizeable percentage of world food production and consumption occurs in developing countries, where the interaction of subsistence, village-oriented agriculture and aggregate food demand/production goals, creates a third possible configuration of market structures.

The challenge in measuring the international consequences of pollution on agriculture is to fold these diverse economic structures into a framework that provides some common measure of social welfare. Both the market-orientated and planned economies have a somewhat common goal, namely to produce efficiently a socially optimal level of agricultural output. Market-oriented economies work towards this goal through the market signals created by the interaction of supply and demand forces. Planned economies rely on indirect approaches based on planning goals, such as perceived consumption needs and associated allocations of resources to meet these goals. Interactive, iterative decision processes are frequently used to measure the contribution of alternative resource allocations to the attainment of the specified goals. However, a fundamentally different decision problem confronts subsistence growers, who both produce and consume their output. Numerous behavioural models have been advanced to capture this dual role but in general it is believed that such growers are risk-averse and thus produce first for family or local consumption, using historically proven farming practices. Production moves into domestic or international markets only after the primary goal of subsistence has been achieved. In this producer/consumer setting, pollution affects the individual in ways distinct from the cases heretofore treated. Specifically, while the growers' quantity (yield) and the quality of commodities may be reduced by pollution, as in any other economic setting, the effects on the grower may be more severe in terms of individual welfare. First, any productivity loss reduces marketable surplus and possibly even the subsistence component of production. Second, changes in quantity and quality of production may affect labour productivity if the grower's health is influenced, given that the grower (and immediate family) provide the major input, labour, into the agricultural production process. Third, pollution-induced yield restrictions may intensify risk-averse behaviour, thereby reducing the likelihood of the grower adapting new, yield-increasing technologies.

These unique dimensions of subsistence agricultural economies suggest that measuring the economic effects of pollution on developing countries will require economists to incorporate some of these producer/consumer/labour/supplier linkages into the assessment framework. Available literature on economic development may provide some guidance. In addition, the concept of the 'household production function' proposed by Tinbergen (1956) and expanded by Lancaster (1966) offers a possible framework within which to explore the effects of pollution on the decision problem facing producers and consumers.

However desirable it may be to develop complete economic representations of subsistence and other behavioural models in the short term, easily manageable research on the global consequences of pollution may be limited to gross evaluations of changes in world trade flows in agricultural commodities. Since world trade in agricultural commodities tends to respond to market (price) signals, the competitive models used in some existing trade flow studies may be appropriate. Specifically, a general equilibrium framework that encompasses agricultural production and consumption in each country, and which then captures the net effect of each country on world trade, is required. In practice, existing econometrically based world trade models of agriculture, such as those of the United States Department of Agriculture, may be one means of evaluating aggregate effects of air pollution and other environmental changes.

The size and complexity of trade models can be overwhelming. Thus an idealised model that would integrate a careful and accurate representation of resource allocation/consumer choice issues for diverse market structures, with a general equilibrium specification of world trade, may be beyond the research resources of most institutions or countries. However, acquiring summary information on world producers and trade with which to implement some components of a trade flow approach is a research task that could currently be accomplished. Possible alterations in trade arising from global climate change could then be imposed on current flows to obtain crude estimates of the global consequences of such environmental change. Whatever approach is taken, it may be time for economic and biological researchers to adopt a more global perspective regarding the effects of pollution from other sources upon agriculture.

Wherever they have been done, prospective and retrospective assessments of pollution impact upon agriculture provide convincing evidence of substantial economic effects having plausible policy relevance. Nevertheless the weight of the evidence cannot be interpreted as a call for a proliferation of biological and economic studies unique to each site, time and food and fibre allocation system. Research funds and talents in any country are too valuable to permit this.

As earlier noted, greater emphasis in the appropriate biological sciences upon axiomatic methods rather than mere replication has the potential to conserve research resources. The development of generic dose–response relations will enhance model robustness, where robustness can be defined as the domain of circumstances where the model can be applied without undergoing structural revisions (Adams and Crocker, 1982). The development of generic models will take a long time, however, as no grand model exists which carries across nearly all sites, times and allocation systems. One can then systematically exploit the exchangeability (transferability) concepts of Bayesian statistics set forth in Lindley and Smith (1972). These concepts test the existence of a common structure which generates random samples drawn from a number of distinct groups. They allow one to draw systematic and communicable inferences about the value of a parameter in one group from the observations on all groups. If exchangeability is complete, the identical model would apply to each group and one could then pool the data from all studies and transfer without revision the results for one group to any other group. At the other extreme, if exchangeability is utterly absent, then each group has its own unique structure, and no transfer of results or pooling of data would be justified. Each group would be totally idiosyncratic.

Between these two extremes, the Lindley and Smith (1972) formulation produces a weighted average of the data from other samples and data from the group (time, place and allocation system) of concern, where the weights are determined by the precision of the pooled versus the individual sample estimates. Thus, rather than having to conclude that dose–response functions and economic consequences across times, locations and allocation systems either are totally similar or are totally dissimilar, one can evaluate the exact extent to which they are similar. For many policy problems the results from one setting, though not derived from a structure identical to that in the pollution setting of immediate concern, may nevertheless be close enough to allow full or at least adequate discrimination among relevant policy alternatives. Exchangeability allows the research planner to assess the degree of closeness as well as implications of degrees of closeness for estimating the pay-offs of the policy alternatives.

Concluding comment

This chapter has dealt with a range of theoretical and empirical issues involved in measuring the economic consequences for agriculture of pollution from other sources. Some fairly sharp answers have been obtained about the nature and magnitude of these effects from the studies performed to date. Nevertheless, complex challenges confront economists and their associates in the plant sciences as they move from the static, partial equilibrium site and regional results of current assessments to a more encompassing dynamic, general equilibrium, less site- and pollution-specific framework. No matter how useful the economic and biological research done to date has been, only the easy tasks have been done. Addressing more complex environmental issues such as global climate change will sometimes require a willingness in the near term to compromise disciplinary integrity in order to provide policy-relevant information. Policy relevance in the near term may also demand a tolerance of fairly simple models. Although it is naive to view simplicity as desirable *per se*, researchers should be prepared to offer strong proof that proposals to build ever more complex biological and economic models will ultimately yield information for which the wait will be worth while in terms of reducing otherwise unavoidable arbitrary and incoherent elements in policy choices. More detailed analyses do not guarantee more perfect economic assessments, whatever the theoretical and empirical faults in existing work.

References

Adams, R.M. (1986), 'Agriculture, forestry, and related benefits of air pollution control: a review and some observations', *American J. Agricultural Economics*, 68, 464–72.

Adams, R.M., Callaway, J.M., and McCarl, B.A. (1986), 'Pollution, agriculture and social welfare: the case of acid deposition', *Canadian J. Agricultural Economics*, 34, 3–19.

Adams, R.M., and Crocker, T.D. (1982), 'Dose–response information and environmental damage assessment: an economic perspective', *J. Air Pollution Control Assoc.*, 32, 1062–7.

Adams, R.M., and Crocker, T.D. (1984), 'Economically relevant response information and the value of information: the case of acid rain', in T.D. Crocker, ed., *Economic Perspectives on Acid Deposition Control*, Boston, Mass., Butterworth, 35–64.

Adams, R.M., Crocker, T.D., and Katz, R.W. (1984), 'Assessing the adequacy of natural science information: a Bayesian approach', *Review of Economics and Statistics*, 66, 568–75.

Adams, R.M., Crocker, T.D., and Katz, R.W. (1985), 'Yield–response data in benefit–cost analyses of pollution-induced vegetation damage', in W.E. Winner, H.A. Mooney and R.A. Goldstein, eds., *Sulfur Dioxide and Vegetation*, Stanford, Calif., Stanford University Press, 56–74.

Adams, R.M., Crocker, T.D., and Thanavibulchai, N. (1982), 'An economic assessment of air pollution to selected annual crops in southern California', *J. Environmental Economics and Management*, 9, 42–58.

Adams, R.M., Hamilton, S.A., and McCarl, B.A. (1986), 'The benefits of pollution control: the case of ozone and U.S. agriculture', *American J. Agricultural Economics*, 68, 886–93.

Adams, R.M., and McCarl, B.A. (1985), 'Assessing the benefits of alternative oxidant standards on agriculture: the role of response information', *J. Environmental Economics and Management*, 12, 264–76.

Atkinson, S.E., Adams, R.M., and Crocker, T.D. (1985), 'Optimal measurement of factors affecting crop production: maximum likelihood methods', *American J. Agricultural Economics*, 67, 414–18.

Atkinson, S.E., and Crocker, T.D. (1987), 'A Bayesian approach to assessing the robustness of hedonic property value studies', *J. Applied Econometrics*, 2, 27–45.

Brown, D., and Smith, M. (1984) 'Crop substitution in the estimation of economic benefits due to ozone reduction', *J. Environmental Economics and Management*, 11, 327–46.

Boyer, J.S. (1982), 'Plant productivity and environment', *Science*, 218, 443–7.

Cameron, C.A. (1974), *Garden Chronicle*, 1, 1257–78.

Crocker, T.D. (1982), 'Pollution-induced damages to managed ecosystems: on making economic assessments', in J.S. Jacobson and A.A. Miller, eds., *Effects of Air Pollution on Farm Commodities*, Arlington, Va., Izaak Walton League of America, 103–24.

Crocker, T.D. and Horst, Jr., R.L.(1981), 'Hours of work, labor productivity, and environmental conditions: a case study', *Review of Economics and Statistics*, 63, 361–8.

Crocker, T.D. and Regans, J.L. (1985), 'Acid deposition control: a benefit–cost analysis', *Environmental Science and Technology*, 19, 112–15.

Day, R.H. (1965), 'Probability distributions for field crops', *J. Farm Economics*, 47, 713–41.

Diervert, W.E. (1974), 'Applications of duality theory', in M.D. Intriligator and D.A. Kendrick, eds., *Frontiers of Quantitative Economics*, II, Amsterdam, North Holland, 3.

Diner, A., and Yaron, D. (1986), 'Optimisation of municipal waste water and reuse for regional irrigation', *Water Resources Research* , 2, 331–8.

Dixon, B.L., Garcia P., and Mjelde, J.W. (1985), 'Primal versus dual methods for measuring the impact of ozone on cash grain farms', *American J. Agricultural Economics*, 67, 402–6.

Feinerman, E., and Yaron, D. (1983), 'The value of information on the response function of crops to soil salinity', *J. Environmental Economics and Management*, 10, 72–85.

Garcia, P., Dixon, B.L., Mjelde, J.W., and Adams, R.M. (1986), 'Measuring the benefits of environmental change using a quality approach: the case of ozone and Illinois cash grain farms', *J. Environmental Economics and Management*, 13, 69–80.

Hamilton, S.A., McCarl, B.A., and Adams, R.M. (1985), 'The effect of aggregate response assumptions on environmental impact analyses', *American J. Agricultural Economics*, 67, 407–13.

Heck, W.W., Taylor, O.C., Adams, R.M., Bingham, G., Miller, J., Preston, E., and Weinstein, L. (1982), 'Assessment of crop loss from ozone'. *J. Air Pollution Control Association*, 32, 353–61.

Heck, W.W., Cure, W.W., Rawlings, J.O., and Zaragosa, L.G., *et al.* (1984), 'Assessing impacts of ozone on agricultural crops', *J. Air Pollution Control Association*, 34, 729–35, 810–17.

Howitt, R.E., Gossard, T.W., and Adams, R.M. (1984), 'Effects of alternative ozone levels and response data on economic assessments: the case of California crops', *J. Air Pollution Control Assoc.*, 34, 1122–7.

Kopp, R.J., Vaughn, W.J., Hazilla, M., and Carson, R. (1985), 'Implications of environmental policy for U.S. Agriculture: the case of ambient ozone standards', *J. Environmental Management*, 20, 321–31.

Lancaster, K.J., (1966), 'A new approach to consumer theory', *J. Political Economy*, 74, 132–57.

Lau, L.J. and Yotopoulous, P.A. (1979), 'The methodological framework', *Food Research Institute Studies*, 17, 11–22.

Lindley, D.V. and Smith, A.F.M. (1972) 'Bayes estimate for the linear model', *J. Royal Statistical Society*, Series B, 34, 1–41.

Matthews, P.J. (1984), 'Control of metal application rates from sewage sludge applications in agriculture', *CRC Critical Reviews in Environmental Control*, 14, 220–49.

McGartland, A.M. (1987), 'The implications of ambient ozone standards for US agriculture: a comment and some further evidence', *J. Environmental Management*, 20, 139–46.

Mjelde, J.W., Adams, R.M., Dixon, B.L., and Garcia, P. (1984), 'Using farmers' actions to measure crop loss due to air pollution', *J. Air Pollution Control Assoc.*, 34, 360–3.

Ott, S.L., and Forester, D.L. (1978), 'Landspreading: an alternative for sludge disposal', *American J. Agricultural Economics*, 60, 555–8.

Page, W.P., Arbodgast, G., Fabian, R.G., and Ciecka, G. (1982), 'Estimation of economic losses to the agricultural sector from airborne residuals in the Ohio river basin', *J. Air Pollution Control Ass.*, 32, 151–4.

Rawlings, J.D., and Cure, W.W. (1985), 'The Weibull function as a dose–response model for studying air pollution effects', *Crop Science*, 48, 423–39.

Roback, J. (1982), 'Wages, rents, and the quality of life', *J. Political Economy*, 90. 1257–78.

Seitz, W.D. (1974), 'Strip-mined land reclamation with sewage sludge: an economic simulation', *American J. Agricultural Economics*, 56, 799–804.

Shortle, J.S., Dunn, J.W., and Phillips, M. (1986), *Economic Assessment of Crop Damage due to Air Pollution: the Role of Quality Effects*, staff paper 118, Department of Agricultural Economics, Pennsylvania State University, State College, Pa.

Shriner, D.A., Cure, W.W., Heagle, A.S., Heck, W.W., Johnson, S.W., Olson,

R.J., and Skelly, J.M. (1982), *An Analysis of Potential Agriculture and Forestry Impacts of Long-range Transport Air Pollutants*, ORNL-5910, Oak Ridge, Tenn., Oak Ridge National Laboratory.

Silberberg, E. (1978), *The Structure of Economics*, New York, McGraw-Hill.

Stanford Research Institute (1979), *An Estimate of the Nonhealth Benefits of meeting the Secondary National Ambient Air Quality Standards*, report prepared for the National Commission of Air Quality, Washington, D.C.

Strauss, J. (1986), 'Does better nutrition raise farm productivity?', *J. Political Economy*, 94, 297–320.

Takayama, T., and Judge, G.G. (1971), *Spatial and Temporal Price and Allocation Models*, Amsterdam, North Holland.

Tinbergen, J. (1956), 'On the theory of income distribution', *Welwirtschaftliches Archiv*, 77, 155–75.

Tingey, D.T., Thutt, G.L., Gumpertz, M.L. and Hogsett, W.E. (1982), 'Plant water status influences ozone sensitivity of bean plants', *Agriculture and Environment*, 7, 243-54.

Weibull, W. (1951), 'A statistical distribution of wide applicability', *Applied Mechanics*, 18, 293–7.

Young, C.E. and Epp, D.J. (1986), 'Land treatment of municipal waste water in small communities', *American J. Agricultural Economics*, 62, 238–43.

11. The impact of sewage sludge on agriculture

M. Linster

Introduction

Background

In the past fifteen years the percentage share of population linked up to a sewage treatment facility has steadily increased in OECD member countries. This has led to progressively larger quantities of sewage sludge, which must then be treated and disposed of as effectively as possible.

Between 1970 and 1980 the volume of sludge produced rose by about 15 to 30 per cent, depending on the country concerned, and current output in OECD countries is estimated at approximately 12 million tonnes of dry matter per year (table 11.1). In view of the trends in water quality standards and improved treatment systems, it may be assumed that the output will increase even further during the next ten years.

Sewage sludge can be of domestic or mixed (industrial plus domestic) origin, depending on the type of network feeding the effluents to the treatment plant. Thirty to forty per cent of the sludge is currently used in agriculture or in similar activities such as forestry, wasteland reclamation or the creation of urban green spaces and landscaping, the remainder being sent to landfill, incinerated or dumped at sea (table 11.1). The choice often goes to the cheapest and most practical solution, depending on the specific local conditions and relevant investment and operating costs.

Out of all these potential disposal methods, agricultural spreading is the only one which takes advantage of the nutrients contained in sludge and often seems to be the cheapest method. It is only with large-scale treatment plants which cater for more than 300,000 inhabitant equivalents that other systems such as incineration are viable. It was therefore chiefly on economic grounds that the agricultural use of sludge was promoted in many countries during the 1970s. The European Community is encouraging and giving priority to this solution. The promotion of the agricultural use of sludge has been accompanied by the demonstration of the fertiliser value of sludge and its beneficial effects on crop yields. Above all, it provides an attractive alternative to inorganic fertilisers, since it produces similar effects at lower cost.

However, it was discovered that although sludge may have positive effects on agriculture, it also entails hazards for agriculture, for the environment in general and man in particular. The presence of toxic substances (heavy

Table 11.1 Annual production of sludge and disposal methods, OECD countries, early 1980s

Country	Volume of sludge produced (000 tonnes dry matter/year)	Sludge disposal methods (% of total)				
		Soil spreading	Landfill	Incineration	Dumping at sea	Not specified
Canada	500	42	18	40	—	—
USA (a)	4500	42	15	27	4	12
Japan (b)	na	14	42	na	35	8
Australia	na	na	na	na	na	na
New Zealand	na	na	na	na	na	na
Austria	140	47	—	20	—	33
Belgium	70	10	80	10	—	—
Denmark	130	45	55	10	banned	—
Finland	130	41	37	—	—	22
France (c)	840	20	46	20	1	—
Germany	2100	32	56	10	2	—
Greece	3	—	100	—	—	—
Iceland	na	na	na	na	na	na
Ireland	20	4	39	—	47	—
Italy (d)	500	30	50	20	—	—
Luxembourg	6.5	90	10	—	—	—
Netherlands	230	58	27	1	9	5
Norway	55	40	40	—	5–10	10–15
Portugal	na	na	na	na	na	na
Spain	45	60	20	—	20	—
Sweden	210	60	30	—	—	10
Switzerland	150	71	—	29	—	—
Turkey	na	na	na	na	na	na
UK	1500	41	27	4	28	—
Yugoslavia	na	na	na	na	na	na
North America	5000	42	..	..	..	..
Europe–CEC (e)	6000	29	45	7	19	—
OECD Europe (e)	7000	..	..	..	..	..
OECD total (e)	12000	..	..	..	..	..

(a) Twenty-six per cent in 1976.
(b) Dumping at sea covers coastal discharges.
(c) An additional 13 per cent are treated for dumping and agricultural use.
(d) Data for incineration include non-specified.
(e) Estimates.
na not available

Source: CEC (Cost 68), OECD, national reports.

metals) and pathogens present appreciable potential hazards. The range of impact of these substances is wide, they affect many environments and interact in complex ways which make the size of the hazard they present difficult to quantify and control. These considerations have led to research programmes aimed at identifying the processes concerned and thereby encouraging the improved use of sludge. In particular, the European Communities developed the COST 68 programme covering research projects by member States and non-member countries such as Canada, the United States and Finland.

In many countries sewage sludge is used preferentially in agriculture provided the harmful effects (frequently due to excessive and uncontrolled spreading) can be curbed and farmers can be convinced of the value of sludge. Many member countries have adopted regulations to encourage the improved use of sewage sludge. However, the standards proposed and the basis for this use differ markedly from one country to the next, owing, among other things, to the uncertainties still remaining as to the scientific processes involved and the specific national context.

Characteristics and composition of sludge

Depending on its origin (domestic, industrial) and the type of treatment used, sewage sludge may have very different physical and chemical properties. Analysis shows that it contains appreciable amounts of organic matter and nutrients such as nitrogen and phosphorus, hence its value in agriculture (table 11.2). Other constituents of sludge are as follows:

Table 11.2 Composition of sludge compared with other fertilisers (% dry matter)

Constituent	liquid sludge	Solid sludge	Composted sludge	Farmyard manure	Pig farm effluent
Organic matter	40–60	30–60	88	60	80
Total nitrogen	2·5–6	1·5–3	3	1–3	4–6
Phosphorus (P_2O_5)	3–8·4	2–8·4	1·5	0·5–2·5	2–5
Potassium	0·5–1·5	0·5	0·5	2·5–3·5	3–5

Source: French Ministry of the Environment, Paris.

1 *Main trace elements* such as zinc, copper, manganese and boron: these elements, which are vital at low concentrations for plants, become phytotoxic above a certain threshold, but their concentrations are generally low.
2 *Other trace element* such as lead, cadmium, mercury and arsenic: their contents in sludge widely exceed those in the soil, they have no nutritional value and they may be hazardous at relatively low concentrations.
3 *Pathogens* (bacteria, viruses, parasites): the most representative group of bacteria are Salmonellas; parasites occur in the form of worms, worm eggs or protozoan cysts.

4   *Calcium*: liquid sludge may contain 0·2 to 1·5 per cent of CaO, and solid sludge from 2 to over 20 per cent, but the concentration is too low for sludge to be regarded as calcium fertiliser (apart from limed sludge).
5   *Toxic organic compounds*: except in rare cases, sludge contains very few organic pollutants (pesticides, detergents), since they are mostly destroyed during the different treatment processes.

Benefits relating to the agricultural use of sewage sludge

The benefits relating to the agricultural use of sewage sludge depend on several factors connected with the quality and characteristics of the sludge and its capacity to meet agricultural requirements. The effects aimed at are the nitrogen, organic matter and phosphorous effects. Sewage usually contains the main nutrients required by plants, but its capacity to produce the desired effect and to 'compete' with commercial inorganic fertilisers depends on several factors, including:

1   Origin of the sludge.
2   Treatments undergone.
3   Nature of the host soil.
4   Type of crop.
5   Timing and method used for spreading.
6   Weather conditions.

Nitrogen fertiliser value of sludge
Nitrogen is an essential element for good crop yields but may lower water quality (leaching of nitrates) and harm crops when excess amounts are present in the soil.

Sludge contains nitrogen in two different forms: inorganic, which can be directly taken up by the plants, and organic, which can be taken up only as it turns into inorganic nitrogen in the soil. This gradual conversion may take one to twelve months or even longer and constitute, depending on the type of crop, a drawback or an advantage over commercial fertilisers. For instance, this is advantageous for maize production. Furthermore, nitrogen present in an inorganic form in the soil is better retained by the latter and leaching risks are much lower.

The nitrogen efficiency of sludge depends on its total nitrogen content, the share of nitrogen initially occurring in inorganic form (ammonia), and the organic carbon content, which influences the mineralisation and denitrification processes. It is also governed by external factors, such as temperature, as heat may lead to important nitrogen losses during the first week following spreading through volatilisation of ammonia. Further losses may arise through denitrification and leaching. There are also considerable variations during the year for sludge from the same sewage treatment plant: total nitrogen may vary by a factor of 2 and ammoniacal nitrogen even more.

The biggest difference between sludge and commercial fertilisers is that the latter contain only inorganic nitrogen, which can be directly taken up by plants. Field experiments using different types of crop have shown that the nitrogen efficiency of sludge varies from one crop to the next, depending on the type of sludge used. For instance, the amount of nitrogen taken up by rye grass from stabilised liquid sludge is roughly the same as that taken up after inorganic fertiliser treatment. In other cases, and with other types of sludge, this percentage may be lower but, in general, crop yields following sludge spreading are as good as those obtained following commercial fertiliser spreading, at equal quantities of inorganic nitrogen. To obtain equivalent yields, it is therefore often necessary to apply larger amounts of nitrogen from liquid sludge than of nitrogen from commercial fertilisers.

The wide range of results obtained by different studies highlights how difficult it is to determine the precise nitrogen fertiliser value of the sludge. Many studies are highly specific and merely consider one type of sludge under given conditions. It is therefore very difficult to draw reliable conclusions application to natural conditions.

Organic fertiliser value of sludge

The organic matter content of sludge (on average 50 per cent of the dry weight) is sufficient for its use as organic fertiliser. However, although the factors governing soil fertility such as physical and chemical characteristics and organic matter content are fairly well known, much less is known about the precise effect of sludge on these characteristics. As with any organic fertiliser, the use of sludge improves soil structure, hydraulic characteristics, water retention and the availability of nutrients to plants.

As for the nitrogen and phosphate fertiliser effects, it is difficult to evaluate the precise role of sludge as organic fertiliser. It seems that it is relatively low when moderate amounts are spread, but it may be high in the case of repeated massive applications at regular intervals. The need to apply large quantities may, however, lead to heavy metal accumulation and over-fertilisation, which in turn can result in water pollution.

Phosphate fertiliser value of sludge

The phosphorus content of sludge chiefly depends on the processes used for treating the waste water from which it is produced. Sludge produced by mechanical-biological treatment, which removes 25 per cent of the phosphorus in waste water, does not have a very high phosphorus content, whereas that from physio-chemical treatment contains large amounts. The average phosphorus content may vary by a factor of 2 depending on whether the treatment involved phosphate removal.

Absorption of phosphorus by plants is affected by low temperature, drought, poor soil and deficiencies in other nutrients such as nitrogen.

Opinions differ regarding the phosphate effectiveness of sludge. Some studies show that sludge has almost the same phosphate efficiency as certain inorganic fertilisers such as basic slag and rock phosphates when it is produced by physio-chemical treatment (good solubility in the soil and high availability for plants, the best results being obtained with lime-treated sludge), whereas that from biological processes is much less efficient. However, the average efficiency is lower for phosphates than for nitrogen.

Economic value of sludge

The commercial value of sludge and the economic advantage of its agricultural use are difficult to quantify. A French study recently priced the gross value of sludge at approximately FFr 300 per ton of dry matter. However, this figure takes into account only the initial nitrogen and phosphorus content and not the actual fertiliser efficiency.

Sludge is often supplied free of charge to farmers or sold at prices lower than its actual value in order to promote its use. In some cases the storage and/or transport and even spreading costs are paid by the farmer, or in others by the local authorities, since they are too high to be borne by the users.

Conclusion

The overall fertiliser efficiency of sludge seems good compared with commercial fertilisers. The nitrogen fertiliser effect is more than satisfactory (especially for the stabilised liquid sludge) and the phosphate effect moderate (physio-chemical sludge). Nutrients are no doubt released more slowly than from inorganic fertilisers, but the long-term results are sometimes better than those obtained with inorganic fertilisers. The potassium content alone remains low, and it may therefore be necessary to supplement the sludge with commercial fertiliser. To obtain the same yields, however, higher quantities of nitrogen and phosphorus must be applied than of inorganic fertilisers. Some studies have also shown the existence of a yield threshold beyond which only inorganic fertilisers produce higher yields.

It is therefore difficult to determine the actual efficiency of sludge. It not only depends on the intrinsic qualities of the sludge but is influenced by many different factors, including environment ones such as pH, soil texture, air temperature and climate. Its wide range and, in particular, nitrogen content fluctuations may lead to problems in the agricultural use, since the amount of nitrogen which can be taken up by the crops is specific. To prevent surplus nitrogen spreading it is necessary to analyse the sludge at regular intervals to determine the precise nitrogen content. More information is also required on after-effects or long-term efficiency, which are difficult to quantify but also play an important role compared with commercial fertilisers.

Hazards relating to the agricultural use of sludge

Alongside the positive effects on crops, sludge may have harmful effects on agriculture and the environment as well as on man. These harmful effects may result from surplus application of nutrients and the presence in sludge of substances such as heavy metals, pathogens and organic compounds. The hazards may be summarised as follows:

1 Direct negative effects on agriculture, affecting animals and crops.
2 Human health hazards through contamination of the food chain.
3 Impact on other environments and eco-systems (soil, water, air) through volatilisation, leaching, run-off and accumulation.

These hazards are compounded by the combined effect of several different pollution sources, such as sewage sludge plus atmospheric pollution (acid rain, heavy metal deposits, etc.). The most hazardous elements are heavy metals, which may act as a limiting factor in the agricultural use of sewage sludge.

Heavy metals

Origin of heavy metals
Heavy metal contamination of waste water, and hence sludge, arises from domestic activities, run-off water (leaching from roads, rainwater, etc.) and industrial discharges. Conventional treatment systems remove 60 to 72 per cent of the cadmium, 28 to 73 per cent of the chromium, 45 to 70 per cent of the copper, 20 to 70 per cent of the nickel, 54 to 73 per cent of the lead and 40 to 74 per cent of the zinc present in effluents. A practical study conducted in Switzerland found that run-off water plays a considerable role in the contamination of sludge by heavy metals. Its heavy metal content, especially cadmium, chiefly originates from industrial releases into the atmosphere. The share of domestic pollution varies on average from 10 to 30 per cent, except for zinc and cadmium (up to 50 per cent). In the case of zinc the large share of domestic origin is due to the corrosion of drinking water supply systems and the presence of the metal in certain detergents. At the industrial level the study found massive contamination in the vicinity of 20 per cent from the incineration of domestic waste and composting (storage areas, slag cleaning, stack gas scrubbing). The importance of diffuse sources in the overall heavy metal balance is therefore far from negligible, and in this context monitoring of industrial effluents would be only partly effective.

Impact of heavy metals
The problems arising from heavy metals are chiefly due to the fact that their half-life is very long (approximately a thousand years in most cases). They therefore tend to accumulate and represent very long-term hazards. The scale of heavy metal effects on crops varies, depending on the type of soil, the natural and artificial content of heavy metals in the soil, the nature of the crops and the effect of other polluting sources.

Phytotoxic effects
It is generally recognised that heavy metals are a limiting factor in plant growth. Their toxic effect depends not only on the total amount present in the soil but above all on their concentration in the soil solution. The metals

present in the liquid phase are those which can be directly taken up by the plant roots. The effect of soil characteristics (pH, organic matter content, etc.) is paramount in this respect. The more acid a soil the less it retains heavy metals, which enter the liquid phase and thereby become more available to plants. It is therefore very important to have good knowledge of the soil characteristics, the metals present and the mobility of the metals in the sludge–soil–plant–food chain system.

The element presenting the greatest toxicity hazards is cadmium, owing to its high mobility through the food chain. It is known to be highly available for absorption by plants, unlike other elements, such as lead, which are difficult to absorb.

There is no doubt that the spreading of sludge increases heavy metal concentration in the soil solution. The combined supply of organic matter may, however, limit (at least initially) the transfer of heavy metals to the liquid phase. In general, heavy metals from inorganic sources are thought to be more available to plants than those from organic sources.

The phytotoxic effects of certain metals are cumulative, as with zinc, copper and nickel. This is why some standards are expressed in zinc equivalents.

Comparative studies of yields obtained following the application of sludge with low heavy metal concentrations and of sludge with high concentrations have shown a slight decrease in crop yields for highly contaminated sludge. A French study of crops that had received massive amounts of sludge polluted by cadmium and nickel showed yield losses of 20 per cent for maize crops. The lower yields were accompanied by yellowing of the older leaves, increased heavy metal concentration in leaves, and a general decrease in the phosphorus content. On the other hand, nickel accumulated chiefly in seed. For other types of crops such as lettuce, yields were not affected in spite of a higher heavy metal concentration in the leaves. The effects observed for average doses were fairly similar to those observed with massive ones. It therefore seems that plants growing on soil with a high heavy metal content rapidly reach a maximum level beyond which further sludge applications have little effect.

Soil contamination
The difference in concentration between soil and sludge highlights the hazards arising from agricultural use of sludge. In the United Kingdom the average cadmium concentration in sludge is 29 mg/kg of dry matter (1980) whereas basic concentrations in uncontaminated soil are less than 1 mg/kg of dry matter.

The level of accumulation of heavy metals from sludge in the soil depends on their availability for absorption by plants and for leaching to ground water. For soils with a pH of at least 6 the risk of leaching is extremely slight. Such soils have a good metal retention capacity and the elements contained in the sludge may on the contrary accumulate in the topsoil layer. This is why the repeated application of sludge containing heavy metals may lead to progressive and irreversible contamination of agricultural land. In the past, excessive spreading of sludge has made land unusable for farming purposes.

Food chain contamination

The accumulation of heavy metals in the topsoil creates a hazard not only for plants but also for grazing animals, which may ingest the metals directly. Entry into the food chain therefore occurs through both crops and contaminated animals. The most sensitive crops to the accumulation of heavy metals are fruit and vegetables.

Impact on human health

The greatest problems arise from the toxic elements accumulating in the body, such as cadmium and mercury. Usually sludge contains only low concentrations of such elements, but since they are highly toxic it is necessary to monitor closely the amount present in sludge for agricultural spreading.

Conclusion

There can be no doubt as to the hazards relating to the presence of heavy metals in sludge used in agriculture, and much research has been conducted in this field. However, there are still several information gaps which, combined with the many factors involved, make them difficult to evaluate precisely. The actual impact depends not only on the metal content of sludge but also on the initial concentration in the soil, the effect of other pollution sources, including acids and heavy metal deposits on leaves, soil acidification by acid rain or the application of certain inorganic fertilisers, and also metal mobility through different environments.

Crop yield data are also difficult to interpret, since the beneficial effects of nutrients and organic matter cannot be dissociated from the harmful effects of heavy metals. In addition, the few studies showing yield reductions were conducted following the spreading of highly contaminated sludge. It is also difficult to predict plant response accurately and to extrapolate from laboratory studies. Caution is therefore required when interpreting the findings, although the impact should not be minimised.

Pathogens

Origin of pathogens

The nature of the pathogens present in sludge varies from one country to the next and their concentration depends on the quantity of animal waste present in effluents and the infection level of the local population. In European temperate climates, approximately 10 per cent of parasite eggs present in waste water are thought to be of human origin. The main sources of pathogens are tanneries, slaughterhouses, certain microbiological laboratories and hospitals.

Effects of sludge treatment on the pathogen content of sludge

Untreated raw sludge shows the highest pathogen level of all sludges, higher than that of treated waste water. The next highest is stabilised or slightly composted sludge. The number of vegetative bacteria is considerably reduced, but the effects on cysts, eggs and spores are thought to be only temporary and to depend on temperature. Lime-treated sludge or sludge irradiated at low doses contain no vegetative bacteria with the exception of parasites and bacterial spores and a few other pathogens. The most effective treatment processes are pasteurization, high-dose irradiation, high-temperature composting, aerobic thermophil stabilisation and high-temperature treatment.

This account of the effect of treatment processes is fairly sketchy and their actual effectiveness is a mater of debate. An effective treatment for one type of organism is not necessarily effective for another. It should also be remembered that parasite eggs or cysts can be highly resistant, even after fairly long periods. Furthermore, sludge may be recontaminated by equipment infected during storage, transport and even spreading operations.

Fate of pathogens in the soil

Sludge applies to the soil undoubtedly contains viable human pathogens. However, these are not all hazardous to livestock or human health: they may either be present in too few numbers or, depending on the environmental conditions, die rapidly following spreading. The factors involved are as follows: initial numbers of organisms, weather conditions (temperature, frost, humidity. sunshine), concentration of hydrogen ions and mineral salts in the soil, soil permeability, the presence of available organic matter and that of predator micro-organisms. Owing to these many factors, it is difficult to evaluate survival time, and the results vary from one study to the next. In the soil the survival time for Salmonellas varies from less than thirty days to over one year. In grass the survival time is shorter and ranges between ten days and three weeks. The shortest survival times are usually observed at the tips of grass blades which are exposed to the sun, while longer survival time are observed on the lower parts. The risk is therefore not the same, depending on the type of grazing animal: for instance, sheep graze grass tips and pigs graze the lower parts, including roots and soil.

Hazards relating to the presence of pathogens

HAZARDS TO AGRICULTURE

The presence of pathogens in sludge is chiefly a hazard to grazing animals and has no effect on plant growth and crop yield. Plants are regarded above all as the 'vectors of pathogens.

The risk of infecting grazing animals decreases with time; and in most cases, provided that animals are put out to graze only some time following spreading, the risk is limited. Studies have shown that exposure of animals to contaminated pastures does not necessarily lead to infection of the herd. In spite of this, in some countries, such as Switzerland, Germany and the Netherlands, cases of salmonellosis have been directly linked with the spreading of infested sludge. A comparison between the infection rate among cattle grazing on land treated with sludge and that on untreated land has revealed significant differences. In the United Kingdom, however, research on this aspect has not revealed any evidence of a direct link between sludge and diseases. The problem mostly arises in areas where land available for sludge spreading is limited, and animal population density is high and the pathogen content, especially that of Salmonellas, of sludge is high because untreated raw sludge is used. Apart from Salmonellas, no other species of bacteria seem to be transmitted by sludge. As for parasites, it has been shown that sludge can act as a vector for worms such as thread and tape worms. For other species, the role of sludge is much less certain and would require additional research and more systematic epidemiological monitoring. Also, very little work has been done concerning viruses.

HAZARDS TO MAN

Pathogens may be transmitted to man in different ways. The principal methods are as follows:

1 Contamination of persons working or playing on land treated with sludge.
2 Contamination of treatment plant personnel.
3 Contamination of fruit and vegetables grown on land treated with sludge
4 Contamination of surface water through run-off.
5 Contamination of animals grazing on land treated with sludge.
6 Transmission by domestic and wild animals, including birds and insects.

Nevertheless, the actual hazards to human health remain slight and, in most cases, can be controlled through good practice guidelines. The few known cases of infection were probably caused by the consumption of infected milk or raw vegetables before standards were laid down in this field.

Conclusion

There is still insufficient information available on the actual danger posed by the presence of pathogens in sludge, and in particular about its identification. It is difficult to draw a direct link between the spreading of sludge and the onset of diseases in animals and man. The problems raised by the identification of this cause-to-effect relationship are chiefly due to the many different factors involved and to the difficulty of grading these factors. Sludge is not the only vector of pathogens, and where any disease has been identified the other transmission modes should also be taken into account. Furthermore, in man, diseases, since the symptoms are not always specific, are not always diagnosed. The only known cases of infection were due to the application of raw sewage on crops eaten raw and gathered too soon after sludge spreading.

The effectiveness of the different treatment methods is also a matter of debate and there is still insufficient information on the subject to provide a sound basis for any preventive measures to be taken.

In any case it is accepted that the presence of pathogens in sludge is not a limiting factor for agricultural use. Much more concern is shown about the hazards to man than to agriculture as such. Potential harmful effects on animals and man may be kept to a minimum by complying with application and storage instructions and prohibiting the use of raw sewage.

Toxic organic compounds

The least known pollution from sludge concerns organic compounds, especially since there have not been any clear indications so far on any harmful effects on plants, soil or man following agricultural spreading.

The main source of organic pollutants in sludge is industry. Sludge concentrations are usually very low (ranging from 0·01 to 0·23 p.p.m. for PCBs), the highest concentrations occurring in sludge from large urban treatment plants.

Organic pollutants are not taken up by plants as readily as heavy metals. Most of the research on the subject concerns PCBs. These are less degradable in soil than other organic micropollutants, and their half-life ranges from two months to five years, depending on the number of chlorine atoms present. They may therefore accumulate in living tissues throughout the food chain.

In the long term, organic pollutants and especially PCBs might pose a real problem but there is still insufficient information on the subject.

Pollution transfer to other environments

Under certain conditions, nutrients and other substances contained in sludge may pollute other environments, such as surface water, ground water and even air through run-off, leaching, volatilisation. These impacts mostly result from the application of excess quantities of sludge and hence the substances concerned, and from an unsuitable application procedure not taking into account the spreading period, type of soil and crop and weather conditions. Nitrogen supplied in this way may increase air pollution through volatilisation of ammonia and the nitrogen dioxide produced by denitrification, surface-water pollution through run-off of organic nitrogen and ammonium, and ground-water pollution through nitrate leaching.

Leaching
Leaching of nutrients and other substances is governed by several factors, including:

1 The amount of water percolating through the soil, depending on rainfall, evapotranspiration and soil water retention capacity.
2 The quantity of substances present in the soil, regardless of whether or not their origin is natural or supplied through fertilisers.
3 Soil type: losses are usually higher for sandy soils and lower for clay soils.
4 Crop type: losses are higher for fallow land and land used for crops with short growth period, and lower for permanent grassland.
5 Type of substance concerned: it seems that nitrogen losses through leaching can be considerable, whereas leaching of phosphorus and heavy metals is usually much lower.

Research conducted in Switzerland and Denmark has shown that, in the case of nitrates, losses through leaching may vary, depending on the amount applied and environmental conditions. It is strongly affected by the soil's water retention capacity. On the other hand, phosphorus leaching does not seem to be influenced by the quantities applied. Heavy metal leaching is usually very low and even insignificant on lime-treated soil or soil with a high organic content.

Volatilisation

Nitrogen losses through volatilisation in the form of ammonia and nitrogen dioxide produced through denitrification occur during surface spreading of sewage sludge and over the period immediately following spreading. Very few papers have been published on losses through volatilisation. Available results show that losses may be very high, depending on atmospheric conditions, the spreading period and the method of application. The losses may total up to 50 per cent of the nitrogen applied and have economic implications for the farmer and environmental ones for air quality. They may be kept to a minimum by complying with a few application rules. For instance, it is preferable to apply the sludge to soil on which crops are being grown as opposed to bare soil. Best results are obtained through injection of the sludge into the soil, immediately followed by ploughing.

Conclusion

Pollution transfer risks through the agricultural use of sewage sludge show that this hazard affects several media and it is necessary to take into account existing interactions when drawing up preventive measures. However, it seems that such risks, including leaching, are no higher than those connected with other types of fertiliser such as inorganic fertilisers and animal waste.

Measure taken to reduce the hazards

Most of the measures taken to reduce the hazards relating to the agricultural use of sewage sludge concern the accumulation of heavy metals and transmission of pathogens. They are based on the principle that the pollution sources are the sludge and not the waste water. Many countries have adopted regulations including standards and user instructions. Unfortunately the basis of the regulations varies markedly from one country to the next, owing to the scientific information gaps remaining in this field. In the case of heavy metals, for instance, there are four different basic principles:

1 Concentration in sludge should not exceed certain limits.
2 The quantities of heavy metals supplied to the soil should not exceed certain limits.
3 Soil concentrations should not exceed certain limits.
4 Heavy metals raise fewer problems if the sludge is supplied through small successive applications, as opposed to a single large one.

Table 11.3 outlines the application of these principles in selected countries.

For pathogens, also, restrictions on the use of sludge vary from one country to the next. The United Kingdom and other countries authorise the

Table 11.3   Heavy metal regulation in selected OECD countries

Country	(1)	(2)	(3)	(4)
Canada	X		X	
Denmark	X			
Finland	X			X
France	X	X		
Germany	X	X		
Netherlands	X			X
Norway	X		X	X
Sweden	X			X
Switzerland	X			X
UK		X		
USA		X		

(1)   Heavy metal in sludge should not exceed defined limits.
(2)   Heavy metal in soil should not exceed defined limits.
(3)   Heavy metal loading to agricultural land should not exceed defined limits.
(4)   Heavy metals are less likely to cause problems if added to the soil in several small increments over an extended period rather than in one or few large increments.

Source: CEC (Cost 68).

application of raw sewage to grassland provided a waiting period is observed between the application and putting the animals out to graze. In Sweden the use of non-pasteurised sludge is prohibited on grassland and cultivated land. Most of these measures, however, are not aimed at removing the hazard, which would be too costly, but rather at reducing it.

European Community measures to control sewage sludge

An initial effort to standardise and harmonise rules for the use of sewage sludge was recently made by the European Communities, which on 6 June 1986 adopted a directive concerning the application of sewage sludge to soil. It is aimed chiefly at preventing harmful effects on soil, vegetation, animals and man, and will enter into force in 1989.

The main objective of the directive is to prevent the accumulation of heavy metals in soil. It provides for:

1   A limitation of the heavy metal content of sludge for agricultural use.
2   A limitation of the amount of heavy metals which may be applied to agricultural soil through sewage sludge.
3   A limitation on the heavy metal content in soil to which sludge is to be applied.

The use of sludge is prohibited whenever the heavy metal content of the soil already exceeds the proposed limits. To prevent the accumulation of metals in soil beyond the proposed limits, the Community member States have two options at their disposal. First, they may lay down the maximum quantities of sludge which may be applied to the soil per unit of area per year while observing the national limit values for heavy metal concentration, provided these limits are selected within the range given by the directive. Second, they may lay down limit values for the quantities of heavy metals introduced into the soil per unit of area and unit of time (directive standards). The other provisions are as follows:

1   A waiting period of three weeks must be observed between the application of the sludge and grazing.
2   The limit values of heavy metals in soil may be exceeded by 50 per cent if the soil pH is higher than 7.
3   The application of sludge is prohibited on soil in which fruit and vegetable crops are growing, with the exception of fruit trees, and a waiting period of ten months must be observed before harvesting the fruit and vegetables in contact with the soil and normally eaten raw.
4   The application of sludge to soil with a pH of less than 6 is not prohibited (contrary to what the Commission had proposed), but member States may lay down stricter limits for the heavy metal content of soil where necessary and must take into account the increased mobility and availability to the crop of heavy metals caused by any lowering of the pH.
5   Member States must report to the Commission at regular intervals on the agricultural use of the sludge, namely every four years as from 1991.
6   Farmers using sewage sludge must have regular access to analytical results concerning heavy metals and other parameters.

Conclusions

The presence of heavy metals in sludge is clearly a worrying problem for the agricultural use of sewage. It constitutes an appreciable hazard to the environment (accumulation in the soil, leaching to ground water under certain conditions) as well as to man (contamination of food chains) and may affect plant growth (lower crop yields). However, several studies suggest that the environmental hazards of metals in sludge are less important than those from other sources such as inorganic fertilisers and atmospheric pollution. Known cases of cattle poisoning by heavy metals have nearly all been connected with industrial pollution. In the case of cadmium, for instance, the contribution of sewage sludge is obviously a cause for local concern, but at national level it is minimal compared with the quantities supplied through air pollution.

Pathogens are also an appreciable potential hazard, although less worrying than for heavy metals and perhaps more easy to control. In any case, this does not constitute a limiting factor for agriculture.

Hazards arising from the presence of other chemicals are not yet well known and do not seem to raise any problems so far. There is therefore a need for more precise information concerning all these hazards and, especially, more comparable data. The balance between beneficial and harmful effects is not always clear, and efforts should be made to evaluate the actual value of sludge for agriculture more accurately.

Conclusions and outlook

It seems clear that sewage sludge must be managed just like any other type of waste and that the solution chosen should be both economically viable and environmentally acceptable. All countries seem to agree that sludge contains a certain amount of nutrients which can be effectively exploited only through agricultural use. This solution is not only thought to be the most economically efficient but also meets the general principles of action determining waste management policy.

Although sewage sludge unquestionably has value as fertiliser, there is some doubt as to its actual effectiveness. Its use places greater constraints on the farmer than commercial fertilisers, for which application rules are fairly straightforward and the fertiliser element content remains constant. Farmers are therefore reluctant to use sludge as fertiliser, and sludge producers are mostly obliged to supply the sludge free of charge. Some countries such as Sweden apply a commercial fertiliser tax in order to promote sludge spreading.

In spite of this situation and the uncertain benefit–impact balance, agricultural use remains the best solution in many instances. The other disposal techniques are often more expensive and pose environmental hazards. For instance, incineration may cause pollutant transfer to the atmosphere, dumping at sea pollutes the marine environment and landfill entails leaching runoff and volatilisation risks.

For the agricultural use of sewage sludge to develop under the best possible conditions, it must not only represent the most effective waste management solution but also be of benefit to agriculture. Several requirements need to be met:

1   Agricultural use should be the best possible solution, taking local conditions into account, including soil quality, the nature of the crops, climate, distance between the treatment plant and final location, transport costs, essential treatment process costs, size of the treatment plant and nature of the effluents treated.
2   The sludge should be of sufficient quality to meet the main requirements of agriculture without the need for additional large quantities of inorganic fertilisers.
3   The concentration of heavy metals and other potentially toxic substances must not exceed certain thresholds beyond which the environment and man might be contaminated and crop yields reduced.
4   The volume of sludge to be applied in order to meet fertiliser requirements should not lead to hazardous levels of toxic substances.
5   The fertiliser content of sludge should be determined more precisely and as soon as possible, using rapid, easy and inexpensive measurement methods.

For these requirements to be met, it is essential to take a number of measures.

With regard to regulatory measures, strict standards should be laid down in order to protect the environment and guarantee maximum agricultural effectiveness. In order to limit downstream risks, the measures should take into account concentration of hazardous elements in sludge, initial concentrations in the soil and their trends over time. Further measures should be taken in parallel in order to act at the source and restrict pollutant emissions into waste water. In view of the diffuse nature of certain sources, in particular heavy metal, the effectiveness and viability of the latter type of measure remains to be proved.

With regard to financial and/or tax incentive measures, the use of sludge in agriculture may be subsidised and/or the purchase of commercial fertilisers made subject to a special tax.

Educational measures are also required to convince farmers of the value of sludge and to change and adapt farming practices so that the sludge can be used effectively. Such changes include establishing a waiting period between spreading and crop harvesting or grazing, compliance with standards and limits on the maximum amounts of sludge to be spread. All soil analysis and monitoring following spreading should be carried out in co-operation with the farmer. Scientific research in order to close existing gaps is also needed.

To be effective, the above measures should not be confined to a sectoral approach but should cover all aspects and take into account other relevant policies. For instance, stricter standards for industrial emission of atmospheric pollutants might lead to pollution transfer to water, thereby affecting the pollutant content of sludge and its effects on agriculture. There is therefore a practical need to integrate all these aspects and the relevant policies at the level of both environmental policy as such (integration of waste, water and air policies) and also other policies such as agricultural and environmental. Effective management from the economic, agricultural and environmental standpoints should never be sectoral.

12. Overview: the integration of agricultural and environmental policies

M.D. Young

In the past agriculture has had a long positive association and a beneficial impact on the environment but, with improving technology and increasing support for agriculture, some agricultural practices have come into conflict with environmental objectives. As demonstrated throughout this book, one of the reasons for this conflict has been the lack of integrated policies which make appropriate trade-offs between competing agricultural and environmental objectives.

The essential theme of the chapters in this book is the integration of agricultural and environmental policies. An integrated approach requires environmental considerations to be taken fully into account at an early stage in the development and implementation of agricultural policies leading, through improved decision-making and policy implementation, to improved environmental quality and ultimately a more prosperous agricultural sector. Successful integration requires conscious trade-offs between competing objectives.

In developing new agricultural and related regional development policies, consideration needs to be given to a trilogy of three factors: (1) the need to enhance the positive contribution which agriculture can make to the environment; (2) the need to reduce agricultural pollution; and (3) the importance of adapting all agricultural policies so that they take full account of the environment (figure 12.1).

First, the positive role of non-polluting forms of agriculture can be enhanced through the introduction of management agreements which:

1 Are of sufficient length that the expected environmental benefits can be realised.
2 Compensate farmers for the lost value of production.
3 Reimburse farmers for extra management costs.

The aim of management agreements should be to improve the environment, especially landscape amenity and conservation value in the areas where agricultural pollution is not a dominant problem.

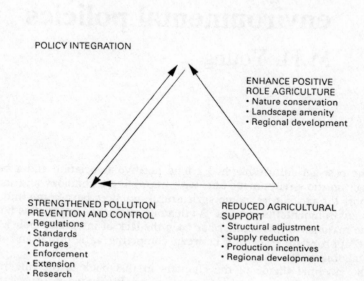

ADMINISTRATIVE INTEGRATION
– AGREED ENVIRONMENTAL
   QUALITY OBJECTIVES
– RECOGNISE RESPONSIBILITY
– REGIONAL INTERGRATION

POLICY INTEGRATION

ENHANCE POSITIVE
ROLE AGRICULTURE
• Nature conservation
• Landscape amenity
• Regional development

STRENGTHENED POLLUTION
PREVENTION AND CONTROL
• Regulations
• Standards
• Charges
• Enforcement
• Extension
• Research

REDUCED AGRICULTURAL
SUPPORT
• Structural adjustment
• Supply reduction
• Production incentives
• Regional development

Figure 12.1 Opportunities for the integration of agricultural and environmental policies

Second, the reduction of agricultural pollution will be most efficiently achieved through the use of regulations, standards, charges, enforcement and advisory procedures in a manner which is consistent with the 'polluter-pays' principle. Efforts should be made to overcome the perceived difficulties associated with applying this principle to the control of agricultural pollution from diffuse sources.

Third, policies principally designed to achieve agricultural and other sector objectives, such as the reduction of agricultural surpluses, can and should be adapted in a manner which produces environmental benefits.

With regard to the third factor it is recommended that surplus reduction programmes be targeted so that they simultaneously reduce surpluses and agricultural pollution and enhance environmental quality. Successful targeting mechanisms include financial incentives to set aside land which is a source of pollutants and the restriction of the programmes to areas which are the source of environmental problems. In the interests of achieving better integration, it is also recommended that targeting mechanisms should not be constrained by the 'polluter-pays' principle, which requires that the cost of agricultural pollution prevention and control is met by the farmers who cause the pollution.

Administrative considerations

Successful integration requires policy-makers to give full consideration to, and accept responsibility for, the effects of their policies on the objectives of all other sectors. This is as true for the effects of environmental policies on agriculture as it is for the effects of agriculture on the environment.

One necessary pre-condition to this concept of 'responsibility' is that policy-makers must jointly agree to the objectives of interacting and interdependent sectors. Greater progress will be made when these objectives are clear, measurable and time-specific. Progress will also be more likely when there is a clear commitment, at Ministerial level, to the integration of environmental and agricultural policies.

Because of the regional disparities in physical, social and economic conditions, successful integration will usually require the development of strategies which rely upon a mixture of policies. Some of these will be national in character while others, particularly in larger countries, will be regional. Experience in some of these larger countries is revealing that the opportunities for integration are often greater at a regional level.

Current opportunities

Current opportunities for the better integration of agricultural and environmental policies include:

1 Changing institutional arrangements so that there is a real dialogue between those responsible for protecting the environment and those responsible for agriculture.
2 Introducing administrative procedures which encourage collaboration between agencies, such as the mandatory referral of policy proposals to other departments.
3 Applying environmental impact assessment procedures to agricultural activities and policies.
4 Further harmonising the standards and procedures used in regulating the use of agricultural inputs.
5 Controlling the use of agricultural inputs that damage the environment, modifying agricultural practices, particularly through education and extension.
6 More stringently enforcing existing regulations.
7 Requiring farms to prepare management plans which indicate how they will use inputs and adopt practices in a manner which does not have adverse effects on the environment.
8 Introducing charges on inputs such as fertilisers and practices such as the spreading of animal manure to finance the cost of advisory and other activities designed to prevent and control agricultural pollution.
9 Removing impediments to the adoption of environmentally favourable practices such as integrated pest management which remain in the eligibility criteria for the receipt of financial assistance.

10 Changing product quality standards so that they reduce incentives to use pesticides.

11 Making taxation policy neutral to agricultural and environmental objectives.

12 Entering into management agreement and other arrangements with farmers to improve landscape amenity and nature conservation value.

As illustrated throughout this book, many OECD countries have already begun to take many of the above opportunities for integration. In particular, policies which further develop the role of the farmers in maintaining and improving landscape amenity and furthering nature conservation objectives are advocated. At the same time, in implementing pollution prevention and control measures it is suggested that more account needs to be taken of the effects of environmental policies on agriculture in general and agricultural incomes in particular.

A major problem in many countries is the degree to which current land-use regulations are poorly defined and, even where well defined, not enforced. There is a need, particularly in areas where there is significant agricultural pollution, for the more precise definition of non-polluting agricultural practices.

There is strong pressure, spread over many if not all member countries and, indeed, world-wide, for prices to reflect the true cost to society of the provision of goods and services. In the context of agriculture and the environment this means that farmers should pay for the full marginal cost of any natural resources used, particularly those which are in scarce supply, such as water. This concept applies to all industries and is just as relevant to agriculture as it is to the use of all other resources.

Pollution can be reduced and biological diversity on modern farm land can be increased if farmers are encouraged to use artificial fertilisers and pesticides more sparingly and more precisely. This can be achieved with new technological advances backed up by sound advice and appropriate management guides.

There are also a large number of opportunities to improve the environment at modest cost and little inconvenience to the farmer. Leaving field margins unfertilised, abandoning small pockets of land, planting trees and retaining hedges can work wonders for wildlife.

Emerging opportunities

New opportunities for integrating agricultural and environmental policies are emerging as current agricultural policies are reformed to overcome existing problems. In particular, as measures to lower production levels and, particularly, to limit budgetary expenditure on price support are realised, invaluable opportunities to achieve these objectives and simultaneously improve the environment can be expected to emerge.

Major changes to agricultural policy which may occur in the near future and offer opportunities for the development of integrated policies include changes to price support policies and tariff arrangements, the introduction of cross-compliance requirements, direct income support to farmers, the establishment of land set-aside programmes and the introduction of quotas on inputs and outputs.

One opportunity which is being implemented in the United States is the attachment of conservation-compliance conditions to government pro-grammes which increase farm income. This approach requires farmers to comply with a set of pre-specified land-use conditions and practices in return for the receipt of government payments. In some countries this shift from market intervention policies to those which involve direct deficiency pay-ments to farms would be required before cross-compliance provisions could be introduced. The same concept can also be used to reduce the tendency of production incentives to encourage production to expand into wetlands and other environmentally fragile areas.

Land set-aside policies have the potential to improve agriculture, budge-tary and environmental goals simultaneously. Opportunities for realising these benefits will be greatest when:

1 The period for set-aside is sufficient to induce farmers to diversify into environmentally favourable activities such as forestry.
2 The title given to the land that is withdrawn from production encourages farmers to enhance their role in conserving, maintaining and improving the environment.
3 Set-aside programmes are targeted to areas with environmental problems, including key production areas.
4 Land is selected on a tender basis which includes an assessment of the likely off-site impact of continued production and the potential budget savings from setting the land aside.
5 The programmes do not include short-term rotational set-aside.

As discussed in several chapters, the suggested period for the set-aside of land to achieve conservation benefits needs to be in excess of ten years. Opportunities to set aside highly productive land in filter strips along the edge of streams and in ground-water areas should not be discounted. In some situations it is noted that such an approach can produce substantial environ-mental benefits and budgetary savings from reduced surplus production.

The reduction of support measures which induce production or the intro-duction of quantitative restrictions on production can be expected to decrease agricultural pollution in areas where the intensity of agricultural production is high. The effect of reduced price support, however, is likely to be different in environmentally sensitive areas where landscape amenity, cultural heri-tage, wildlife habitat and species diversity values are important. In these areas the provision of financial assistance to encourage farmers to continue with environmentally favourable practices, such as the repair of stone walls and terraces, leaving unsprayed headlands and edges around fields, is a recommended strategy.

If the foreshadowed changes in agricultural policies do occur, then sub-stantial structural changes can be expected in the agricultural sector. These changes could result in short-term economic pressures which could result in substantial environmental losses. Consequently, to obtain net environmental and social benefits from a reduction of support measures, there will probably be a need for additional expenditure on the maintenance of landscape amenity, nature conservation and community development objectives. Additional expenditure will be particularly needed in environmentally sensitive areas.

Index

Abandoned land, 55, 229, 230, 292
Acid deposition, 12, 13
  and animal manure, 156
Administrative
  efficiency, 289
  procedures, 339
Aerial spraying ban, 58
Agriculture
  positive environmental effects of, 248
  positive role of, 256, 337
Air pollution, 14, 295–319, 309
  cost of, 304
  cost to consumers, 307
  effects on workers, 308
  gains from, 304
Animal manure, 64
  ammonia from, 168
  disposal costs, 167
Austria, 221–252

Best management practice, 210, 214–5
Bureaucracy, 21

Cadmium, 57
Cancer, risk from pesticides, 96
Chemicals, substitute for labour, 96
Clean Water Act, 190
Co-operatives, 274
Common Agricultural Policy, 9, 20,
    147, 185, 256, 259
Comparative advantage, 309
Compensation, 258, 268, 280, 283, 285,
    337
  for lost rights, 255
  for ongoing costs, 244
Conservation
  adviser, 274–5
  compliance, 341
  ethic, 288

Conservation Reserve Programme, 92,
    94, 101, 102, 103, 105, 106, 108,
    199, 200
Contingent valuation, 214
Costs, 197
Credit
  interest-free, 176
  long-term low-interest-rate, 178–9
Crop-edge programme, 20, 29, 36
Cross compliance, 199, 201, 202, 204,
    341
Cultural reasons for expansion, 122

Dairy waste, 74
Decoupling, 30
Direct income transfers, 43
Direct payments, 221, 240, 243, 249,
    289, 341
Diversity, 51
  of cropping, 12
  of species, 14
Drinking water pollution
  area of, 156
  by nitrates, 156
Drinking wate, 14, 57, 126
  Denitrification of, 133
  EC guidelines, 126
  nitrate in, 123
  purification cost, 43
Drinking water standard, 214

Economic instruments, see fertilizer tax,
    manure levy, input tax, pesticide
    levy, livestock feed levy, research
    levy
Energy, forestry, 71
Environmental impact assessment, 136,
    339
Environmentally sensitive area, 73, 259,
    272–3, 277–287, 342
Erosion rates, 182, 297

Eutrophication, 13, 55, 74, 209, 213–4
Export subsidies, 82
Extensification, 20, 45
Extension, 30, 88, 339
Externalities, 138

Fertilizer, 46
  Fertilizer, application rate, 26, 29, 84
  cadmium in, 57
  optimal application rate, 144
  profit from reduced use, 73
  restrictions, 282
  tax, 61–4, 87
Filter strips, 102, 103, 106, 341
Food quality, 13
France, 115
Free trade, 155
Future generations, costs imposed on, 193

Game hunting, 235
Germany, 9, 38
Green fallow, 20, 30, 40
Groundwater pollution, 129, 156
  from fertilizer, 97
  from pesticides, 97
Gully control programme, 178

Headage payments, 280
Heavy metals, 57
  animal feed standards, 170
  in sewage sludge, 320–21, 326
  regulation, 333
Hedgerows, 16

Income transfers, 30, 31
Input tax, 31, 40, 46
Insurance schemes, 42
Integrated pest management, 29
Integration, 1, 41, 86, 91, 138, 194,
    216, 288, 290, 337–342
  administrative, 195, 212, 265, 339
  substantive, 194, 199, 211, 213
Intensification, 44
Intensive animal production, 64, 122
  consequences for environment, 155
  distribution of, 123
  expansion restrictions, 168

Land set-aside, 341
Land-use planning, 177, 236, 255, 268, 290
  zoning, animal production, 134

Landscape
  amenity value, 160–1
  change, 246
  definition, 254
Leaching, effects of price on, 83
Legislation, 17, 63–64, 133, 162, 236, 237
Less favoured areas subsidies, 281
Levy on inputs, 339
Liability, 136
  see also pollution
Livestock feed, levy on, 165
Loans, see credit

Management
  agreement, 258, 271, 282, 340
  plans, 339
Manure
  bank, 132, 167
  export, 132, 151
  levy, 165
  levy, exemption, 166
  liquid, 30
  ploughing in, 73
  processing, 132
  production, accounting for, 165
  required storage capacity, 79, 83
  spreading regulations, 55, 78, 136,
    151, 162, 163–4, 166–7, 169
  storage, subsidy, 80
  see also pollution; sewage sludge
Methaemoglobinaemia, 57
Minimum standards, 136
Modelling, 98
Mountain areas, 221–252

Nature conservation, 237
Netherlands, 38, 147–172
Nitrates, 55
Nitrogen
  leaching, 79
  effect of price on production, 71
Non-government organisations, and
    conservation, 20
Non-point source pollution, 198
Nutrient leaching, 88

Objectives, 339
Orderly agriculture, 17, 21, 46
Organic fertilisers, 26
Ozone, 310

Perched water tables, 186

Pesticides, 340
  application controls, 66
  application rates, 24, 26
  ban, 59
  dependence, 51
  effects on fauna, 59
  in ground water, 213–4
  poisoning of fish, 209
  proportion used in US farm
    programme, 91
  regulations, 195
  residues, 57
  risks to farm workers, 96
  spray drift, 57–8
  tax, 61–4, 87
Phosphates, 55
Pig production, 50
Plans, 290
Plant yield, 297–8
Polluter pays principle, 21, 39, 45, 56,
    138, 141, 145, 172, 203, 204, 338
Pollution
  drinking water, 24
  estimation of costs, 314
  from agriculture, 228, 249, 338, 341
  heavy metal, 13, 26, 123, 126, 160
  liability for, 19
  liquid manure, 123
  monitoring, 297
  nitrates in ground water, 213–4
  off-farm costs of, 189, 193, 197, 198
  optimal, 141
  phosphate, 156
  sensitive areas, 73
  tax, 141–4
  transboundary effects, 309
  water, 26, 55
  see also air pollution; sewage sludge;
    water pollution
Portugal, 175–188
Price
  control, 52
  elasticity, 46
  maintenance, 240
Price reduction
  and income, 33
  and land use intensity, 33
  vs input tax, 33
Price support, 17, 30, 60, 86, 94, 176,
    180, 239, 256, 267, 341
  effect of reduced, 341
  effects on environment, 53
  environmental consequences of, 60–1

  impact of reduced, 185, 292
  influence on production, 148
  reduced, 42
Property rights, 139
Public access, 260
Public health, 96

Re-afforestation, 178, 229, 234
Recreation, 206, 209, 214, 223, 230,
    245–6
Regulations, 73, 211, 339
  enforcement, 31
  standards, charges, 338
Research, 88
  levy, 66, 165
  potential of, 67
Rotation
  importance of, 184
  loss of, 92
Rural Clean Water Programme, 196,
    212–4

Self-sufficiency, need for, 148
Set aside, 40, 68, 94, 195
Sewage sludge, 320–336
  economic effects, 310
  fertiliser value, 323–5
  food contamination, 328
  hazards of, 329–30
  regulations on use, 332–3
  treatment of, 329
  use of, 320
Shell fish contamination, 129
Sites of Special Scientific Interest,
    256–7, 272, 283, 286, 287
Slurry bank, 132
Sodbuster, 92
Soil drainage, 179, 183
Soil erosion, 74, 182, 189, 197
  impacts of, 184
  T values, 191
Soil pollution, see sewage sludge
Specialization, 86
Species, rare plants, 58
Standards, 14, 43, 339, 340
Straw, use of, 176, 181
Structural adjustment, 53, 55, 276
Subsidy, 139
Surface drainage, 184
Surplus, 68, 249
  production, 82, 312
  reduction, 44
Swamp buster, 92, 199–200

Switzerland

T values, 192, 193, 198
Targets, 21, 91, 135, 156, 198, 199,
        202, 338
    phased, 163
Tariff barriers, effects of, 225
Tax, *see* economic instruments
Tourism, 223
    income from, 260
Trade-offs, 337
Training, 66
Transaction costs, 40

United Kingdom, 253–292
United States, 189–218

Universal Soil Loss Equation, 203
US Farm Bill, 190, 199

Vegetables, nitrate in, 36
Voluntary agreements, 271

Water, phosphorus in water, 13
Water pollution
    effect on recreation, 72
    from sewage, 72
    pesticides in water, 14
Wetlands, 230
Wildlife and Countryside Act, 258, 268,
        271, 272, 287
Willingness to pay, 301
Woodland management, 275

SOCIAL SCIENCE LIBRARY

Oxford University Library Services

WITHDRAWN

Marston Road

Oxford OX1 3UQ

Tel: (2)71093 (enquiries and renewals)

http://www.ssl.ox.ac.uk

This is a NORMAL LOAN item.

We will email you a reminder before this item is due.

Please see http://www.ssl.ox.ac.uk/lending.html
for details on:

- loan policies; these are also displayed on the notice boards and in our library guide.

- how to check when your books are due back.

- how to renew your books, including information on the maximum number of renewals. Items may be renewed if not reserved by another reader. Items must be renewed before the library closes on the due date.

WITHDRAWN

- level of fines; fines are charged on overdue books.

Please note that this item may be recalled during Term.